Analog Circuits and Signal Processing

The *Analog Circuits and Signal Processing* book series, formerly known as the *Kluwer International Series in Engineering and Computer Science*, is a high level academic and professional series publishing research on the design and applications of analog integrated circuits and signal processing circuits and systems. Typically per year we publish between 5-15 research monographs, professional books, handbooks, and edited volumes with worldwide distribution to engineers, researchers, educators, and libraries. The book series promotes and expedites the dissemination of new research results and tutorial views in the analog field. There is an exciting and large volume of research activity in the field worldwide. Researchers are striving to bridge the gap between classical analog work and recent advances in very large scale integration (VLSI) technologies with improved analog capabilities. Analog VLSI has been recognized as a major technology for future information processing. Analog work is showing signs of dramatic changes with emphasis on interdisciplinary research efforts combining device/circuit/technology issues. Consequently, new design concepts, strategies and design tools are being unveiled. Topics of interest include: Analog Interface Circuits and Systems; Data converters; Active-RC, switched-capacitor and continuous-time integrated filters; Mixed analog/digital VLSI; Simulation and modeling, mixed-mode simulation; Analog nonlinear and computational circuits and signal processing; Analog Artificial Neural Networks/Artificial Intelligence; Current-mode Signal Processing; Computer-Aided Design (CAD) tools; Analog Design in emerging technologies (Scalable CMOS, BiCMOS, GaAs, heterojunction and floating gate technologies, etc.); Analog Design for Test; Integrated sensors and actuators; Analog Design Automation/Knowledge-based Systems; Analog VLSI cell libraries; Analog product development; RF Front ends, Wireless communications and Microwave Circuits; Analog behavioral modeling, Analog HDL.

Carl D'heer • Patrick Reynaert

THz and Sub-THz CMOS Electronics for High-Speed Telecommunication

Architectures and Circuits for Future 6G Transceivers

Springer

Carl D'heer
Huawei Technologies
Heverlee, Belgium

Patrick Reynaert
ESAT-MICAS
KU Leuven
Heverlee, Belgium

ISSN 1872-082X ISSN 2197-1854 (electronic)
Analog Circuits and Signal Processing
ISBN 978-3-031-64438-2 ISBN 978-3-031-64439-9 (eBook)
https://doi.org/10.1007/978-3-031-64439-9

This Springer imprint is published by the registered company Springer Nature Switzerland AG
The registered company address is: Gewerbestrasse 11, 6330 Cham, Switzerland

To those brave enough to attempt CMOS circuit design at high frequencies. May your coffee be strong and your simulations converge.

Preface

The demand for high-speed telecommunication systems has grown rapidly in response to the emergence of advanced data-intensive applications and increased global connectivity. To support the need for higher data rates, the sub-THz (100–300 GHz) and THz (0.3–3 THz) frequency bands have attracted considerable attention due to their large available bandwidths. These frequency bands could potentially support tens to hundreds of gigabits per second in various application domains, ranging from short-distance data transfer to immersive VR/AR and long-distance point-to-point links. Sub-THz and THz transceivers will not only play a key role in the development of next-generation wireless communication (6G), but also enable high-speed links through dielectric waveguides, which could prove vital for the advancement of fully autonomous vehicles and energy-efficient data centers. However, the large propagation loss and poor signal quality above 100 GHz limit the range over which these high-speed (sub-)THz links can operate effectively.

CMOS technology, known for its superior digital circuit performance, low-cost manufacturing, and high production yield, presents an attractive solution for implementing transceiver integrated circuits. By leveraging CMOS technology, transceivers can be integrated with digital processing blocks while harnessing the benefits of its large-scale production capabilities. However, despite recent efforts to scale down transistors and enable high-frequency operation, the maximum operation frequency of transistors in CMOS has saturated at around 300 GHz. Additionally, passive components in deeply scaled CMOS suffer from degraded quality factors, and a lower supply voltage limits the dynamic range of signals. These factors lead to poor system performance and efficiency when the circuits at (sub-)THz frequencies are implemented in standard CMOS.

This book focuses on the development of innovative architectural and circuit-level solutions to enable high-speed (sub-)THz communication in CMOS technologies. A comprehensive numerical model of high-speed links is developed to identify critical performance bottlenecks and provide system-level design guidelines for higher speed and efficiency communication systems. Based on the modeled trends, various transceiver architectures are analyzed for their feasibility. Multiplier-last THz transmitters are discussed to enable high data rate transmission above the transistor's maximum frequency. A mixed-signal transceiver architecture is proposed, which achieves high-speed operation at sub-THz frequencies with improved energy efficiency compared to conventional architectures. This transceiver relies on direct-digital (de)modulation, supporting high-order modulation schemes without the need for high-performance, power-consuming data converters. This book also presents a practical design methodology with several circuit-level innovations to ensure efficient CMOS circuit implementation at high frequencies.

The proposed solutions are validated through the implementation of several integrated circuits and the realization of a complete high-speed sub-THz link.

1. A tripler-last 390 GHz transmitter is designed, supporting 28 Gb/s with the non-conventional spectrally efficient BPOOK modulation scheme.
2. A 135 GHz direct-digital modulation 16-QAM transmitter is presented, capable of generating 16-QAM signals up to 36 Gb/s without the need for a high-speed multi-bit DAC or offline signal processing.
3. A 135 GHz direct-digital demodulation 16-QAM receiver is proposed, demonstrating 16-QAM demodulation up to 32 Gb/s directly to binary outputs, eliminating the need for a high-resolution ADC.
4. A complete 16-QAM link operating at 135 GHz is established over a wireless channel and a dielectric waveguide, achieving a maximum data rate of 32 Gb/s with on-chip modulation and demodulation, without any offline digital signal processing, while achieving state-of-the-art energy efficiency. All designs were implemented in 28 nm bulk CMOS technology and validated through extensive measurements.

Leuven, Belgium Carl D'heer
April 2024 Patrick Reynaert

Contents

List of Abbreviations

6G	Sixth Generation.
AC	Alternating Current.
ADC	Analog-to-Digital Converter.
AP	Access Point.
APSK	Amplitude and Phase-Shift Keying.
ASK	Amplitude-Shift Keying.
AWG	Arbitrary Waveform Generator.
AWGN	Additive White Gaussian Noise.
BB	Baseband.
BEOL	Back End of Line.
BER	Bit-Error Rate.
BiCMOS	Bipolar Complementary Metal-Oxide-Semiconductor.
BJT	Bipolar Junction Transistor.
BPF	Band-Pass Filter.
BPOOK	Binary-Phase On-Off Keying.
BPSK	Binary Phase-Shift Keying.
BSF	Band-Stop Filter.
BW	Bandwidth.
CDR	Clock Data Recovery.
CM	Common-Mode.
CMOS	Complementary Metal-Oxide-Semiconductor.
CMR	Common-Mode Rejection.
CPFSK	Continuous-Phase Frequency-Shift Keying.
CPW	Coplanar Waveguide.
CR	Carrier Recovery.

CTLE	Continuous-Time Linear Equalizer.
CW	Continuous Wave.
DAC	Digital-to-Analog Converter.
DC	Direct Current.
DE	Drain Efficiency.
DEC	Decoding.
DEMOD	Demodulation.
DFE	Decision Feedback Equalizer.
DM	Differential-Mode.
DPFSK	Discontinuous-Phase Frequency-Shift Keying.
DPSK	Differential Phase-Shift Keying.
DSP	Digital Signal Processing.
DWG	Dielectric Waveguide.
ECC	Error Correcting Code.
EIRP	Equivalent Isotropic Radiated Power.
EM	Electromagnetic.
EMI	Electromagnetic Interference.
ENC	Encoding.
ENOB	Effective Number of Bits.
EQ	Equalization.
EVM	Error Vector Magnitude.
FEC	Forward Error Correction.
FEOL	Front End of Line.
FFE	Feed-Forward Equalizer.
FinFET	Fin Field-Effect Transistor.
FIR	Finite Impulse Reponse.
FS	Full-Scale.
FSK	Frequency-Shift Keying.
FSO	Free Space Optical.
FSPL	Free-Space Path Loss.
GAA	Gate All Around.
GBW	Gain-Bandwidth.
GCPW	Grounded Coplanar Waveguide.
GDE	Group-Delay Equalizer.
HEM	Hybrid Electromagnetic.

HPF	High-Pass Filter.
IC	Integrated Circuit.
IF	Intermediate Frequency.
IIP3	Input-ReferredThird Order Intercept Point.
IM	Intermodulation Distortion.
IoT	Internet of Things.
IR	Infrared.
ISI	Intersymbol Interference.
ITU	International Telecommunication Union.
LAN	Local Access Network.
LNA	Low-Noise Amplifier.
LO	Local Oscillator.
LOFT	Local Oscillator Feedthrough.
LPF	Low-Pass Filter.
LSB	Least-Significant Bit.
MIM	Metal-Insulator-Metal.
mm-Wave	millimeter-Wave.
MOD	Modulation.
MOM	Metal-Oxide-Metal.
MOS	Metal-Oxide Semiconductor.
MOSFET	Metal-Oxide Semiconductor Field-Effect Transistor.
MSB	Most-Significant Bit.
MSK	Minimum-Shift Keying.
NMOS	N-Type Metal-Oxide Semiconductor.
NPDP	Neutralized Pseudo-Differential Pair.
NRZ	Non-Return-to-Zero.
OIP3	Output-Referred Third Order Intercept Point.
OOK	On-Off Keying.
OP1dB	Output-Referred 1-dB Compression Point.
OTA	Over-the-Air.
P1dB	Input-Referred 1-dB Compression Point.
PA	Power Amplifier.
PAE	Power Added Efficiency.

PAM	Pulse Amplitude Modulation.
PAPR	Peak-to-Average Power Ratio.
PB	Passband.
PBO	Power Back-Off.
PCB	Printed Circuit Board.
PD	Predistortion.
PDF	Probability Density Function.
PLL	Phase-Locked Loop.
PMOS	P-Type Metal-Oxide Semiconductor.
PN	Phase Noise.
PP	Power Penalty.
PRBS	Pseudo-Random Binary Data Sequence.
PS	Pulse Shaping.
PSD	Power Spectral Density.
PSK	Phase-Shift Keying.
PSR	Power-Supply Rejection.
PVT	Process Voltage Temperature.
QAM	Quadrature Amplitude Keying.
QPSK	Quadrature Phase-Shift Keying.
RC	Raised-Cosine.
RF	Radio-Frequency.
RHP	Right-Half Plane.
RL	Return Loss.
RMS	Root Mean Square.
RRC	Root Raised-Cosine.
RX	Receiver.
SNR	Signal-to-Noise Ratio.
SOI	Silicon-on-Insulator.
SPDT	Singe-Pole Double-Throw.
SQNR	Signal-to-Quantization-Noise Ratio.
SRF	Self-Resonance Frequency.
TE	Transverse Electric.
TEM	Transverse Electromagnetic.
THz	Terahertz.
TM	Transverse Magnetic.
TX	Transmitter.

UV Ultraviolet.

VCO Voltage-Controlled Oscillator.
VSWR Voltage Standing Wave Ratio.

WLAN Wireless Local Access Network.

List of Symbols

t	Time
c	Speed of Light in Vacuum
λ	Wavelength
f	Frequency
C	Channel Capacity
C_S	Spectral Capacity
Q	Q-Factor
ω	Angular Frequency
L	Inductance
C	Capacitance
d	Distance
L	Length
W	Width
H	Entropy
I	Mutual Information
T	Period
B	Bandwidth
A	Amplitude or Area
R	Data Rate
D	Symbol Rate
f_c	Center/Carrier Frequency
T_s	Sampling Period
f_s	Sampling Frequency
dB	Decibel
T_s	Symbol Period
T_b	Bit Period
ϕ	Phase
θ	Phase
η	Energy Efficiency
η	Spectral Efficiency
h	Modulation Index

T	Temperature
k	Boltzmann Constant
N_0	Thermal Noise Floor
E	Energy
P	Power
F	Noise Factor
NF	Noise Figure
dBc	Decibels Relative to Carrier
dBm	Decibels Relative to 1 mW
dBi	Decibels Relative to Isotropic Radiator
α	Raised-Cosine Roll-Off Factor
G	Gain
E	Electric Field
D	Electric Flux Density
B	Magnetic Flux Density
H	Magnetic Field
ρ	Charge Density
J	Current Density
ϵ	Permittivity
μ	Permeability
σ	Conductivity
k	Wave Number
v_p	Phase Velocity
η	Wave Impedance
S	Poynting Vector
γ	Propagation Constant
α	Attenuation Constant
β	Phase Constant
n	Optical Index
V	Voltage
I	Current
Q	Charge
R	Resistance
G	Conductance
G_m	Transconductance
Z	Impedance
Y	Admittance
B	Susceptance
X	Reactance
ρ	Resistivity
Z_c	Characteristic Impedance
Γ	Reflection Coefficient
V_T	Threshold Voltage of a MOSFET
μ	Charge Carrier Mobility
f_T	Transition Frequency

f_{max}	Maximum Oscillation Frequency
δ	Skin Depth
k	Magnetic Coupling Coefficient
K	Rollet Stability Factor
μ	Geometric Stability Factor
V_E	Vertical Eye Height
τ	Time Constant
e	the natural number (2.7183)
π	Pi (3.14159)

Chapter 1
Introduction

1.1 High-Speed Telecommunication: Evolution and Trends

This section will provide a brief overview of the history of telecommunication and its relationship with the evolution of the semiconductor industry. Recent trends in the main branches of telecommunication will be discussed, and an outlook is provided on the future communication requirements of the world.

1.1.1 A Brief History of Electronic Telecommunication

1.1.1.1 From Smoke Signals to Radio

The word "telecommunication" is a combination of the Greek prefix "tele-" ($\tau\eta\lambda\epsilon$), meaning "far off, at a distance," and the Latin noun "communicare," meaning "to share" or more literally "to make common" [1]. Telecommunication, often referred to as just "communication" throughout this book, is thus defined as the transmission or sharing of information by various types of technologies over a certain distance [2]. It originated from a desire to be able to communicate over a distance greater than the human voice is capable of, but with a similar level of practicality. Over the last centuries telecommunication technology has seen an extraordinary evolution and has been a prominent driver of human innovation. An overview of the evolution in the last two centuries is depicted in Fig. 1.1.

Telecommunication has a rich history that spans thousands of years. Early forms of communication involved smoke signals and drumbeats to signal the presence of a military camp. Pigeon post was first introduced by the Persians and later adopted by the Romans for military purposes. During the Middle Ages, beacons were often used to signal the approach of enemy troops. In the late eighteenth century, Claude

C. D'heer, P. Reynaert, *THz and Sub-THz CMOS Electronics for High-Speed Telecommunication*, Analog Circuits and Signal Processing,
https://doi.org/10.1007/978-3-031-64439-9_1

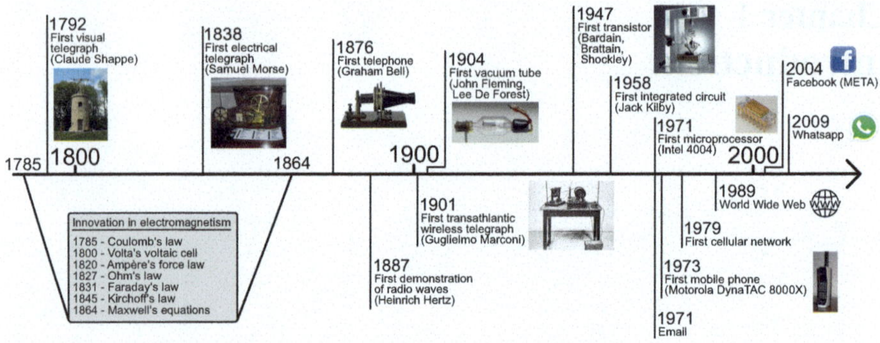

Fig. 1.1 Overview of the evolution of telecommunication technology

Chappe invented the first visual telegraph, which used a series of towers with movable arms or paddles to spell out letters and numbers, enabling messages to be transmitted over long distances.

The pioneering work of Alessandro Volta, André-Marie Ampère, Carl Gauss, Michael Faraday, and others in the field of electricity and magnetism paved the way for the development of the first electrical telegraphs. Most notably, in 1838, Samuel Morse invented an electrical telegraph that revolutionized communication due to his creation of the Morse Code, a highly efficient method of encoding messages. Further improvements on the electrical telegraph lead to the invention of the electric telephone by Graham Bell in the 1870s.

Another significant breakthrough occurred in 1864 when Scottish physicist James Clerk Maxwell formulated a set of equations, known as Maxwell's equations, that established the relationship between electricity and magnetism. Maxwell's equations consolidated the previously distinct fields of electricity and magnetism, and predicted the existence of electromagnetic waves, which became the foundation of radio communication. Heinrich Hertz provided the first demonstration of these electromagnetic waves, or radio waves, in 1887. Then, in 1901, Guglielmo Marconi sent the first wireless telegraph message across the Atlantic Ocean, igniting the radio revolution. His first transmitter made use of an elevated copper sheet to function as a monopole antenna. Additionally, the invention of the vacuum tube in 1904 enabled the generation of stronger and more reliable radio waves which facilitated longer range radio transmissions and significantly enhanced the quality of audio transmissions. The development of wireless communication then quickly progressed to global radio and television broadcasting, marking a significant turning point in the history of telecommunications.

1.1.1.2 The Start of the Semiconductor Era

A major breakthrough occurred in 1947 at Bell Labs when John Bardeen, Walter Brattain, and William Shockley invented the first transistor, a solid-state device capable of amplifying and switching electronic signals [3]. Bardeen and Brattain developed the first point-contact transistor, while Shockley introduced the improved bipolar junction transistor (BJT) in 1948, which entered production in the early 1950s and marked the beginning of the semiconductor era. Transistors quickly replaced vacuum tubes for most applications, as they were smaller, more reliable, and consumed less power. In 1958, Jack Kilby of Texas Instruments proposed the idea of the integrated circuit (IC), which combined multiple transistors and other electronic components onto a single piece of silicon, enabling the creation of much more complex and powerful electronic devices in a much smaller form factor. The maximal potential of the IC was realized with the invention of the metal-oxide-semiconductor field-effect transistor (MOSFET) in 1959 by Mohamed Atalla and Dawon Kahng at, you can already guess where, Bell Labs. MOSFETs consumed even less power than BJTs and were more scalable, leading to incredible miniaturization and mass-production of MOSFET-fueled ICs that kick-started the digital revolution. The MOSFET fabrication process was named CMOS (complementary metal-oxide semiconductor), as it uses complementary and symmetrical pairs of p-type and n-type MOSFETs for logic functions. This technique enables the implementation of compact, fast, and extremely energy-efficient digital circuits, driving the semiconductor industry to innovate and develop newer and better CMOS processes.

Since the invention of the transistor and the integrated circuit, the semiconductor industry has undergone an unimaginable evolution. The continuous scaling of a single transistor has led to an increasing number of transistors on a single integrated circuit over time. In 1965, Gordon Moore, the former CEO of Intel, predicted early on that the number of transistors on an integrated circuit would double every year [4], which he later revised to approximately every 2 years. Since 1975, his prediction has been surprisingly accurate, as illustrated in Fig. 1.2. In just 50 years, the first microprocessor, Intel 4004, fabricated in a 10 μm process with 2300 transistors has transformed into a processor containing billions of transistors with a feature length of only 7 nm. While traditional transistor scaling has plateaued in recent years, other innovations such as chiplets and 3D integration are keeping Moore's law alive, at least in terms of performance.

The unprecedented scaling of the transistor size brought about significant advancements in various industries, including telecommunications. The increasing transistor density enabled the realization of increasingly complicated systems at a reducing cost per component, spurring economic growth. Furthermore, as will be discussed in Chap. 3, smaller transistors enable faster and more energy-efficient operation. This leads not only to more powerful and efficient processors but also to a steady increase in achievable communication speeds. The latter is exemplified by the exponential growth (similar to Moore's law) in data rates observed in

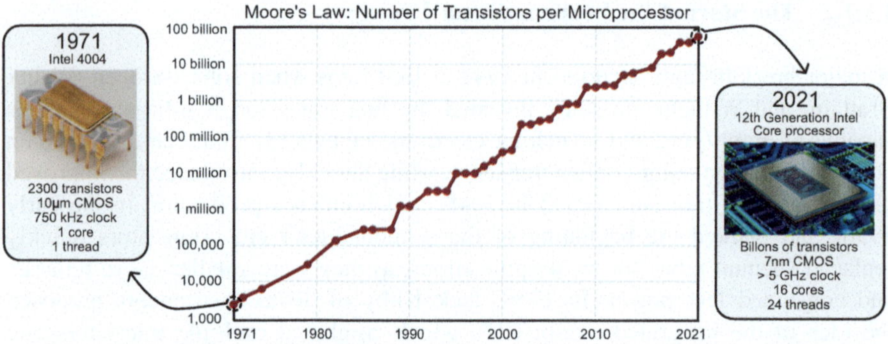

Fig. 1.2 Semi-log plot of the transistor counts for microprocessors against dates of introduction, showing a nearly doubling every 2 years as predicted by Gordon Moore. (Data is from [5])

telecommunication networks, which is often referred to as Edholm's law [6], named after Phil Edholm who predicted this pattern.

Upon reflection, it is remarkable to acknowledge the transformative impact that the invention of the transistor and integrated circuit had on the fields of telecommunication and computing, transforming our world and paving the way for many subsequent technological advancements. While the history lesson stops here, a more detailed and entertaining description of the history of electronics, the radio, and telecommunications can be found in the beginning of Thomas Lee's legendary textbook [7].

1.1.2 Classification of Telecommunication Methods

Telecommunication networks can be broadly classified into two categories: wireline and wireless telecommunication, with five distinct technologies as depicted in Fig. 1.3.

1.1.2.1 Wireline Communication

Firstly, wireline communication includes all communication between devices using a wired connection, which can either be an electrical conductor (such as a copper cable) or a dielectric fiber (with optical or radio waves).

Copper-based wireline communication has been a reliable technology for many applications over the past decades, including USB cables, coaxial cables, and printed circuit board (PCB) traces. While capable of efficiently transferring data over short distances, advancements have enabled copper technology to support high data rates up to 100 Gb/s over several meters. However, the high frequency-dependent loss of the copper channel causes significant attenuation and distortion

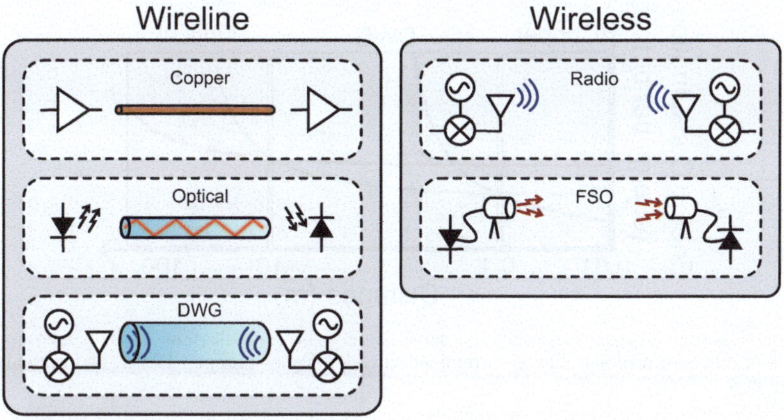

Fig. 1.3 Overview of the most prominent telecommunication technologies: wireline (copper, optical, and DWG) and wireless (radio and FSO)

of high-speed signals, limiting link performance. Expensive cables and intense signal processing are required to mitigate this distortion, making these links costly and power consuming at longer distances and/or higher data rates. Additionally, copper incurs extra material costs and weight, which can be significant for some applications. Furthermore, the conducting copper wires can act as antennas picking up noise and undesired signals from the environment, called electromagnetic interference or EMI, potentially corrupting the signal.

In contrast, optical dielectric fibers have significantly lower losses than copper channels, allowing high-speed communication over much longer distances. For example, optical fiber technology is employed in transatlantic communication between America and Europe, supporting enormous data rates over a distance of more than 6000 km. However, a crucial part of an optical link is the interface between the electronics and the optical channel, necessitating a conversion from the electrical signal to an optical signal and vice versa. Although there have been significant advancements in this area, enabling high-speed and low-loss conversion, these technologies are still expensive and consume large amounts of power. Additionally, the interface between the optical connector and fiber requires tight mechanical tolerances and precise alignment and is sensitive to vibrations. While optical fiber communication is a great solution for long-distance applications, it is not as efficient as copper for short distances due to the costly electrical/optical conversion and sensitive alignment.

Finally, dielectric waveguide (DWG) communication is an emerging wireline technology that combines dielectric fibers with radio waves instead of optical signals, providing unique advantages over copper and optical links. Compared to copper cables, DWGs exhibit significantly lower losses for propagating high-frequency radio waves, leading to larger communication distances. DWGs can be made from cheap plastic, making them more cost-effective than copper channels.

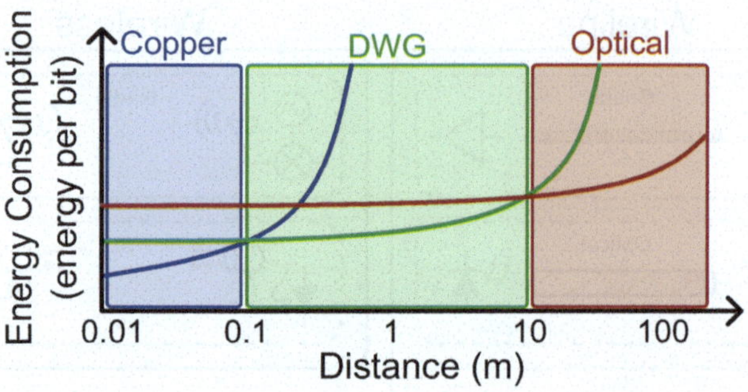

Fig. 1.4 Comparison of wireline communication technologies (copper, DWG, and optical) over link distance

DWGs are also resistant to EMI and weigh significantly less than copper channels. Compared to optical links, DWGs require lower mechanical tolerances and are more robust against misalignment. As DWG communication is compatible with the radio frequency (RF) circuits used for wireless radio communication, this technology will be further examined in this book.

Figure 1.4 compares wireline communication technologies in terms of energy consumption per bit across link distance. For short distances, copper links are the most efficient due to their simplicity. However, as the communication range increases, they quickly become more power hungry due to the lossy channel. For long distances, optical links are the best solution thanks to the low loss characteristics of optical fibers. However, the high overhead in power makes them less effective for short distances. For medium distances, DWG communication emerges as the optimal technology, providing a balanced solution between copper and optical links.

1.1.2.2 Wireless Communication

The second category is wireless communication, which is defined as the transfer of information between two points without the use of a guided medium, typically employing antennas to transmit and receive radio waves. Wireless networks are often further divided into two types: Wireless Local Access Networks or WLANs and cellular networks. While WLANs provide wireless connectivity over a fixed, smaller area, cellular or mobile networks offer coverage over a larger area with high user mobility. Wireless networks enable large area connectivity without the need for physical cables or connections. Although wireless links may lack in reliability due to varying operating conditions, they gain in convenience and cost-effectiveness over wireline links since installing cables is often expensive and impractical. An in-depth analysis of the specific properties and challenges associated with wireless radio communication will be presented later on.

To complete the discussion on wireless communication, it is important to mention free-space optical (FSO) communication, which uses electromagnetic waves at optical frequencies to wirelessly transmit data through free space, rather than through an optical fiber. FSO is a useful technology when high-speed data transmission is required, but physical connections are impractical or impossible to implement. However, FSO links are heavily influenced by atmospheric factors, such as rain, fog, dust, and heat, as optical wavelengths are more easily absorbed compared to radio waves. This makes FSO links less reliable compared to other wireless communication methods, but they can be useful for certain niche applications, such as communication in space where there is no atmosphere to absorb the signal.

1.1.3 Recent Trends in the World of Telecommunications

Figure 1.5 demonstrates the data rate trend in wireline networks, WLANs, and cellular networks, showcasing the exponential growth observed in Edholm's law. The most notable trends in each of these categories will now be discussed.

1.1.3.1 Wireline Communication

Wireline communication links play a critical role in many different application domains, providing highly reliable and high-speed data transfer capabilities over

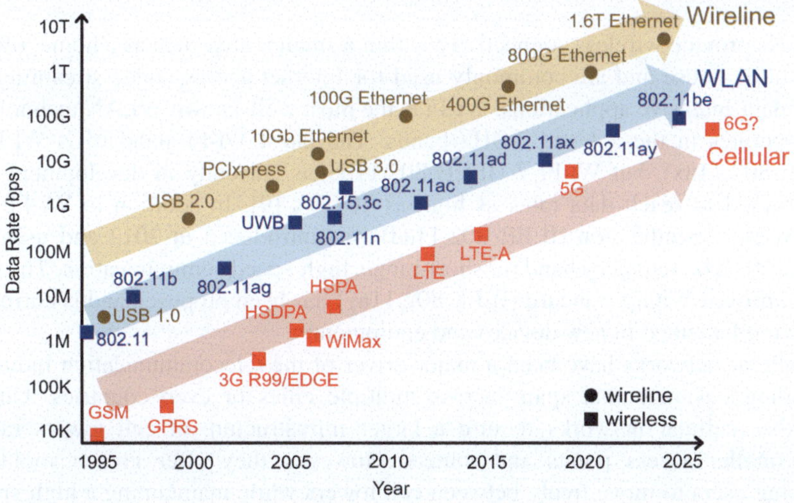

Fig. 1.5 Data rate trends in wireline networks, wireless local area networks (WLAN) and cellular networks (figure expanded from [8])

a large range of distances. Local area networks or LANs are among the most notable applications, connecting multiple devices within a relatively small area, with Ethernet being the most prominent standard. For instance, a house or an office may have multiple devices like computers, laptops, and other smart devices connected in order to communicate with each other and share resources such as printers, storage devices, and Internet connectivity. Such interconnects can be realized using copper-based wireline links with speeds up to 10 Gb/s. In contrast, data centers use high-performance LANs that employ a variety of copper and optical links to connect servers and storage systems, supporting speeds up to 400 Gb/s and distances up to 100 m. Ethernet standards for LANs are continuously evolving, with 400 Gb/s Ethernet already established and further developments underway for 800 Gb/s and even 1.6 Tb/s Ethernet standards.

Wireline links are not limited to short distances and are also utilized to connect buildings within a city or for long-haul networks that span thousands of kilometers. Fiber-optic links are typically used for these longer distances, and the development of advanced technologies like wavelength division multiplexing (WDM), dense coherent optical communication, and various optical network components have enabled data rates up to 400 Gb/s over these extensive distances.

Finally, RF DWG communication has gained a lot of popularity recently and is improving rapidly as this technology can be a great complement to short-range copper links and long-range optical links. As will be discussed later, RF DWG communication has shown to be highly efficient for short-to-medium distances from 1 to 10 m already achieving data rates up to 100 Gb/s.

1.1.3.2 Wireless Communication

WLANs provide wireless connectivity within a smaller area such as a home, office, and coffee house and are commonly used for Internet access, video streaming, or other data-intensive applications. Wi-Fi is the most well-known WLAN technology and operates in the 2.4 and 5 GHz bands. The latest Wi-Fi standard is Wi-Fi 6 (IEEE 802.11ax), but Wi-Fi 7 (IEEE 802.11be) is currently in development and is expected to reach data rates as high as 40 Gb/s [9]. In addition to Wi-Fi, the first WiGig specification (IEEE 802.11ad) was introduced in 2012 and uses the higher 60 GHz frequency band for short-range, high-speed communication. The new and improved WiGig standard (IEEE 802.11ay) has been proposed and is currently being implemented in new devices and equipment.

Cellular networks have been a major driver of the telecommunication industry, providing coverage that spans across multiple cities or even countries. Unlike WLANs, cellular networks require a larger infrastructure of cell towers rather than smaller access points and routers. However, they offer greater mobility, allowing users to move freely between cell towers while maintaining a high-speed connection. When first introduced in the 1980s, mobile networks only supported analog voice transmission in its first generation (1G). However, the advent of digital communication in 2G and mobile Internet in 3G gave rise to mobile

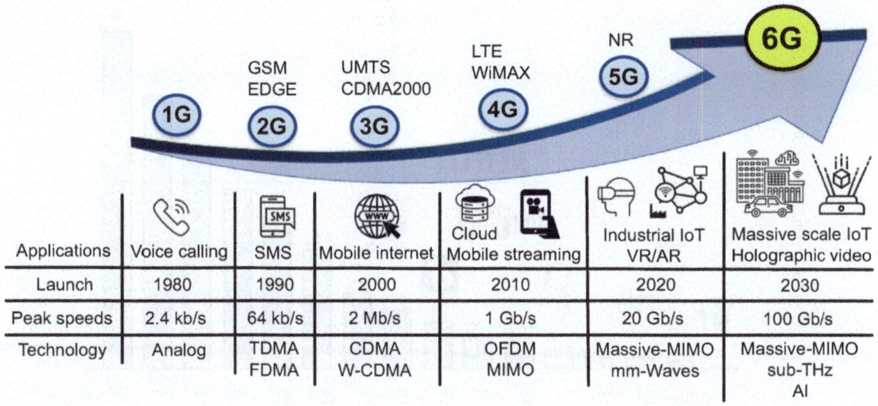

Fig. 1.6 Evolution of cellular telecommunication networks in the last several decades

global connectivity and transformed cell phones into smartphones. The significant innovations introduced in 4G made high-definition video streaming possible and brought cloud technology to the forefront. Currently, 5G cellular networks are being deployed worldwide, supporting data rates up to 20 Gb/s in optimal conditions. This will fuel the growth of data-intensive applications such as UHD video streaming, online gaming, virtual/augmented reality (VR/AR), and industrial Internet of Things (IoT). Research efforts have already begun for the next (sixth) generation of cellular networks (6G), which could potentially support data rates of hundreds of Gbps and enable futuristic applications such as holographic video conferencing and truly immersive VR/AR experiences. The evolution of mobile networks is illustrated in Fig. 1.6 [10].

In addition to WLANs and cellular networks, ultra-wideband (UWB) and Bluetooth (BT) are two other wireless technologies commonly used for short-range communication purposes. These technologies offer additional functionality, such as location tracking, device pairing, and data transfer between devices. Due to their limited range compared to mobile networks, they can be considered as WLAN technologies.

1.1.4 The Never-Ending Need for Speed and Efficiency

The rapid advancements in telecommunication and computing technologies have given rise to numerous emerging data-intensive applications, resulting in an ever-increasing amount of data traffic in our highly connected world. Over the past decade, the number of Internet users worldwide has more than doubled, and global Internet traffic has expanded by a factor of 20 [11]. This exponential growth in data

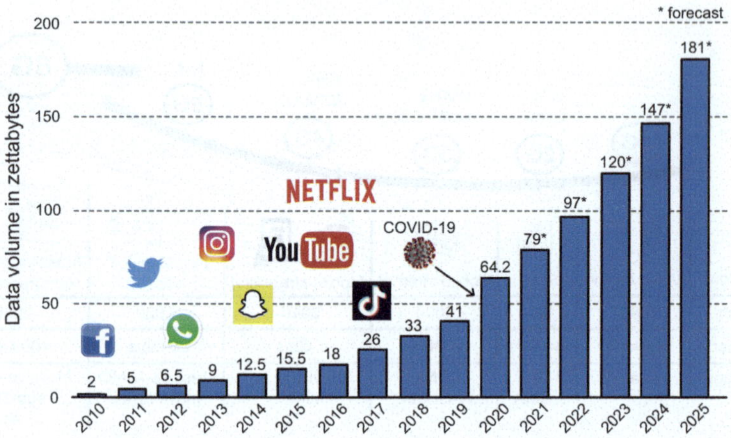

Fig. 1.7 Volume of data created, captured, copied, and consumed worldwide from 2010 to 2020, with forecasts from 2021 to 2025 [12]

creation is vividly illustrated by Fig. 1.7. According to projections, the world will consume a staggering 181 zettabytes[1] of data in 2025.

In order to meet the demands of these future data-hungry applications, continuous advancements toward higher capacity communication networks are essential. A great deal of effort is being devoted to the development of faster Ethernet standards, such as 800G and 1.6T Ethernet, to support increased wireline connectivity, particularly in data centers that have to store, process, and transmit all of the data generated by today's digital economy. To ensure that all mobile users can stay connected and utilize their data-intensive applications, cellular networks must continue to evolve as well. As a result, significant resources are being dedicated to the deployment of emerging 5G networks and research into future 6G network technologies, with the goal of enabling global broadband, low-latency connectivity.

While speed is undoubtedly a crucial factor in developing the next generation of telecommunication technology, it is not the only one. The International Energy Agency estimates that data centers and data transmission networks each consumed approximately 1–1.5% of the world's electricity (equivalent to around 250–350 TWh) in 2021 [11]. Even with the tremendous yearly growth in data usage, rapid improvements in energy efficiency have helped to moderate this increase in energy demand. Reducing the energy consumption further and minimizing the environmental impact of the information industry are crucial for the sustainable development of our world. As such, energy efficiency (often expressed in energy per bit) remains a critical specification in designing future telecommunication systems and is an important theme emphasized throughout this book.

[1] 1 zettabyte = 10^{21} bytes or 1 billion terabytes.

Finally, it is important to note that financial considerations play a crucial role in driving innovation. In addition to energy efficiency, the cost efficiency ($ per bit) of the complete system is a top priority for semiconductor industries. This involves carefully selecting the most appropriate technology, packaging, system architecture, and many other details that will be discussed in depth throughout this book.

1.2 The THz and Sub-THz Spectrum

1.2.1 Overview of the Electromagnetic Spectrum

Terahertz and millimeter waves are part of the electromagnetic (EM) spectrum which is a range of different types of EM radiation. Figure 1.8 provides an overview of the EM spectrum, highlighting the different types of radiation and their most notable applications. Each type of radiation has a distinct wavelength and frequency, which are related according to the following equation:

$$c = \lambda \cdot f \ , \tag{1.1}$$

where c is the speed of light (equal to approximately 300,000 km/s in vacuum), λ the wavelength, and f the frequency of the EM wave. Different wavelengths or frequencies are employed for various purposes, and as a result, the EM spectrum can be categorized into various regions. The boundaries between certain regions are often somewhat arbitrary and can vary depending on the particular field of study.

Radio waves (<300 MHz) have the longest wavelength in the spectrum (>1 m) and are mostly used for radio and television broadcasting. Higher frequency microwaves[2] (300 MHz–30 GHz) are extensively used in cellular and satellite

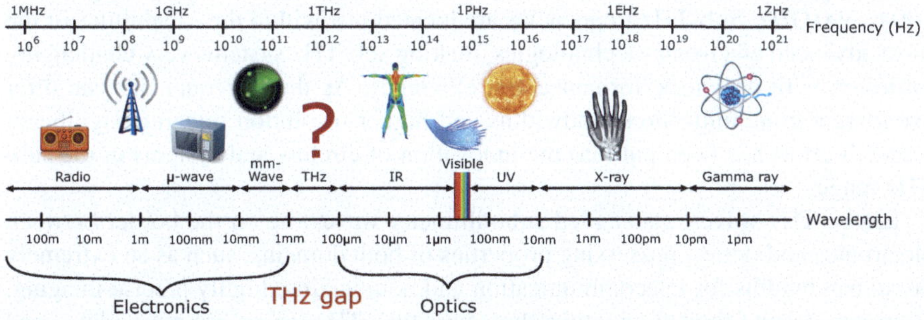

Fig. 1.8 Overview of the electromagnetic spectrum with the most important bands and applications

[2] Interesting fact: The prefix micro- is not meant to suggest a wavelength in the micrometer range, but rather that the wavelengths are "small" (i.e. shorter) compared to radio waves.

telecommunications, as well as for microwave heating. Microwaves are also commonly used for radar and imaging applications. Millimeter-wave (mm-Wave) and terahertz (THz) radiation falls between 30 and 300 GHz and between 300 GHz and 3 THz, respectively. These relatively unexplored regions are used for advanced radar, imaging, and telecommunication applications, as will be discussed later. The infrared (IR) region (wavelengths between 700 and 1 mm) is primarily used for heating (e.g., IR heaters and ovens) and remote sensing (e.g., thermal imaging and weather satellites). The visible light spectrum is the range of wavelengths (between 400 and 700 nm) that we can see with our eyes and is used for illumination and visual communication. Ultraviolet (UV) radiation (wavelengths between 10 and 400 nm) is used for sterilization, fluorescent lighting, and, of course, tanning. Lastly, X-ray and gamma ray radiation occupies the region with the shortest wavelengths and is utilized for medical imaging and treatment. Even smaller wavelengths are used in the realm of high-energy particle physics but are not included in this discussion. This book will mainly focus on two regions: the mm-Wave and THz regions.

1.2.2 THz, Sub-THz, and mm-Wave

The mm-Wave band, which derives its name from its wavelength in the millimeter range (1–10 mm), has gained significant attention recently due to its potential for high-speed communication (thanks to the large available bandwidth) and high-resolution imaging and sensing (mostly due to the short wavelength). Thanks to the continuous scaling of IC technologies, transistors have become progressively faster and can now provide sufficient gain in the lower end of the mm-Wave range, enabling commercial mm-Wave products at high volumes, such as 77 GHz automotive radar.

The higher end of the mm-Wave band, also called the sub-THz band, ranges from 100 to 300 GHz. Sub-THz frequencies are located just within the capabilities of the most advanced electronics technologies, making sub-THz systems very challenging to design with decent performance and efficiency. As these frequencies can offer even larger communication bandwidths and higher resolution imaging, significant research effort has been put into the innovation of circuits and systems in the sub-THz range.

Lastly, THz waves, also called submillimeter waves, lie on the border between electronics and optics, possessing properties of both domains, such as an extremely broad bandwidths for telecommunication and nonionizing, highly precise imaging. However, despite their fascinating characteristics, THz waves are not widely used due to a lack of available THz sources and detectors. Specifically, THz waves are too high in frequency for electronics and too low in frequency (or too long in wavelength) for optical components. As a result, the range where THz power can be insufficiently generated or detected is referred to as the "THz gap." THz systems are therefore practically nonexistent in commercial products and are only found in high-end laboratories as expensive, bulky, and fragile instruments. Nonetheless, recent research aims to overcome this THz gap to exploit the unique properties of THz radiation.

1.3 Telecommunication at (Sub-)THz Frequencies

Communicating at frequencies above 100 GHz offers some very promising opportunities, as well as several inevitable challenges. Despite the vast amount of available bandwidth and the potential for a highly compact system size, designing efficient (sub-)THz circuits remains a daunting challenge due to the high propagation losses and inadequate performance of active devices. This section discusses these benefits and challenges and their potential impact on various applications.

1.3.1 Benefits

Communication systems above 100 GHz can benefit from the large available bandwidths and scalability of passive components to realize extremely high data rates with a compact form factor.

1.3.1.1 Modulation Bandwidth

The main motivation to operate at higher frequencies is to take advantage of the large absolute bandwidths available at those frequencies. To better understand why, it is important to introduce an equation formulated by Claude Shannon [13]. This equation, which will be discussed more extensively in the next chapter, essentially defines the theoretical upper limit of the data rate that can be achieved (also known as the capacity C) over a communication channel:

$$C = \mathrm{BW} \cdot \log_2(1 + \mathrm{SNR}) \,, \tag{1.2}$$

where BW denotes the bandwidth of the channel and SNR the available signal-to-noise ratio over that channel, expressed as follows:

$$\mathrm{SNR} = \frac{P_{\mathrm{signal}}}{P_{\mathrm{noise}}} \,. \tag{1.3}$$

The signal-to-noise ratio is a measure of the signal quality and can be enhanced by increasing the signal power P_{signal} and/or by reducing the noise power P_{noise}. However, there are limitations to both methods. The maximum signal power that can be transmitted is strictly restricted by regulations for health and safety reasons and by technological constraints. Additionally, noise is an inevitable and persistent aspect of the physical world and can only be reduced to a certain extent. Furthermore, since the relationship between SNR and capacity is logarithmic, there are only diminishing returns to be gained from these limited improvements in SNR.

A more popular strategy to improve the capacity of communication channels is to focus on increasing the available bandwidth, which is directly proportional to the capacity and thus offers more opportunities for improvement. This drive for more bandwidth is a key motivation for exploring telecommunication at higher frequencies, such as the THz and sub-THz bands. There are two main reasons why higher carrier frequencies can provide larger bandwidths.

Firstly, the usage of the different bands in the electromagnetic spectrum is strictly regulated to avoid interference between multiple users and is coordinated at an international level by the International Telecommunication Union (ITU). Not only is the allocation of each frequency band strictly determined, the transmitted signal must also comply with a detailed frequency emission specification within and around the band, which is called the spectral mask. At lower frequencies, the spectrum is very crowded with shared applications such as mobile, satellite, meteorological, maritime, and military, which leaves only limited stretches of bandwidth available for high-speed wireless communication. Current cellular and Wi-Fi networks, which use sub-6 GHz bands, only have channel bandwidths of 10s–100s of MHz. Although new 5G bands have opened in the lower mm-Wave spectrum, supporting larger bandwidths up to several GHz, significant amounts of bandwidth become available above 100 GHz. Recently, the ITU has allocated tens of GHz toward fixed and mobile wireless communication from 92 GHz up to 275 GHz [14], as visualized in Fig. 1.9. Particular interest has been given to the available spectrum within the W-band (92–114.25 GHz) and D-band (130–174.8 GHz) by ETSI [15, 16] and ECC [17, 18] for fixed service broadband systems. Above 275 GHz, the spectrum is yet to be allocated, presenting great potential for ultra-high speed links at THz frequencies. The increased attention to this band led to the introduction of the IEEE802.15.3d standard that focuses on frequencies from 252.72 to 321.84 GHz [19].

As the operation frequency increases, the dangers of interference become less problematic due to larger propagation loss and higher directivity of the links. Nevertheless, all regulations set by the ITU must be adhered to rigorously. After all, even though a large amount of spectrum has been allocated toward high-

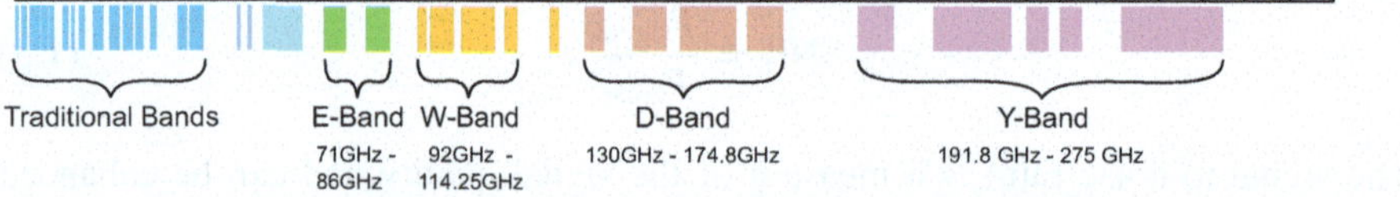

Fig. 1.9 Overview of the different frequency bands allocated to fixed and mobile wireless communication by the ITU [14]

capacity wireless communication, the fragmentation of these bands (see Fig. 1.9) limits the maximum continuous bandwidth to 12.5 GHz (from 151.5 to 164 GHz). As a result, commercial high-capacity wireless links will have to be implemented using multiple bands rather than one large, continuous band. To avoid interference between neighboring bands and limit the disturbance of sensitive applications operating within these bands, the frequency allocations remain crucial in this part of the spectrum. Note that the spectrum regulations do not apply to guided DWG communication at these frequencies, as signal emission and thus interference is very limited outside of the waveguide.

Second, in addition to the increased available bandwidth at higher frequencies, achieving the necessary bandwidths is easier for circuits. After all, the same fractional bandwidth, which generally dictates the design complexity and efficiency, leads to more absolute bandwidth as the frequency increases. For resonant radio frequency (RF) circuits, this fractional bandwidth is typically represented by the quality factor (Q) of the resonator, resulting in the following linear relationship between bandwidth (BW) and carrier frequency (f_c):

$$\mathrm{BW} = \frac{f_c}{Q} . \tag{1.4}$$

For instance, a circuit with 10% fractional bandwidth at 10 GHz has an absolute bandwidth of only 1 GHz, while a similar design at 100 GHz produces 10 GHz of absolute bandwidth. This explains the motivation behind exploring higher frequencies, which offer large bandwidths for high-speed links.

1.3.1.2 Form Factor

Going to higher frequencies also offers the benefit of decreasing the form factor of realizable (sub-)THz systems, which can be explained by two observations. First, all electromagnetic structures that scale with the wavelength, such as antennas and transmission lines, will shrink as the frequency increases since the wavelength is inversely proportional to the frequency (as shown in Eq. 1.1).

Second, most RF circuits are based on resonator circuits that resonate at or around the desired operation frequency. To resonate at a carrier frequency ($\omega_c = 2\pi f_c$), the inductance (L) and capacitance (C) of the resonator need to satisfy the following relation:

$$\omega_c = \frac{1}{\sqrt{L \cdot C}} . \tag{1.5}$$

While resonator-based circuits will be more elaborately discussed later on, one can already observe that for a fixed capacitance (often determined by the size of the transistor components) the necessary inductance lowers as the frequency increases. Since most of the chip area is often consummated by inductor coils to attain the desired inductances, the footprint of the chip decreases substantially as the operation frequency increases. Both of these observations are valid not only for on-chip passive components (antennas, inductors, and transmission lines) but for off-chip components as well (like filters or antennas on a printed circuit board). The possible compactness of (sub-)THz solutions opens the door to a new range of applications, as will be discussed later.

1.3.2 Challenges

Next to the promising benefits, working at very high frequencies unfortunately comes with numerous drawbacks. The biggest bottleneck of (sub-)THz systems is the limited signal-to-noise ratio (SNR) which is caused by multiple factors.

1.3.2.1 Propagation Loss

The first reason for the low system SNR in (sub-)THz systems is the significant propagation loss at higher frequencies. The primary factor responsible for this is the so-called free-space path loss (FSPL), which refers to the signal attenuation of a wireless link in free space (without obstacles) with omnidirectional antennas.[3] The FSPL is not a real "loss" in the sense that part of the signal power gets attenuated during transmission. Rather, since the transmitter radiates power in all directions, the power density of the electromagnetic wave reduces as it propagates further away from the transmitter. At the receiver, only a small part of this power is captured by the antenna. The amount of received power scales with the area of the antenna, which is proportional to the wavelength and inversely proportional to the frequency. The FSPL thus increases with increasing distances (d) and higher frequencies (f), as demonstrated by the expression of the FSPL:

$$\text{FSPL} = \left(\frac{4\pi d}{\lambda}\right)^2 = \left(\frac{4\pi d f}{c}\right)^2. \tag{1.6}$$

Figure 1.10 provides a visualization of the propagation loss caused by the FSPL in a wireless, or over-the-air (OTA), link for various frequencies over distance. As shown in the figure, the FSPL increases rapidly with distance for all frequencies. However, at higher frequencies, the loss is significantly higher compared to lower frequencies. For instance, a 10 m wireless link at 100 GHz suffers from 40 dB more loss than at 1 GHz. Consequently, higher frequency wireless communication links are limited in range compared to their lower frequency counterparts due to the large propagation loss.

[3] An omnidirectional antenna is an antenna that radiates the same in all directions.

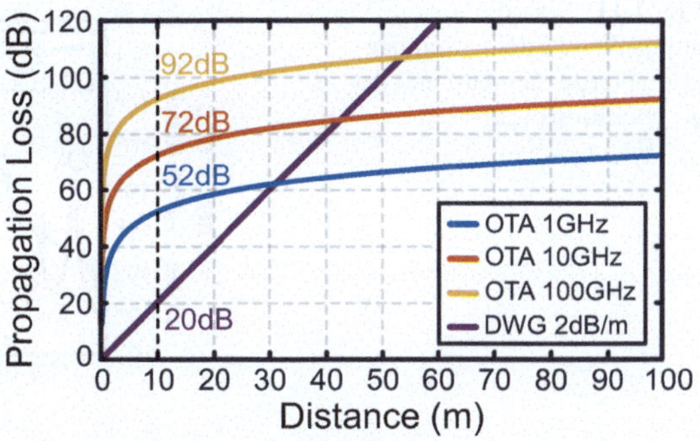

Fig. 1.10 Propagation loss versus distance of wireless (OTA) links at 1 GHz, 10 GHz and 100 GHz, as well as of a DWG link with a loss of 2 dB/m

One potential solution to mitigate the issue of propagation loss at higher frequencies is to utilize more directive antennas, e.g., by constructing a large array of antenna elements. This approach is especially attractive in the (sub-)THz frequency range, where the size of a single antenna element is relatively small. By using antenna arrays, it is possible to construct large directive antennas with a limited total area, which can help to increase the communication range of (sub-)THz wireless links. However, using antenna arrays also comes with its own set of challenges and opportunities, which will be discussed in more detail later on.

Another approach to mitigate the power loss issue in free space is to confine the signal within a waveguide. By transmitting a (sub-)THz signal through a dielectric waveguide (DWG), the power loss increases exponentially with the waveguide length. The theoretical principles of DWG propagation will be discussed later, but Fig. 1.10 shows that a DWG with a loss of 2 dB/m presents a lower propagation loss than wireless communication over a distance of tens of meters. Although a DWG transmission lacks the flexibility of wireless operation, it enables the utilization of the vast bandwidth available at high frequencies for wireline networks.

It is worth noting that wireless links on the Earth propagate through the Earth's atmosphere, which contains various gas molecules (primarily oxygen and water vapor) that can cause absorption resonance lines at certain frequencies. When electromagnetic waves propagate at these frequencies, they are partly absorbed by the molecules and thus attenuated. The width and height of the absorption lines depend on atmospheric properties such as temperature, air pressure, and water vapor density. To ensure that excessive signal loss does not occur during transmission, it is important to avoid the absorption peaks caused by oxygen (O_2) and water vapor (H_2O). Figure 1.11 illustrates the specific attenuation of the atmosphere per kilometer over frequency at sea-level pressure, 15 °C, and a water vapor density of 7.5 g/m^3 (standard) and zero density (dry air) [20]. Although atmospheric attenuation can cause signal loss, it is not a significant concern for most (sub-)THz wireless links since they typically cover distances below 10 m. The free-space path loss is usually the main factor that leads to low signal-to-noise ratios.

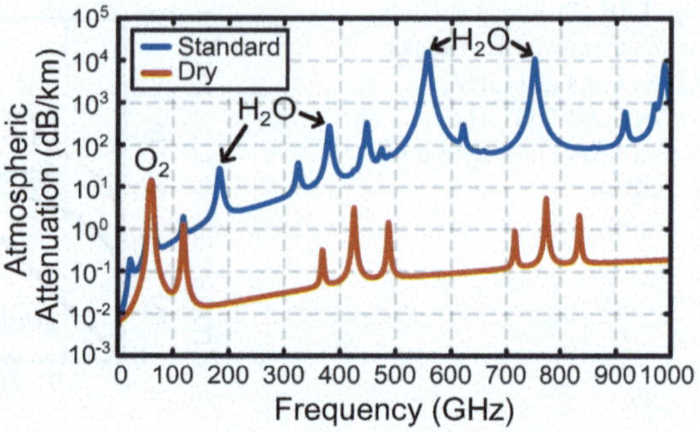

Fig. 1.11 Specific atmospheric attenuation per kilometer in standard and dry air conditions [20]

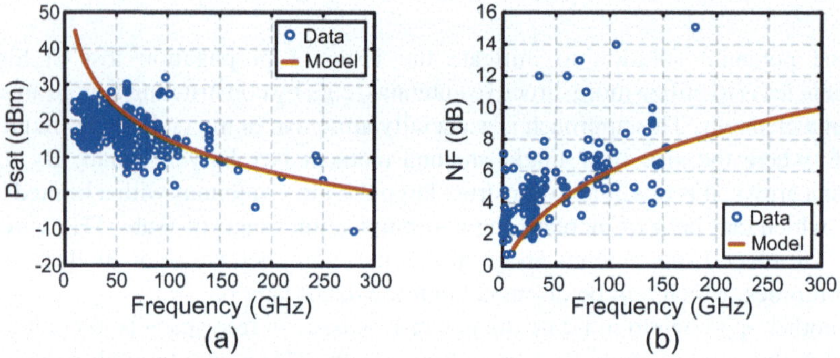

Fig. 1.12 Performance of state-of-the-art mm-Wave CMOS transmitters (output power) [21] (**a**) and receivers (noise figure) [22] (**b**)

1.3.2.2 Transmitter and Receiver Performance

Another reason for low SNR values above 100 GHz is the performance of the transmitter and receiver chips, which is impacted mainly by the behavior of the transistors. After all, transistors play a critical role in amplifying the signal to generate high output power for the transmitter and minimizing the noise in the receiver. Despite the advancements in scaling that have enabled the operation of transistors in the mm-Wave spectrum, even in CMOS technology, their performance, such as power gain, deteriorates rapidly beyond 100 GHz. This trend is clearly depicted in Fig. 1.12, which illustrates the frequency-dependent performance of state-of-the-art mm-Wave transmitters [21] and receivers [22] in CMOS.

For the transmitter, the most important specification influencing the SNR is the output power. Figure 1.12a shows the saturated (maximal) output power (P_{sat}) of mm-Wave transmitters over frequency, indicating a steep decrease over frequency. Particularly in the THz range, transistors will not be able to provide gain at all, leading to an even more disastrous drop in output power.

The receiver SNR performance is typically captured using its noise figure. The noise figure or NF quantifies the amount of noise the circuits will add to the signal.

As this number steadily increases with frequency, illustrated by Fig. 1.12b, more noise will inevitably be added to the signal as the frequency is increased.

It is obvious that the decrease in signal power and increase in circuit noise will seriously degrade the signal-to-noise ratio and thus the signal quality as the carrier frequency rises above 100 GHz, especially when approaching the THz range. Why this happens will be explained extensively later on, but it is clear that this puts a big limit on the attainable communication range at high frequencies. In order to attain optimal system performance and energy efficiency despite these limitations, innovative circuit design techniques will be necessary.

1.3.3 Applications

1.3.3.1 Considerations for (Sub-)THz Telecommunication

As previously discussed, telecommunication systems operating at frequencies above 100 GHz present a potential for achieving high data rates in a compact form factor; however, the low SNR values limit their communication range. This trade-off between bandwidth and SNR becomes more pronounced as the frequency of operation increases. For instance, while sub-THz links at 100 GHz can deliver very high data rates over a decent range, THz links operating at frequencies above 300 GHz may provide ultra-high data rates but only over very short distances. Bearing this trade-off in mind, a variety of different applications can be envisioned.

Wireless communication offers a range of applications, ranging from high-speed and compact links over short distances to highly directive links for high-speed point-to-point connections over longer distances. The directivity of a link can be improved by constructing a large array of antenna elements, which is feasible due to the small size of each antenna at high frequencies. However, the size of the array may be impractical for consumer applications, limiting high-directivity links to larger point-to-point networks, such as base stations. Such directive links have the added advantages of being more secure against eavesdropping and more resilient to interference due to the large attenuation of undesired signals. However, the narrow beam width associated with high-directivity antennas significantly complicates beam alignment and locking, especially when one of the antennas is in motion.

Communicating above 100 GHz also proves to be promising for wireline communication, as using a directive DWG medium avoids the large propagation losses associated with wireless links. Compared to wireline copper and optical links, (sub-)THz DWG links can provide high-speed communication over short-to-medium distances with high efficiencies. The most promising applications of (sub-)THz links in both wireless and wireline communication will now be discussed.

1.3.3.2 Wireless Communication

At distances below 10 cm, (sub-)THz links can provide high-density, high-speed wireless interconnects with a small footprint, making them a promising replacement for copper-based wireline links to transfer large amounts of data between chips within a large system or package, such as between a high-performance processor (think AI) and off-chip memory, which is often a bottleneck in high-performance computing [23].

Wireless (sub-)THz links could also serve as a wireless connector for data transfer between computers, storage devices, and other high-speed peripherals, replacing short USB cables or Ethernet connections. For instance, the Apple Watch series 7 supports 60.5 GHz wireless data transfer for internal firmware updates without the need for a physical cable or interface [24].

Moreover, data centers can also benefit from wireless (sub-)THz links by augmenting the primarily wired network with adaptable wireless links to handle large bursts in data traffic [25].

Sub-THz and THz links can also significantly enhance mobile networks (6G) by complementing existing infrastructure to form a more efficient and heterogeneous network. This means that older, reliable cellular networks (4G, 3G) with wide coverage will be combined with small, short-range, and high-speed cells, known as data kiosks or information showers, which can handle large data transfers to off-load the network. These broadband access points are strategically placed in areas with high human flow, such as train station gates, public building entrances, or shopping malls. Whenever a mobile device is in proximity to such a terminal, heavy and delay-tolerant data can be transferred in a matter of seconds, for example, software updates or high-quality multimedia content, like movies. By deploying small (sub-)THz cells, the macro-scale network (cellular or WLAN) can be significantly off-loaded, thereby making all global connectivity more efficient. However, it is important to note that these small cells will not replace the entire mobile network, but rather work in tandem with existing infrastructure [23, 26].

(Sub-)THz links can enable ultra-high-speed WLANs with medium-range (1–10 m) coverage, building upon the current 60 GHz WiGig standards to support futuristic applications such as ultra-high definition (UHD) multimedia streaming, virtual and augmented reality (VR/AR), and holographic video conferencing. For instance, to stream 8K UHD video content with a resolution of 7680 $\times$ 4320 pixels, 24-bit color depth, and a refresh rate of 120 frames per second (fps), a raw data rate of 96 Gb/s is required. Meanwhile, for truly immersive wireless VR/AR systems, data throughputs of several hundred Gbps and latencies below 10 ms are necessary to avoid motion sickness [27]. Looking even further ahead, the use of (sub-)THz links could enable the projection of realistic displays in holographic video conferencing, which will require Tbps of data [27].

For long distances ($>$100 m), directive (sub-)THz links can be used for high-speed point-to-point telecommunication. Due to the large propagation losses, large antenna arrays are necessary to achieve the desired antenna directivity, making these

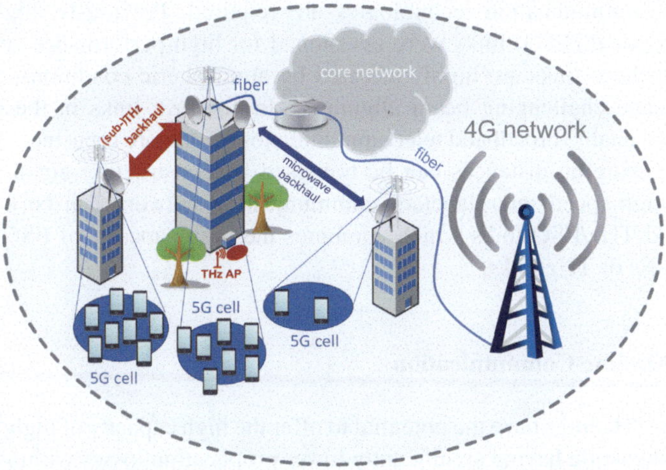

Fig. 1.13 Conceptual diagram of a heterogeneous network with various types of cells

systems impractical for consumer products but suitable for larger systems such as backhaul networks. As the volume of cellular data traffic continues to grow, not only does the data volume between mobile devices and base stations increase but also between different cells in the network and the data centers. Since the data throughput of many devices is combined into backhaul links, the capacity of these links must be several times higher than the data rates of individual users to ensure reliable and timely data delivery from/to multiple users. To achieve these extremely high data speeds over long distances, directive point-to-point (sub-)THz links are the perfect candidate. Recent advancements show that backhaul links beyond 100 GHz, such as the W- and D-bands, will be able to support capacities of up to 100 Gb/s over several kilometers [28].

The diagram in Fig. 1.13 illustrates a heterogeneous network that incorporates various types of cells to provide connectivity. At the center of this network is a low-frequency cellular tower (4G, 3G) that covers a vast area and is linked to the core global network. This large cell contains smaller cells that offer higher speed network access, such as 5G cells with more limited coverage and (sub-)THz access points (APs), commonly referred to as data kiosks, which provide an information shower to users within a specific area, like the entrance to a public building. These high-speed cells are linked to the core network via backhaul links, which can be either microwave links or (sub)-THz links for higher capacity, as well as long-distance optical fiber links, allowing them to off-load the macro-scale cellular network.

Finally, the exploration of (sub-)THz links for wireless communication in space is gaining attention due to the increasing demand for space-related explorations in the coming decades [29]. In order to support services like video feeds, high-resolution scientific data streaming, and virtual reality for controlling rovers and other machines, real-time links with higher capacity than can be handled by current microwave communication technologies are required. Previously, high-capacity free-space optical (FSO) links were envisioned for high-performance space applications, but these links are highly affected by atmospheric conditions, are costly, and experience challenging beam alignment. In contrast, links in the (sub-)THz spectrum can enable broadband telecommunication with more robustness to weather effects, and extreme distances can be traversed by constructing gigantic antenna arrays. A high-speed interplanetary communication network can be constructed using hybrid THz/FSO links which combines the high capacity of FSO links and the robustness of THz links.

1.3.3.3 Wireline Communication

DWG (sub-)THz links have the potential to offer the high capacity of high-frequency wireless links while having significantly lower propagation losses within a range of tens of meters. Unlike wireless links, DWG communication is a wireline technology, which limits its flexibility. However, for many applications that do not require this flexibility, DWG links can provide extensive benefits.

At the smallest scale, DWGs can be implemented as silicon ribbons, which are ideal for chip-to-chip interconnects or high-speed communication over a backplane, offering high-speed, high-density connectivity over very short distances [30]. These tiny DWGs can be employed as a highly broadband and miniaturized interconnect solution for chip-to-chip communication at THz frequencies, as discussed earlier for wireless chip-to-chip links.

DWG links also offer a promising alternative to short-range copper and optical interconnects in data centers for intra-rack (within racks) and inter-rack (between racks) communication, covering distances of 1–10 m. As discussed earlier, copper-based links become increasingly power consuming at higher speeds over longer distances, while optical links are not cost-effective for such short ranges. Moreover, aligning the fiber with the optical laser (transmitter) and photodiode (receiver) can be challenging due to the fiber's tiny cross-section and the optical signal's short wavelength, making it highly sensitive to misalignment. In contrast, DWG links offer the desired speeds while being more energy and cost-efficient in this range, with greater robustness to fiber misalignment and lower link latency.

Finally, with the advent of fully self-driving cars, the need for robust and high-speed interconnects has grown significantly. These autonomous vehicles use a multitude of sensors, such as lidar, radar, ultrasound, and high-resolution cameras to detect traffic lights, road signs, and other vehicles or pedestrians which produce enormous amounts of raw data—an estimated 19 terabytes per hour for a fully autonomous vehicle [31]. This data must be transferred to the central processor in real time to manipulate steering, acceleration, and braking of the car. DWG links can be an energy-efficient and effective solution for high-speed data transfer within the car with minimal latency. Additionally, DWG channels are cost-effective compared to copper, insensitive to EMI,[4] and robust against mechanical misalignment caused by vibrations. Furthermore, they present little additional weight to the car, which is particularly valuable considering the current wiring harnesses of vehicles can weigh up to 70 kg [32]. Reducing this weight can lead to significant cost savings in fuel or battery life.

1.4 Why CMOS Electronics?

As previously mentioned, CMOS technology has rapidly progressed at an exponential pace, driven by significant innovation and economic incentives. Since the introduction of the first MOSFET in 1959, the feature length of individual MOSFETs has scaled down at an incredible rate, reaching a size of just 5 nm[5] in 2019.

In the process, the classical MOSFET structure has undergone significant innovation to combat short-channel effects and improve gate control, resulting in the introduction of 3D transistor structures such as the FinFET and more recently the Gate All Around (GAA) structure. Although the gains in logic gate density from scaling have become increasingly marginal, the ability to build progressively more complex systems has maintained high levels of interest and investment from industry. The remarkable growth of the semiconductor industry is largely driven by the continuous push for more advanced technology nodes to improve the performance of ICs for digital systems (e.g., processor, memory, and machine learning accelerators), which generate the majority of revenue.

The potential advantages of utilizing CMOS technology for telecommunication circuits are numerous. Given that cost is a significant driving force in our capitalistic world, CMOS offers the lowest cost per chip at mass production with the highest attainable yields. Additionally, integrating high-frequency RF circuits with digital logic on a single die enables communication systems to benefit greatly from high-performance digital signal processing and high-density memories, resulting in a significant reduction in costs and improved energy efficiency.

In the early 2000s, transistor scaling enabled GHz frequency operation, which led to the rise of RF CMOS and a boom in the mobile phone market. However, as scaling continued into the realm of nanoscale CMOS, the benefits for digital systems persisted, while RF performance began to suffer. Firstly, the maximal operation frequency of transistors has saturated just below 300 GHz, resulting in poor gain

[4] EMI or electromagnetic interference is the disturbance of a system by external interfering EM signals. Copper wires act as antennas picking up noise and undesired signals in its environment, while DWGs remain largely undisturbed.

[5] Actually, the feature size of the most advanced technology nodes is just a marketing term and does not refer to the physical dimension of the devices. For example, the smallest transistors in a 5 nm technology actually have a minimal size around 20–30 nm, depending on the manufacturer.

performance in the sub-THz range and nonexistent gain at THz frequencies. Additionally, lowering the supply voltage to save power in digital circuits has reduced output power and dynamic range in RF circuits. Furthermore, dense metal stacks with generally thinner metal layers have been implemented to accommodate more complicated routing in digital systems but have been detrimental to RF performance. Lastly, with progressively scaled and dense systems, designing, modeling, and verifying circuits in these advanced CMOS technologies have become much more complex, especially at higher frequencies. Given these persistent challenges, why should we continue to pursue these systems in CMOS?

After all, there exist other technologies that are better suited for high-performance (sub-)THz circuits than CMOS, which is mostly optimized for digital performance.

For instance, III/V semiconductors such as InP and GaAs offer higher electron mobility, a higher transistor breakdown voltage, thicker metal layers, and higher substrate resistivity than CMOS. These advantages translate into higher operation frequencies, more gain, higher output power, and lower losses in passive components such as transmission lines, inductors, and antennas. However, compared to CMOS, these III/V technologies have much lower yields and a higher cost per chip and are incapable of realizing low-power digital circuits.

BiCMOS technologies, on the other hand, provide a good middle ground between CMOS and III/V technologies. BiCMOS combines bipolar transistors, typically SiGe, with a standard CMOS process to enjoy the benefits of both technologies. SiGe bipolar transistors can operate at higher frequencies, provide more gain, and can deliver slightly more output power than MOSFETs while being compatible with a standard CMOS manufacturing process. Although the SiGe transistors are inferior to III/V transistors (lower f_{max} and supply voltage) and the digital performance is not as advanced as the most advanced CMOS process, BiCMOS technologies may provide a good balance between cost, yield, and performance.

The optimal technology for a given application ultimately depends on the specific requirements of that application. For low-volume RF modules that prioritize performance over cost and efficiency, III/V technologies may be the best choice. These might include systems for specific military or space applications where neither cost nor power consumption is a major concern. BiCMOS technologies, offering a middle ground between III/V and CMOS, are typically used for high-performance applications with moderate volumes, reasonable costs, and acceptable power consumption. Examples might include high-speed backhaul links or cellular base stations. For moderate performance applications where cost and energy efficiency are paramount, CMOS technologies are ideal. This includes consumer products such as mobile phones, computers, and IoT sensors or any other application that requires high-volume production at a low cost. Figure 1.14 provides an overview of the different application domains and the optimal technology based on module cost and production volume.

It is evident from the discussion that there is immense potential in using nanoscale CMOS to develop (sub-)THz systems for high-performance consumer applications. However, given the limitations of scaling in achieving the required

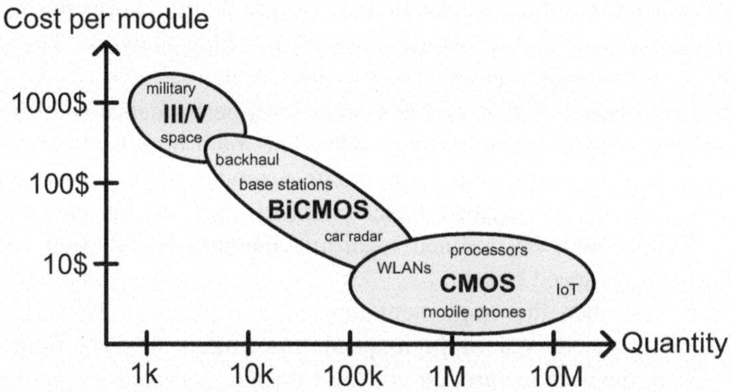

Fig. 1.14 Overview of the different application domains and associated optimal technologies

performance at very high frequencies, there is a need to develop new design techniques that can optimize system performance and efficiency within the constraints of CMOS technology. This noble pursuit forms the primary objective of this book.

1.5 Outline of the Book

This book focuses on the design of high-speed telecommunication systems at sub-THz and THz frequencies in CMOS technologies. The fundamental limitations and design trade-offs will be investigated on various abstraction levels: system level, architectural level, and circuit level. In addition to the general introduction given in this chapter, the following seven chapters dig progressively deeper into the topic and discuss actual implemented communication systems. A brief overview of each chapter is given below.

Chapter 2 discusses the most fundamental theoretical concepts in telecommunication systems. Starting with a brief review of important mathematical ideas, digital modulation is introduced with an overview of the most prevalent modulation techniques and implementations. Then, the general framework of a communication system is discussed along with the most critical factors that affect performance. Finally, the different modulation techniques are compared and the fundamental trade-offs revealed.

Chapter 3 transitions from mathematics toward physics, detailing the basics of electromagnetism and circuit theory to understand the behavior of electronic circuits. The most essential active and passive circuit components are introduced and their behavior is described.

Chapter 4 extends the component-level discussion from the previous chapter toward the design of high-frequency electronic circuits. It outlines the basic theory of impedance matching and stability and extensively details the behavior and design of the two primary building blocks of the circuits in this work: the neutralized pseudo-differential pair and transformer-based matching network. The design of several amplifier-based circuit blocks is also discussed.

Chapter 5 zooms out to analyze the system-level performance and fundamental trade-offs of mm-Wave communication links. The main high-level design aspects are qualitatively discussed, and a quantitative model of a generic mm-Wave communication system is described. Based on the results from this numerical model, several conclusions will be outlined regarding the maximum speed and energy performance of high-speed links.

Chapter 6 describes the implementation of a tripler-last 390 GHz BPOOK transmitter. First, the design of high-speed transmitters at THz frequencies is examined. Then, the architecture, circuit-level design, and measurement results of the this transmitter are outlined.

Chapter 7 presents a 135 GHz direct-digital 16-QAM wireless and dielectric waveguide link. First, the challenges and limitations of traditional transceiver architectures are discussed, and the direct-digital system architecture is proposed as a more energy-efficient solution. Next, the design of all circuit blocks in the 135 GHz system is extensively explained along with the prototype boards with implemented antennas toward the realization of a full communication link. Finally, the measurements are described, and the results are compared to the state-of-the-art.

Chapter 2
Fundamentals of Telecommunication

This chapter introduces important concepts in the world of telecommunication going from information theory to Fourier analyses and digitization. Digital signal modulation is discussed, and some fundamental properties are analyzed. Finally, the performance of communication systems is studied based on various metrics: noise, nonlinearity, and bandlimiting filtering, which are then used to compare the most promising modulation techniques.

2.1 Mathematics

2.1.1 Introduction to Information Theory

Information theory is the mathematical study of the quantification, storage, and communication of information [33]. It originated in the 1920s with the works of Harry Nyquist and Ralph Hartley and in the 1940s with important contributions from Claude Shannon. This section will give an overview of the most important concepts in information theory without too many unnecessary details.

2.1.1.1 What Is Information?

Intuitively, we can understand that information is related to the uncertainty or surprise that is experienced when a message is received. For example, the message "The pope has turned into a frog today." contains more information than "The sky is blue today.". Indeed, a key measure in information theory is entropy, which

C. D'heer, P. Reynaert, *THz and Sub-THz CMOS Electronics for High-Speed Telecommunication*, Analog Circuits and Signal Processing,
https://doi.org/10.1007/978-3-031-64439-9_2

quantifies the amount of uncertainty in the value of a random variable or process. The information or entropy of a random variable or symbol X can be expressed as [34]

$$H(X) = -\sum_{x \in X} p(x) \log_2 p(x) , \tag{2.1}$$

where $p(x)$ is the probability that the variable X takes on the value x. The unit of this particular expression of entropy is "bits," because it uses a logarithm of base 2, as introduced by Shannon and most commonly used today. Using a logarithm of base 10, for instance, yields the entropy in a unit of "nats," named after Hartley who first suggested using the logarithm measure for information in 1928, and thus deserves some credit [35].

Applying this concept to a fair coin toss with equal probabilities for both outcomes yields an entropy of 1 bit.[1] The roll of a fair die, with six equally probable outcomes, provides entropy of 2.6 bits, containing more information than a coin toss since there is more uncertainty in its outcome. It must be noted that the measure of information has nothing to do with the content or usefulness of the message. A completely random message contains a lot of information, even though it might not have any meaning to us. In order to speak of meaningful communication of information, other concepts are needed.

In essence, the fundamental problem of communication can be described as reproducing a message at one point (in space and time) selected at another point. To evaluate the amount of useful information in the received message (i.e., related to the transmitted message), the concept of mutual information can be useful. The mutual information quantifies how much information (in bits) one random variable contains about another random variable. Assuming Y is the received message and X the transmitted message, the mutual information $I(X;Y)$ can be expressed as follows [36]:

$$I(X;Y) = H(Y) - H(Y|X) , \tag{2.2}$$

where $H(Y)$ represents the entropy of the received message Y, while $H(Y|X)$ is the conditional entropy of the received message Y knowing what message X was originally transmitted. While $H(Y)$ contains all information of the output, $H(Y|X)$ only includes the nonmeaningful information (such as random noise). The mutual information $I(X;Y)$ thus only contains the useful information shared between the received output message Y and the transmitted input message X. This concept can be used to theoretically evaluate the rate of information transfer in a communication link and will be discussed more in depth in Chap. 5.

[1] $H = -2 \cdot 0.5 \cdot \log_2(0.5) = 1$ bit.

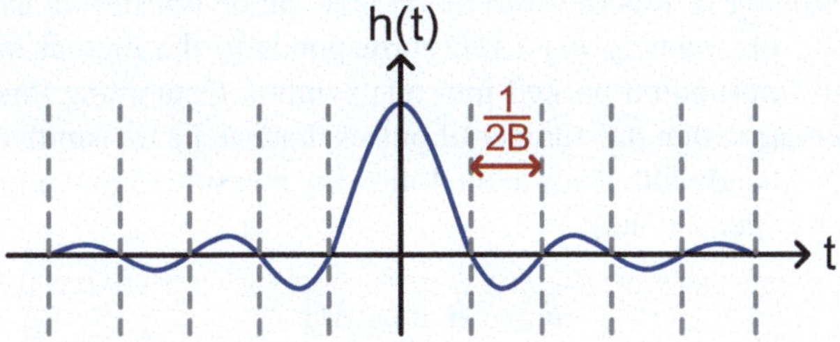

Fig. 2.1 Depiction of a sinc pulse shape where the pulse width is determined by the channel bandwidth

2.1.1.2 Nyquist: Maximum Symbol Rate

Another important contribution came from Nyquist who determined that the number of independent pulses or symbols that could be transmitted over a channel, also known as the symbol rate, is limited to twice the bandwidth of the channel [37, 38]. This can be expressed as follows:

$$T = \frac{1}{2B}, \tag{2.3}$$

where T is the duration of one symbol (in seconds) and B is the bandwidth of the channel (in Hertz). This theory would later be reformulated by Shannon into the famous Nyquist–Shannon sampling theorem, as will be discussed later on in the section on sampling. This upper limit on the obtainable symbol rate is a theoretical limit and requires a pulse shape with specific properties, in particular a sinc pulse shape as illustrated in Fig. 2.1 which has a zero value at fixed symbol period intervals from its maximum. The properties and influence of the pulse shape on the required channel bandwidth will also be discussed later on.

2.1.1.3 Hartley: Quantification of Line Rate

In 1928, Hartley developed a method to quantitatively measure information and its transmission rate (in bits per second). He established that the maximum number of distinguishable pulse levels that can be reliably transmitted and received through a communication channel is determined by the dynamic range of the signal amplitude and the receiver's ability to accurately distinguish these amplitude levels. Assuming the signal amplitude varies over a range of [-A ... A] volts and the receiver has a precision of $\pm\Delta V$ volts, the maximum number of distinct amplitudes M can be expressed as [39]

$$M = 1 + \frac{A}{\Delta V}. \tag{2.4}$$

Note that the number of bits of information that can be transferred using a symbol with M pulse levels, namely $\log_2(M)$, corresponds to the mutual information or maximal useful information packed into one symbol. Combining this insight with Nyquist's observation that the number of pulses that can be transmitted is limited to twice the channel bandwidth leads to the following measure of maximum achievable line rate R (in bits per second):

$$R \leq 2B \log_2(M) . \tag{2.5}$$

Hartley's line rate can be interpreted as the data rate of an error-free channel with $2B$ symbols per second and M signal levels per symbol. How the number of signal levels depends on the noise characteristics of the channel was for someone else to work out.

2.1.1.4 Shannon: Channel Capacity

Building on Nyquist's and Hartley's contributions, Shannon developed the notion of channel capacity in 1948 [13], effectively founding the field of information theory. The channel capacity is the upper limit on the rate at which information can be reliably transmitted (error-free) over a communication channel. In Shannon's noisy-channel coding theorem [40], he states that if information is transmitted at a rate smaller than the channel's capacity, there exist error correcting codes that can make the probability of error at the receiver arbitrarily small. In other words, error-free information transmission is theoretically possible as long as the bit rate is smaller than the channel capacity. Conversely, if the transmission rate lies above this channel limit, all codes will have an error probability greater than zero which increases as the rate increases.

The Shannon–Hartley theorem establishes what this channel capacity is for a finite-bandwidth channel corrupted by Gaussian noise, as illustrated in Fig. 2.2. Essentially, by linking the spread on pulse levels ΔV to the statistical distribution of noise, Shannon specified the effective number of distinguishable pulse levels M in Hartley's law in terms of a signal-to-noise ratio (SNR):

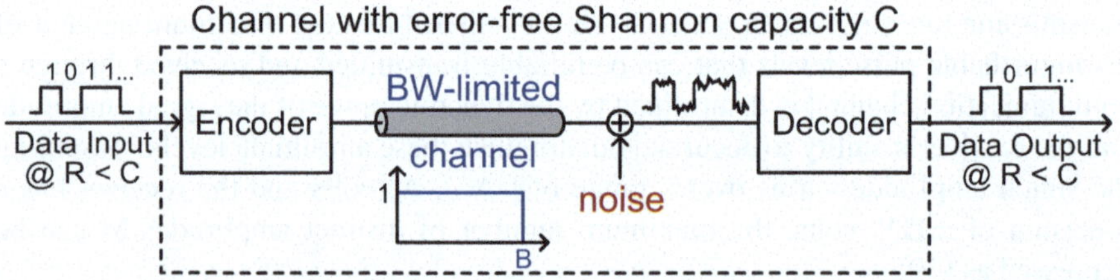

Fig. 2.2 Diagram of Shannon's channel model with capacity C

$$M = \sqrt{1 + \text{SNR}}\,, \tag{2.6}$$

where SNR is defined as the ratio of the average signal power over the average noise power. Plugging this result into Hartley's expression of the line rate yields Shannon's famous equation of the channel capacity:

$$C = B\ \log_2(1 + \text{SNR})\,. \tag{2.7}$$

This expression provides a tight upper limit on the achievable error-free data rate that is supported over a channel with limited bandwidth B and signal-to-noise ratio SNR. The channel capacity should be considered as a theoretical limit as it assumes infinitely sophisticated coding with infinite delay and infinite memory storage and therefore is not achievable in a realistic system. It is, however, an important concept that has served as an important guide in the development of telecommunication and will be extensively studied further in Chap. 5. Having only scratched the surface of what information theory has to offer, the reader is encouraged to wander deeper into the realm of information theory [34, 36].

2.1.2 Welcome to the Frequency Domain

In the field of telecommunication and signal processing, signals are often represented in the frequency domain as it provides a lot of insight into the properties of time-domain signals and eases the calculation of many signal operations. This section briefly summarizes the basic tools needed to become comfortable in the frequency domain.

2.1.2.1 Fourier Transform

The Fourier transform is a mathematical operation that converts a function, often as a function of time, into another complex-valued function that describes the frequency content of the original function. This frequency representation is often called the spectrum of the signal. Given a time-domain waveform $f(t)$, the frequency spectrum of that signal $F(\omega)$ is calculated using the Fourier transform which is defined as follows:

$$F(\omega) = \mathcal{F}\{f(t)\} = \int_{-\infty}^{\infty} f(t)e^{-j\omega t}dt\,, \tag{2.8}$$

where $\omega = 2\pi f$ is called the angular frequency. The reverse operation, calculating the time-domain function from its spectrum, called the inverse Fourier transform, is defined as

$$f(t) = \mathcal{F}^{-1}\{F(\omega)\} = \frac{1}{2\pi}\int_{-\infty}^{\infty} F(\omega)e^{j\omega t}d\omega\,. \tag{2.9}$$

The time-domain signal and its spectrum form a Fourier transform pair, often denoted as follows:

$$f(t) \leftrightarrow F(\omega)\,. \tag{2.10}$$

There are many interesting properties of the Fourier transform, in which you can find any standard mathematics or signal processing textbook, such as [41]. Some of the most relevant ones will be discussed here.

A first important property is the fact that a real-valued, physical signal has a symmetrical spectrum. This means that if $f(t)$ is real, then

$$F(-\omega) = F^*(\omega)\,, \tag{2.11}$$

where the * denotes the conjugate operation. Furthermore, this symmetry means that the magnitude spectrum is even about the origin ($\omega = 0$) and the phase spectrum odd about the origin:

$$|F(-\omega)| = |F(\omega)|\,, \tag{2.12}$$

$$\angle F(-\omega) = -\angle F(\omega)\,. \tag{2.13}$$

For simplicity, typically only the positive frequencies of a physical signal spectrum are depicted without losing any information about the signal. Another interesting property is that adding a time shift to the signal does not alter the magnitude spectrum but only adds a phase shift to the spectrum:

$$f(t - t_0) \leftrightarrow e^{-j\omega t_0}\,F(\omega)\,. \tag{2.14}$$

Conversely, a frequency shift can be attained by the following operation:

$$f(t)e^{-j\omega_0 t} \leftrightarrow F(\omega - \omega_0)\,, \tag{2.15}$$

which is very useful in the context of signal modulation, as will be discussed later. Since the complex exponential $e^{-j\omega_0 t}$ is not a real signal, frequency translation is attained in practice by multiplying by a cosine or sine wave, being its real and imaginary parts.

Very useful in circuit theory is the fact that a time derivative and integration operation translates into a multiplication and division by $j\omega$, respectively, in the frequency domain:

$$\frac{d}{dt}f(t) \leftrightarrow j\omega F(\omega) \tag{2.16}$$

$$\int_{-\infty}^{t} f(\tau)d\tau \leftrightarrow \frac{F(\omega)}{j\omega} + \pi F(0)\delta(\omega) , \tag{2.17}$$

where the impulse function needs to be added for the integration if $F(0) \neq 0$.

Finally, the Fourier transform translates a convolution[2] in the time-domain operation into a multiplication in the frequency domain and vice versa:

$$\mathcal{F}\{f(t) * g(t)\} \leftrightarrow F(\omega) \cdot G(\omega) . \tag{2.18}$$

This is incredibly useful in system theory, where it is common to interpret $g(t)$ as the impulse response of a linear time-invariant (LTI) system, and $(G\omega)$ then represents the frequency response of the system. When $f(t)$ is presented as input to such a system, the output can be calculated as $f(t) * g(t)$ or via a simple multiplication in the frequency domain.

In the end, frequency-domain analysis is often used for electronic circuits and signal processing operations as the frequency domain representation provides useful insight, and many calculations can be simplified considerably. However, calculating the intricate integral of the Fourier transform can be challenging and time consuming. Therefore, Fourier transforms are typically derived using known transformation pairs and properties which you can find all over the Internet [42].

2.1.2.2 Fourier Series

A Fourier series is an expansion of a period function into a sum of sines and cosines. This expansion can give interesting insights into the characteristics of a signal, for instance, to determine the magnitude of the distortion components when doing a harmonic analysis.

The general expression for a Fourier series expansion of a function $f(t)$ periodic with angular frequency ω is given by

$$f(t) = \frac{a_0}{2} + \sum_{n=1}^{\infty} a_n \cos(\omega n t) + \sum_{n=1}^{\infty} b_n \sin(\omega n t) , \tag{2.19}$$

where the coefficients can be easily calculated thanks to the orthogonality property of sine and cosine waves [43]. The Fourier expansion consists of waves with the same frequency as the signal, called the fundamental components, as well as waves at multiples of the fundamental frequency, known as the harmonics. Furthermore, due to the even/odd symmetry of the cosine/sine function, even functions around the origin will result in only cosine components, while odd functions will only contain sine components.

[2] The convolution operation is defined as follows: $f(t) * g(t) = \int_{-\infty}^{\infty} f(x)g(t-x)dx$.

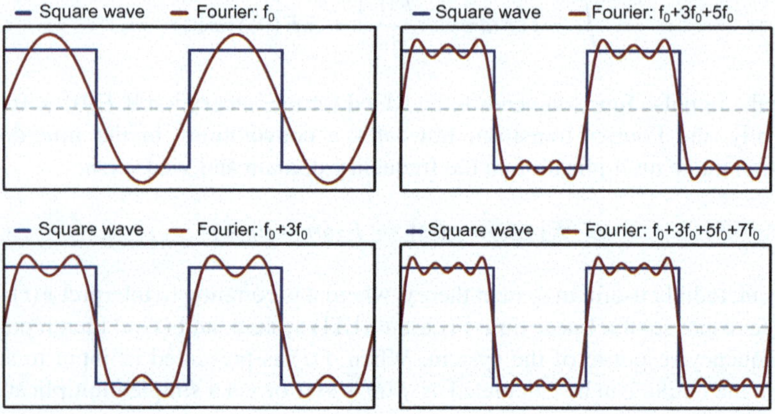

Fig. 2.3 Different Fourier series expansions (red) of a square wave (original in blue)

As an example, a periodic square wave signal (with frequency f_0) can be decomposed into a fundamental sine wave (at f_0) and its odd harmonics (at $3f_0, 5f_0, 7f_0...$) as follows:

$$f(t) = \frac{4}{\pi}\sum_{k=1}^{\infty}\frac{\sin(\omega(2k-1)t)}{2k-1} \tag{2.20}$$

$$= \frac{4}{\pi}\Big[\sin(\omega t) + \frac{1}{3}\sin(3\omega t) + \frac{1}{5}\sin(5\omega t) + ...\Big]. \tag{2.21}$$

Figure 2.3 shows the first four partial sums of the Fourier series for a square wave, showing that as more harmonics are added, the Fourier series converges more to the original square wave signal.

As expected, there is close relation with the Fourier transform discussed before. It can be shown that the frequency spectrum of a periodic signal consists of discrete spectral lines instead of a continuous spectrum [41]. After all, a periodic signal can be represented by a discrete number of waves, its Fourier expansion, each representing a single line in the frequency domain, also called spectral lines. The magnitude and phase of these spectral lines are determined by the coefficients of the Fourier series expansion.[3] Note that a nonperiodic signal would need an infinitesimally small fundamental frequency to construct its Fourier expansion, corresponding to a continuous frequency spectrum.

3 At each frequency, the cosine and sine coefficients a_n and b_n form a vector in the complex plan, determining the magnitude and phase of each spectral component.

2.1.2.3 Power Spectral Density and Bandwidth

Closely related to the frequency spectrum is the power spectral density (PSD) of signal, which describes the normalized power of a waveform versus frequency. Where the spectrum does not give information about the signal power, the PSD is useful to analyze how the power content of signals and noise is affected by filtering circuits or channels.

The PSD is defined as follows [41]:

$$\mathrm{PSD}(f) = \lim_{T\to\infty}\left(\frac{|F_T(f)|^2}{T}\right), \tag{2.22}$$

where $F_T(f)$ represents the spectrum of the waveform $f(t)$ truncated in time over a period T. If the PSD is plotted in decibel units, it yields an identical plot to the magnitude spectrum in decibel units. Hence, in practice, the PSD is often simply replaced by the magnitude spectrum when analyzing the signal power which is more intuitive.

A very important concept in communication and signal processing is the concept of bandwidth (BW), which quantifies over which frequency range the signal occupies energy or power. Many different definitions of bandwidth exist, of which the most useful ones will be discussed. The exact bandwidth value also depends on the nature of the modulated signal. While a baseband (BB) signal has its peak value at a frequency of (almost) zero, a passband (PB) or radio frequency (RF) signal peaks at a nonzero frequency (the center frequency f_c). In general, for most real signals, the PB or RF BW will be twice as large as the BB BW with a similar definition and symbol rate.

The first definition is the fractional in power BW, where the bandwidth is defined as the band where a certain specified fraction of the signal power is concentrated, e.g., 50 % for the half power BW. While this seems like an intuitive way to define BW, it is rather difficult to calculate for real, complex signals. Furthermore, it gives an inaccurate value in case there are undesired spectral components (e.g., spurious components caused by nonlinearities) within the band.

Another definition takes the band to the first null (BB) or between the first nulls around the main lobe (PB) in the signal spectrum. This BW definition is useful for certain theoretical analyses but requires the signal to have a well-defined main lobe, which is not always the case in real signals, and is unrealistic metric if the signal contains significant sidelobes.

Finally, the most popular way to define bandwidth is between the points where the power density has dropped a certain amount below its peak value. Most commonly, a 3 dB decrease is taken as a reference, corresponding to a 50 % drop in power (density) or about a 30 % drop in amplitude. This definition is often used to characterize the bandwidth of a network as well, such as the frequency response (or transfer function) of a filter or channel. The bandwidth cut-off frequency is then defined as the frequency where the magnitude response has dropped with 3 dB. Even though a 3 dB reference is most common, other applications might use different

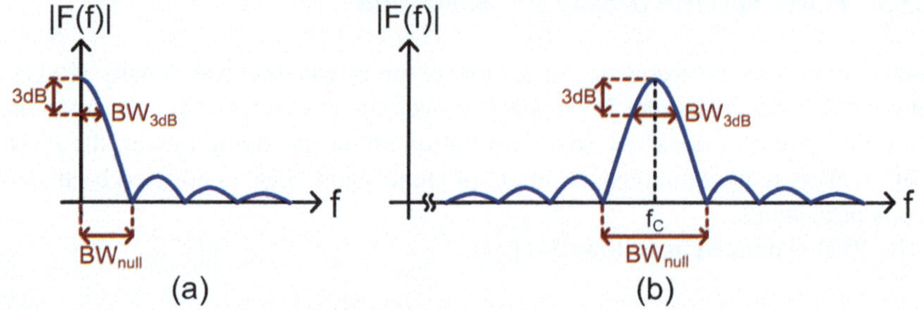

Fig. 2.4 Illustration of two common bandwidth definitions for a baseband (**a**) and passband (**b**) signal

values, such as 6 dB for the return loss of antennas or only 1 dB for transfer functions where flatness is important.

Figure 2.4 shows the null-to-null bandwidth (BW_{null}) and 3-dB BW ($BW_{3\,dB}$) for a typical baseband and passband modulated signal. Since the actual value of the bandwidth heavily depends on the definition, the employed definition should always be clearly communicated.

Lastly, as discussed in Sect. 1.3, strict regulations exist which determine the allocations for each wireless band to avoid interference among them. As this specification is crucial to ensure good operating, regulations typically define a precise spectral characteristic across frequency, called the spectral mask, instead of relying on the rather ambiguous definitions of BW. This spectral mask must be rigorously adhered to limit leakage into neighboring bands.

2.1.3 From Analog to Digital: Sampling and Quantization

All signals in the physical world are analog, meaning they are continuous both in time and in amplitude. Digital signals, on the other hand, are discrete in time and amplitude. Processing and storing signals in the digital domain has multiple advantages, such as higher noise immunity, better accuracy, and reproducibility of results, which is why the majority of communication and computation systems work with digital signals. To convert an analog signal into its digital representation, two discretization operations are required: sampling (time discretization) and quantization (amplitude discretization).

Figure 2.5 shows the conversion of an analog signal into its digital representation with corresponding waveforms. First, the analog signal is uniformly sampled with a period of T_s (the sampling period), resulting in a time-discrete version of the analog signal only updating its value at integer multiples of T_s. Next, the sampled signal is quantized where all analog values are mapped onto a set of uniformly distributed discrete levels. The number of discrete levels is determined by the number of

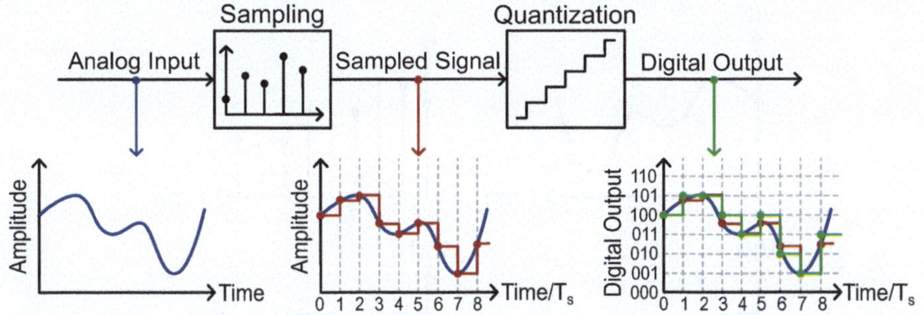

Fig. 2.5 Diagram of the digitization of an analog signal with corresponding waveforms, including sampling and quantization

bits used to represent the digital signal. During both sampling and quantization operations, some information from the original analog signal is inevitably lost. This loss of information can be minimized by increasing the number of samples per unit of time or by using more bits to represent the amplitude of the signal. Both of these operations are now discussed more deeply.

2.1.3.1 Sampling

Sampling is the process of reducing a waveform in continuous time to the discrete time domain. The sampling process can be mathematically described using the "Dirac" function $\delta(t)$ [43]. This function has some interesting properties, most notably that a convolution of a delayed Dirac's function with another function results in the value of that function at the specified delay, or more rigorously

$$\int_{-\infty}^{\infty} f(t)\delta(t-t_0)dt = f(t_0)\,. \tag{2.23}$$

Mathematically, sampling is performed by multiplying the time-continuous signal $v(t)$ with a sequence of Dirac pulses, spaced at T_s, resulting in the following time-discrete, sampled signal $v_s(t)$, also shown in Fig. 2.6:

$$v_s(t) = v(t)\cdot\sum_{n=-\infty}^{\infty}\delta(t-nT_s) = \sum_{n=-\infty}^{\infty} v(nT_s)\,. \tag{2.24}$$

To see what happens in the frequency domain, one can resort to the useful property that a multiplication in the time domain transforms into a convolution in the frequency domain, as discussed in the previous subsection on Fourier transform. First, the Fourier transform of a sequence of Dirac pulses can be expressed as [43]

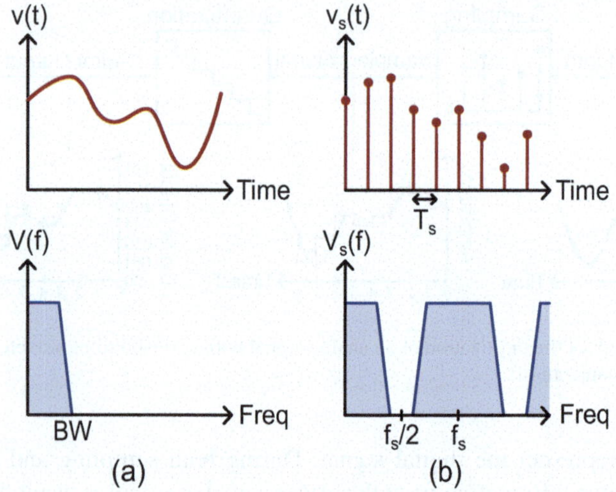

Fig. 2.6 Time-domain waveform of frequency spectrum of an analog signal before (**a**) and after (**b**) being sampled

$$\sum_{n=-\infty}^{\infty} \delta(t - nT_s) \leftrightarrow \sum_{k=-\infty}^{\infty} \delta(f - kf_s) , \tag{2.25}$$

where the sampling frequency is defined by the inverse of the sampling period: $f_s = 1/T_s$. The frequency spectrum of a sequence of Dirac pulses is once again a sequence of Dirac pulses spaced at f_s. The spectrum of a time-discrete signal $V_s(t)$ can thus be expressed as

$$V_s(f) = \frac{1}{T_s} \cdot \sum_{k=-\infty}^{\infty} V(f - kf_s) , \tag{2.26}$$

which shows replicas of the original spectrum at multiples of f_s, as depicted in Fig. 2.6. As the band of the original waveform widens, so do the bands of the replicas, eventually causing overlap between the bands. This overlap, described as aliasing, causes irreversible signal distortion and loss of information. To prevent aliasing, the sample rate must obey the following condition:

$$f_s/2 > BW . \tag{2.27}$$

This condition is called the Nyquist criterion, originating from the Nyquist–Shannon sampling theorem [44], which more generally states that a bandlimited continuous-time signal can be sampled and perfectly reconstructed if the sample rate is more than twice the signal's bandwidth. It was developed by Shannon by linking Nyquist's pulse rate insight to the problem of sampling and reconstruction of continuous signals. To avoid aliasing and its associated information loss, any signal should be band-limited to $f_s/2$ (the Nyquist bandwidth) before sampling, for instance, by a filter, often called an anti-aliasing filter if used for this purpose.

The reconstruction of a sampled signal to its continuous form, called interpolation, is achieved by filtering out all replicas in the frequency domain. In the case where Nyquist's criterion is obeyed, the original signal can be perfectly reconstructed; however when the criterion is violated, the signal will be distorted due to aliasing.

2.1.3.2 Quantization

Quantization is the process of translating a sampled signal with a continuous amplitude range into a digital code. The signal level is compared to a discrete number of reference values, corresponding to a set of digital codes, and truncated to the nearest value, thus discretizing the continuous amplitude as shown in Fig. 2.7a. Most digital amplitude representations use binary coding with a specified resolution or a number of bits N. The total number of levels to represent the signal amplitude is thus equal to 2^N. With the maximum input amplitude defined as the full scale (FS), the difference between two adjacent transition levels is quantified in the analog domain as the least significant bit (LSB) or step width Δ, which is equal to $\Delta = FS/2^N$.

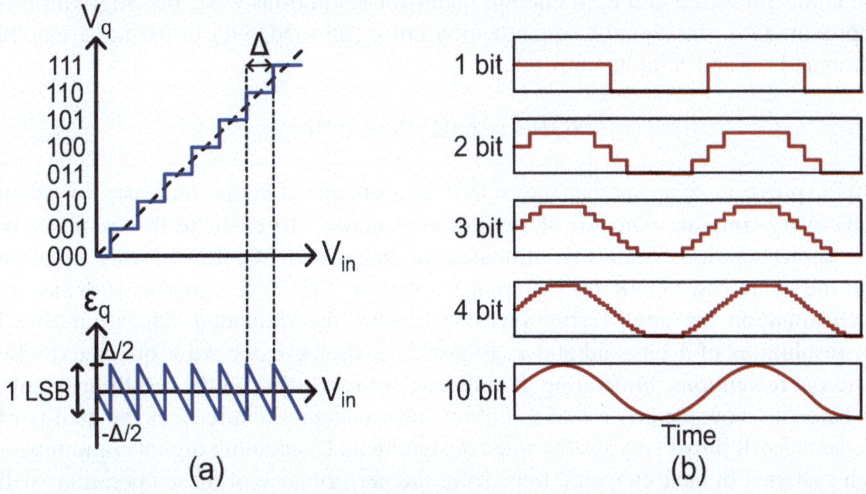

Fig. 2.7 Transfer characteristic of a 3-bit quantizer and associated quantization error (**a**) and waveform of a sine wave quantized with different resolutions (**b**)

Knowing the value of the LSB and the number of bits, the N-bit digital word $[b_0..b_{N-1}]$ can be converted back into a discrete analog amplitude signal as follows:

$$V_q = \Delta \cdot \sum_{i=0}^{N-1} b_i 2^i = \Delta \cdot \left[b_0 + b_1 \cdot 2 + b_2 \cdot 4 + ... + b_{N-1} \cdot 2^{N-1}\right] . \qquad (2.28)$$

Due to the rounding process, this quantized signal has an error compared to the original analog signal, called the quantization error ϵ_q. The quantized signal can be expressed as follows:

$$V_q = V_{\text{in}} + \epsilon_q . \qquad (2.29)$$

In an ideal quantizer, this error has a value within $\pm\Delta/2$ with a linear slope for input signals inside the FS, as demonstrated in Fig. 2.7a.

While the characteristics of ϵ_q depend on the statistical properties of the input signal, making a couple of approximations can lead to interesting conclusions. Specifically, assuming the quantization error has a uniform probability of occurring within its range, the error can be interpreted as white noise added to the input signal with an average power of [43]

$$\overline{\epsilon_q^2} = \int_{-\Delta/2}^{\Delta/2} \epsilon_q^2 d\epsilon_q = \frac{\Delta^2}{12} . \qquad (2.30)$$

This quantization noise approximation is valid for input signals uniformly spanning the entire full-scale and high enough quantizer resolutions [43]. Bearing with this approximation, the signal-to-quantization-noise ratio (SQNR) in dB units can be computed for an FS input sinusoid:

$$\text{SQNR} = 6.02 \cdot N + 1.76 . \qquad (2.31)$$

This expression suggests that the SQNR of a quantized signal increases by about 6 dB every time an extra bit of resolution is added. To evaluate the accuracy of this approximation, Table 2.1 compares the calculated SQNR with this equation and the simulated SQNR for different resolutions [45]. The equation provides an overestimation for small resolutions but matches the simulated values quite well for resolutions of 4 bits and above. Figure 2.7b shows a sine wave quantized with different resolutions, illustrating the influence of resolution on the resulting signal.

Knowing how the process of sampling and quantization influences the quality of the signal will prove very useful when designing and evaluating digital communication systems. In later chapters, improving the performance of these operations will be linked to actual circuit implementations, providing insightful and useful design trade-offs.

Table 2.1 Comparison between calculated and simulated SQNR for different quantizer resolutions N [45]

Number of bits (N)	Calculated SQNR	Simulated SQNR
1	7.78 dB	6.31 dB
2	13.80 dB	13.30 dB
3	19.82 dB	19.53 dB
4	25.84 dB	25.61 dB
5	31.86 dB	31.66 dB
6	37.88 dB	37.73 dB
7	43.90 dB	43.84 dB
8	49.92 dB	49.84 dB

2.2 Digital Modulation

2.2.1 Modulation Fundamentals

Modulation refers to the process of encoding information or data onto one or more properties of a periodic waveform. When information is encoded onto a series of pulses, it is known as baseband modulation or signaling, containing information at frequencies close to 0 Hz or DC. In contrast, passband modulation or signaling refers to the encoding of information on a periodic wave, also known as the carrier, only containing information around the frequency of that carrier. While baseband signals can be transmitted over a conductive channel (i.e., wireline communication), a carrier is needed to make wireless (and DWG) transmission possible as it relies on the propagation of electromagnetic waves. Since wireless and DWG communication are the main topics of this book, passband modulation will be extensively analyzed. However, as passband modulation is fundamentally obtained by upconverting a baseband signal onto a carrier, the properties of the baseband signal have an important influence on the passband signal and must therefore also be well understood.

Furthermore, it is possible to perform analog and digital modulation, determined by how the encoded information is represented. When the information has the form of an analog signal (continuous in time and amplitude), it is called analog modulation. When the information is represented as a set of discrete digital states, it is called digital modulation. As discussed in the previous section, digital signaling has various advantages over analog signaling, one of which being more immune to noise. Indeed, small amounts of noise are removed after quantization as long as it is smaller than the quantization step. In contract, noise added to an analog signal cannot (simply) be removed and will inevitably influence the quality of the modulated signal. For this reason, most reliable high-performance communication systems are digital in nature, and therefore the discussion will mainly focus on digital systems and modulations. However, even though the data is digital, the transmitted and received modulated signals are fully analog in nature, as they are physical signals. A system thus typically contains a digital back end, processing

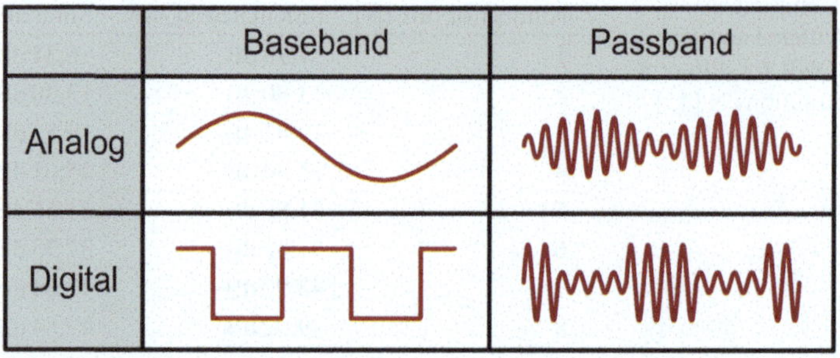

Fig. 2.8 Overview of the different modulation techniques: analog versus digital signaling and baseband versus passband modulation

and storing the digital data, and an analog front-end, which performs the generation and detection of the modulated signals.

Figure 2.8 provides an overview of the discussed modulation techniques, visualizing analog versus digital signaling and baseband versus passband modulation. As can be observed in the figure, a digital passband signal is a continuous signal in nature with certain characteristics (e.g., its amplitude) controlled by digital data. The rest of this chapter will first discuss fundamental properties of baseband signals and then the basics principles of passband modulation and finally analyze various digital passband modulation schemes.

2.2.2 Baseband Signaling

Baseband signaling deals with how digital information is embedded in physical signal pulses. The most straightforward type of baseband modulation is pulse amplitude modulation (PAM), where the digital data is encoded in the amplitude of the signal pulse. In general, if the data has M levels, the modulated signal $s(t)$ is called an M-PAM signal and can be expressed as

$$s(t) = \sum_{n=-\infty}^{\infty} d_n f(t - nT_s) , \tag{2.32}$$

where $\{d_1, d_2, ..., d_n\}$ is the set of data symbols with M possible levels, $f(t)$ the symbol pulse shape, and T_s the duration of one symbol. Information can also be encoded onto other properties of the symbol pulse than its amplitude. For instance, if instead the data is embedded in the width of the pulse, pulse width modulation (PWM) is obtained. Since PAM is the most relevant for passband modulations, other signaling options will be left for the reader to explore.

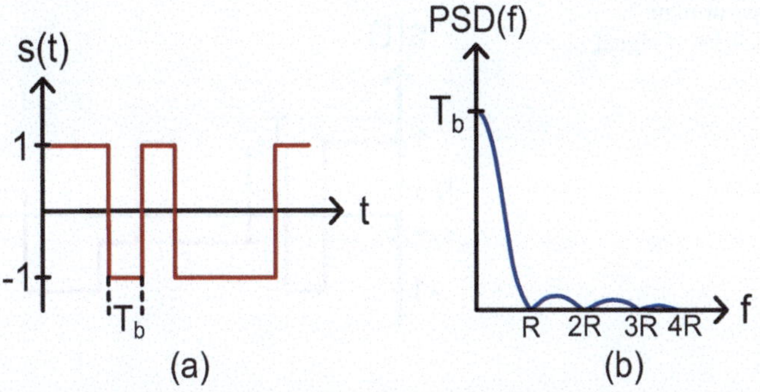

Fig. 2.9 Time-domain waveform (**a**) and PSD (**b**) of a polar NRZ signal with rectangular pulse shape

In the simplest case, the amplitude is encoded with only two possible levels, "0" or "1," resulting in a binary or 2-PAM signal, also known as unipolar non-return-to-zero (NRZ) signaling [41]. More commonly used in differential implementations is polar NRZ signaling, where the amplitude switches between "−1" and "1," as depicted in Fig. 2.9a. This type of polar or balanced signaling has the advantage of not containing any power at DC, resulting in a greater dynamic range for the same signal power compared to unipolar NRZ.

Assuming the data is perfectly random, the PSD of a digital baseband signal can be calculated as follows [41]:

$$\mathrm{PSD}(f) = \frac{|F(f)|^2}{T_s} \sum_{k=-\infty}^{\infty} R(k) e^{j2\pi k f T_s}\,, \tag{2.33}$$

where $|F(f)|$ is the Fourier transform of the pulse shape $f(t)$ and $R(k)$ the autocorrelation of the data. In case a rectangular pulse shape is used, the spectrum can be calculated to be

$$f(t) = \Pi\left(\frac{t}{T_s}\right) \leftrightarrow F(f) = T_s \frac{\sin(\pi f T_s)}{\pi f T_s}\,, \tag{2.34}$$

where $\Pi(t/a)$ represents the rectangle function defined as 1 when $|t| < a/2$ and 0 otherwise. Expression 2.33 can be further evaluated to obtain the PSD of a normalized polar NRZ signal [41]:

$$\mathrm{PSD}_{\text{polar NRZ}}(f) = T_b \left(\frac{\sin(\pi f T_b)}{\pi f T_b}\right), \tag{2.35}$$

where T_b denotes the bit period which is equal to the symbol period in case of binary signal. This PSD is shown in Fig. 2.9b on a linear scale. To encode more bits of information per symbol, an M-level PAM signal contains $\log_2(M)$ bits of

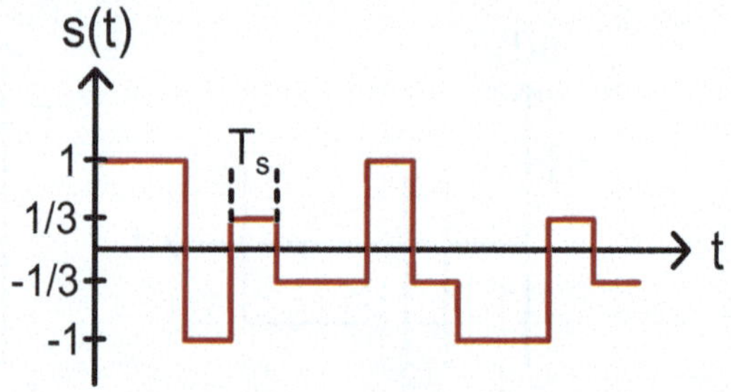

Fig. 2.10 Time-domain waveform of a 4-PAM signal

information per symbol, leading to the following relation between bit rate $R = 1/T_b$ and symbol rate $D = 1/T_s$:

$$R = D \cdot \log_2(M)\,. \tag{2.36}$$

An example of a 4-PAM waveform is shown in Fig. 2.10 with the same symbol period, but four levels per symbol instead of two, leading to twice the amount of bits per symbol. Figure 2.9b shows that the significant signal power is contained in the sidelobes of the spectrum. This is caused by the steep edges, and thus high frequency information, of the rectangular symbol shape. The reduce the sidelobes and power-containing bandwidth of the signal, specific pulse shapes can be used which trade-off steepness in the time domain for bandwidth in the frequency domain. The necessary properties of such pulse shapes and governing trade-offs will be more elaborately discussed later in the chapter.

2.2.3 *Passband Signaling*

For passband modulation, the source data is modulated onto a carrier signal c(t) which can be described as a sinusoid waveform:[4]

$$c(t) = A_c\,\cos(\omega_c t + \phi_c)\,. \tag{2.37}$$

This expression reveals three parameters that can be manipulated for data encoding: the amplitude A_c, the (angular) frequency $\omega_c = 2\pi f_c$ also the called carrier frequency, and the phase ϕ_c. Modulating one or more parameters with binary or multilevel data yields different modulations with their own characteristics. Before diving into the various possible modulation schemes, some basic concepts must be introduced.

[4] Since a passband modulated signal is a propagating electromagnetic wave, i.e. the solution to Maxwell's equation, it has the form of a sinusoid.

2.2.3.1 Modulation, Complex Envelope, and Signal Constellations

Any physical passband modulated signal $s(t)$ can be represented as an upconverted version of a baseband waveform, mathematically expressed as [41]

$$s(t) = Re\left\{g(t)e^{j\omega_c t}\right\} , \tag{2.38}$$

where $g(t)$ is called the complex envelope of $s(t)$. The complex envelope $g(t)$ can be viewed as the baseband equivalent of the passband signal $s(t)$ and is a function of the modulating data signal(s) $m(t)$. The exact mapping function, $g[m(t)]$, can take various forms and depends on the nature of the modulation method. The convenience of the complex envelope representation cannot be overstated, especially for complex modulations where multiple carrier parameters are modulated simultaneously.

The spectrum of a passband signal is directly related to the spectrum of its complex envelope and can be expressed as [41]

$$S(f) = \frac{1}{2}\left[G(f - f_c) + G^*(-f - f_c)\right] , \tag{2.39}$$

where $G(f)$ is the spectrum of the complex envelope $g(t)$. Figure 2.11 demonstrates this upconversion process, represented by a time-domain multiplication by the complex exponential $e^{j\omega_c t}$ (and discarding of the imaginary part), and shows how the baseband spectrum of the complex envelope is translated in the frequency domain to the carrier frequency. Note how the occupied bandwidth around the carrier doubles, in general, compared to the baseband signal.

The complex envelope $g(t)$ has two common representations. First of all, $g(t)$ can be decomposed into a real part, called the in-phase (I) component $x(t)$, and an imaginary part, called the quadrature (Q) component $y(t)$:

$$g(t) = x(t) + jy(t) . \tag{2.40}$$

Alternatively, $g(t)$ can be represented by a magnitude $R(t)$ and a phase $\theta(t)$ in the complex plane:

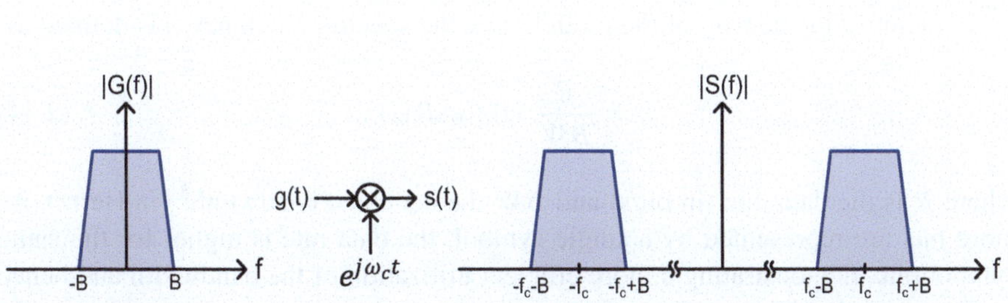

Fig. 2.11 Demonstration of how upconversion leads to a translation in the frequency domain

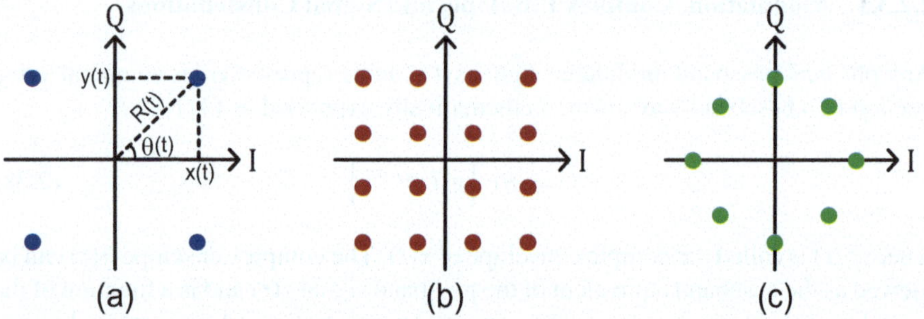

Fig. 2.12 Constellation diagrams of (**a**) a QPSK modulation with annotations, (**b**) a 16-QAM modulation, and (**c**) an 8-PSK modulation

$$g(t) = R(t)e^{j\theta(t)}\,, \tag{2.41}$$

where the conversion between both representations is given by

$$R(t) = |g(t)| = \sqrt{x^2(t) + y^2(t)} \tag{2.42}$$

$$\theta(t) = \angle g(t) = \tan^{-1}\left(\frac{y(t)}{x(t)}\right). \tag{2.43}$$

A passband modulated signal is often visualized using a constellation diagram or plot which is a two-dimensional plot of all the possible signal states as determined by the complex envelope. The horizontal and vertical axes depict the in-phase (I) and quadrature (Q) components, respectively. Figure 2.12 shows several examples of constellation diagrams visualizing different modulation schemes (which will be discussed shortly). For instance, Fig. 2.12a shows the constellation of a QPSK modulated signal with annotated in-phase, quadrature, magnitude, and phase component. The signal consists of four distinct signal states with equal magnitudes spaced $\pi/2$ radians apart. The other constellations in the figure (16-QAM and 8-PSK) contain more magnitude and/or phase levels, presenting a higher number of bits per symbol.

To quantify the density of the modulation, the spectral efficiency η is defined as

$$\eta = \frac{R}{BW} \quad \text{(bits/s/Hz)}\,, \tag{2.44}$$

where R is the data rate (in bit/s) and BW the occupied bandwidth[5] (in Hertz). As more bits are represented by a single symbol, the data rate is higher for the same symbol rate, demonstrating a more efficient utilization of the bandwidth and hence higher spectral efficiency.

[5] Typically the null-to-null bandwidth definition is used in theoretical analyses.

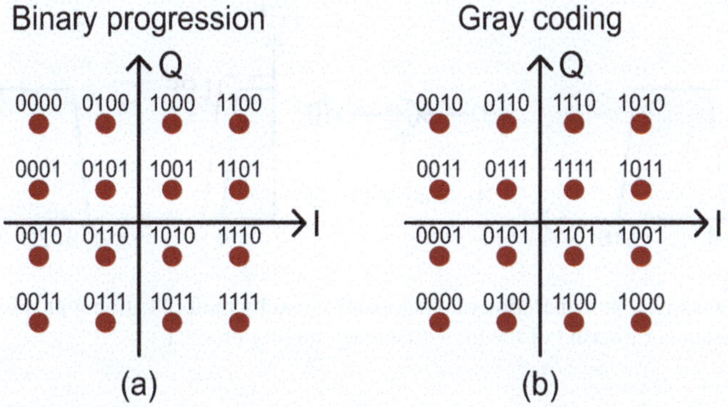

Fig. 2.13 Two different symbol mappers for 16-QAM modulation: binary progression (**a**) and Gray coding (**b**)

To represent multiple bits with a single symbol, the different bit combinations must be mapped to the possible symbols, called symbol mapping. Figure 2.13 illustrates two different symbol mapping methods for a 16-QAM modulation. The first one follows a sequential binary progression, starting in the upper left corner and progressing column-wise down to the lower right corner. While this might look intuitive, a second better option uses a Gray-coded map. Gray coding has the important property that adjacent signal states differ in their symbol assignment by only 1 bit position [46]. This is a very desirable characteristic as mistaking a received symbol for an adjacent symbol only results in a single bit error, as opposed to multiple bit errors in the other mapping method.

2.2.3.2 Demodulation and Eye Diagrams

To extract the original data from the passband modulated signal, called demodulation, the signal should be translated back to baseband. This demodulation, or detection, can be performed in an incoherent or a coherent manner.

During incoherent detection, the original information is extracted by processing only the modulated signal. For example, if the encoded information is contained within the amplitude of the passband signal, this information can be extracted using an envelope detector which outputs a signal that depends on the amplitude of the input. This operation can be expressed as follows:

$$V_{\text{out}}(t) = |A(t)\cos(\omega_c t + \phi)| = A(t) \,. \tag{2.45}$$

Note that all phase information lost during incoherent detection, which limits the application domain of this type of detection.

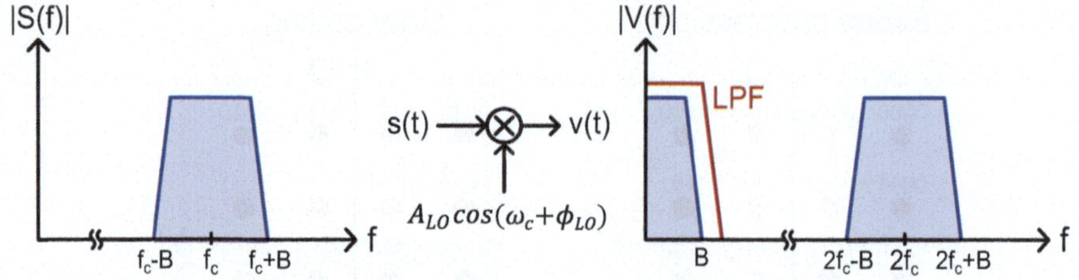

Fig. 2.14 Downconversion operation of a passband signal by multiplication with a local oscillator signal and subsequent filtering of the high-frequency mixing product

In order to retrieve absolute phase information, coherent detection is necessary. During coherent detection, the modulated signal is multiplied by a local reference signal at the carrier frequency, called a local oscillator (LO), which translates the original passband spectrum to baseband and to twice the carrier frequency. After filtering out the high-frequency component at twice the carrier frequency (by a low-pass filter or LPF), the original amplitude and phase information can be extracted from the baseband output. This process is visualized in Fig. 2.14 and can be generally formulated as follows:

$$\begin{aligned} v(t) &= \left[A(t)\cos(\omega_c t + \phi(t))\right] \cdot \left[A_{LO}\cos(\omega_c t + \phi_{LO})\right] \\ &= \frac{A(t)A_{LO}}{2}\left[\cos(\phi(t) - \phi_{LO}) + \cos(2\omega_c + \phi(t) + \phi_{LO})\right] \\ &\downarrow \text{low-pass filter with cut-off} < 2\omega_c \\ &= \frac{A(t)A_{LO}}{2}\cos(\phi(t) - \phi_{LO})\,. \end{aligned} \tag{2.46}$$

As seen from the equation, the amplitude and the phase information can be recovered as long as the LO signal is synchronized (in frequency and in phase) to the carrier of the passband signal. The generation, and especially synchronization, of this LO signal is what makes coherent detection significantly more challenging than incoherent detection.

Regardless of the detection method, the encoded information will be extracted from a binary or multilevel baseband signal by the decision circuits. These circuit blocks essentially guess which symbol was most likely originally transmitted based on the received symbols. To easily visualize the quality of the baseband signal and estimate the performance of the decision circuitry, an eye diagram is often constructed from the waveform. As visualized in Fig. 2.15, the waveform is first broken into sections with a duration of two symbol lengths ($2T_s$) which are then overlayed onto the same time axis and centered on the optimal decision point.

All of these overlapping sections form an "eye" that reflects the quality of the signal, quantified by the vertical eye height and horizontal eye width. When the eye

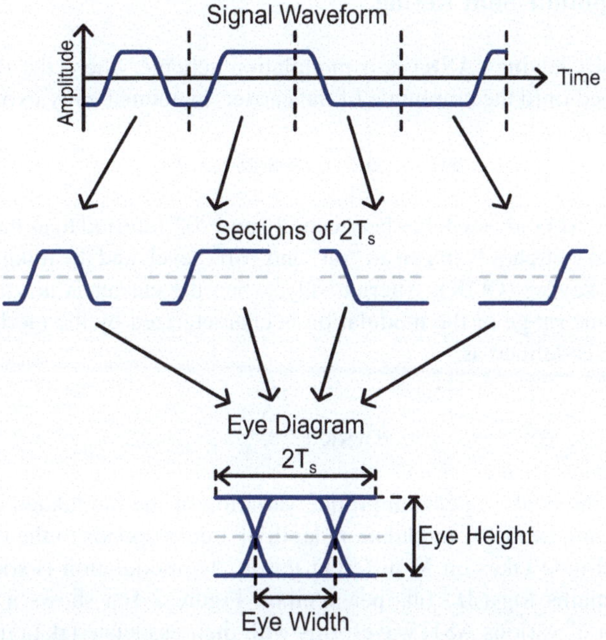

Fig. 2.15 Construction of an eye diagram from a binary NRZ signal waveform

appears open, meaning a large eye height (ideally the maximum amplitude range) and width (ideally one symbol period), the bit decision can easily be made with a low error probability. After all, it is very clear which symbol was transmitted. However, when the eye gets more narrow in height (due to noise or distortion) or in width (due to jitter), it becomes more challenging to reliably determine the symbol that was originally transmitted, and therefore the error of making a wrong bit decision increases. This error probability is the ultimate performance metric of a communication system and will be extensively studied later in this chapter.

2.2.4 *Simple Modulations: ASK, PSK, and FSK*

Simple modulations are defined in this book as modulations where only one property of the carrier wave is encoded with information. The complex envelope can thus be described with a one-dimensional set of signal states $\{d_1, d_2, ..., d_n\}$ or a single data sequence $d(t)$. The baseband data signal can be binary or have multiple levels and can be expressed as a general M-PAM signal like Eq. 2.32.

Three categories of simple digital modulation schemes will be discussed: amplitude-shift keying, phase-shift keying, and frequency-shift keying.

2.2.4.1 Amplitude-Shift Keying

Amplitude-shift keying (ASK) is a modulation scheme where the digital data is directly mapped onto the amplitude of the carrier, mathematically expressed as

$$s(t) = A\,d(t)\,\cos(\omega_c t + \phi)\,. \tag{2.47}$$

When the data $d(t)$ only switches between "1" and "0," referred to as hard switching, the ASK signal switches between an "on" and "off" level, and the modulation is then called on–off keying (OOK). Alternatively, when the carrier is not fully switched off, the dynamic range of the modulation is characterized by the modulation depth h_{ASK}, which is defined as

$$h_{\mathrm{ASK}} = \frac{\Delta A}{2A}\,, \tag{2.48}$$

where ΔA is the peak-to-peak amplitude variation of the modulated carrier and A its maximum amplitude. A modulation depth of 1 corresponds to the OOK case.

If the amplitude takes on M different levels, the modulation is specified as M-ASK and contains $\log_2(M)$ bits per symbol. Figure 2.16a shows a time-domain representation of various ASK waveforms with their associated data inputs, as well as their constellation diagrams showing points on a one-dimensional line.

Assuming the input data is switched at a symbol rate D with a rectangular pulse shape, yielding a baseband null-to-null bandwidth of D (see Fig. 2.9), the resulting ASK signal has a null-to-null passband or RF bandwidth of 2D, as plotted in Fig. 2.16b. Since a symbol rate of D results in a data rate equal to $\log_2(M) \cdot D$, the spectral efficiency of an M-ASK signal is equal to $0.5\,\log_2(M)$ bits/s/Hz.

Furthermore, it must be noted that the spectrum contains power at the carrier frequency, also known as LO feedthrough (LOFT). This is considered as "wasted"

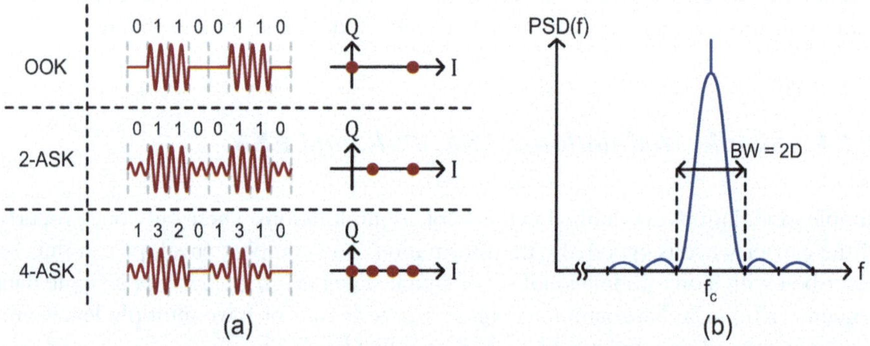

Fig. 2.16 Waveforms and constellation diagrams of various ASK signals with data input (**a**) and PSD of a general ASK signal (**b**)

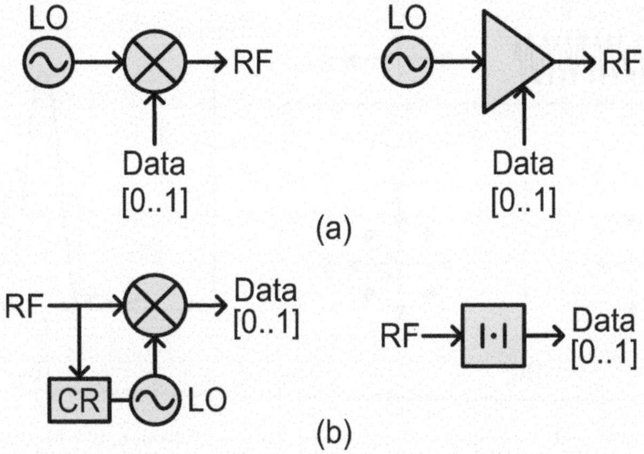

Fig. 2.17 Examples of ASK modulators (**a**) and demodulators (**b**)

power as it does not contribute to the amplitude difference of the useful signal. As the modulation depth increases, the amount of LOFT decreases as does the modulation efficiency. OOK can thus be considered as the most efficient form of 2-ASK. However, in practice, depending on the ASK generator, it can be challenging to turn off the signal completely with fast modulation speeds.

Despite the relatively low spectral efficiency, M-ASK modulations are quite popular, especially at higher mm-Wave frequencies, due to their low implementation complexity. Indeed, an ASK signal can be generated either by upconverting unipolar baseband data onto a carrier or local oscillator (LO) signal or by simply controlling the gain of an amplifier and consequently modulating the amplitude of its output [47], as shown in Fig. 2.17a. Detection can be performed coherently, for instance, by downconverting the signal to baseband with a synchronized LO signal (needing carrier recovery (CR) to recover the frequency and phase from the carrier) or incoherently using an envelope detector or rectifier [47], as depicted in Fig. 2.17b. Especially the incoherent approach is quite straightforward to implement and can lead to very energy efficient implementations.

2.2.4.2 Phase-Shift Keying

When the data is mapped onto the phase of the carrier, the modulation scheme is classified as phase-shift keying (PSK):

$$s(t) = A\cos(\omega_c t + D_p d(t)) , \tag{2.49}$$

where the phase sensitivity $D_p = \pi/2$ if $d(t)$ is assumed to vary between -1 and 1. When the phase is switched between only two states, for example, 0° and 180°, the

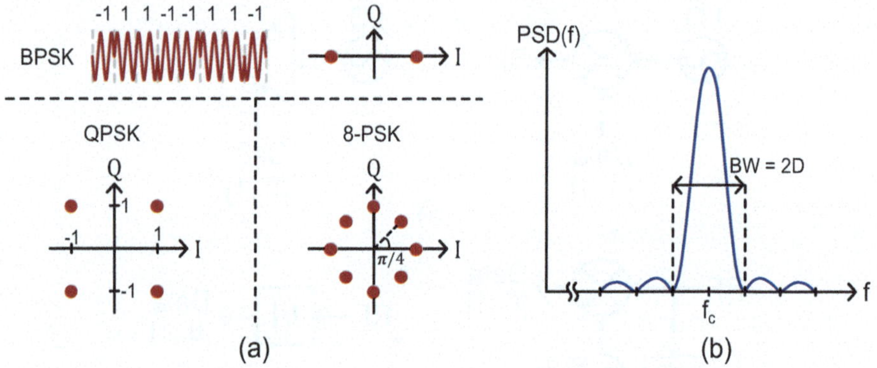

Fig. 2.18 Waveform and constellation diagrams of various PSK signals with data input (**a**) and PSD of a general PSK signal (**b**)

modulation is referred to as binary-phase shift keying (BPSK). A BPSK waveform and equivalent constellation plot is shown in Fig. 2.18a. In general, when M phases are encoded, uniformly spaced at $2\pi/M$ radians apart, M-PSK is established.

The special case where $M = 4$ is also called quadrature-phase shift keying (QPSK), which is especially popular as it can be generated by combining two quadrature BPSK signals, instead of encoding four distinct phases directly onto the carrier. The constellations of a QPSK or 4-PSK (4 phases) and 8-PSK (8 phases) signals are shown in Fig. 2.18a. Despite the complex nature of a generic M-PSK constellation, with a complex-valued envelope, it is still considered a simple modulation since only one parameter, the phase, is modulated.

The spectrum of a general PSK signal modulated at symbol rate D is shown in Fig. 2.18b. This spectrum looks similar to the ASK spectrum, having a null-to-null RF bandwidth of 2D, but does not contain any LO feedthrough. Indeed, no power is wasted to transmit the carrier, which makes the modulation more efficient but makes carrier recovery more challenging. The spectral efficiency is the same as an ASK signal: 0.5 $\log_2(M)$ bits/s/Hz.

BPSK and QPSK are very popular modulation schemes at high frequencies, thanks to their straightforward implementations. The generation can be done in a number of ways, depending on the number of levels, as illustrated in Fig. 2.19a. For BPSK, a carrier can simply be modulated by a polar binary data input, yielding the carrier and its inverted version [48]. Alternatively, this phase inversion can be attained using switches [49]. QPSK modulation is often performed by combining two BPSK signals generated by LO signals that have a relative phase difference of 90° [50]. Finally, a general M-PSK signal (BPSK, QPSK, or higher order M-PSK signals) can be constructed by generating the M necessary LO phases and selecting the correct phase based on the data.

Since all of the information in the M-PSK signal is encoded in the absolute phase of the carrier, coherent demodulation is necessary as shown in Fig. 2.19b.

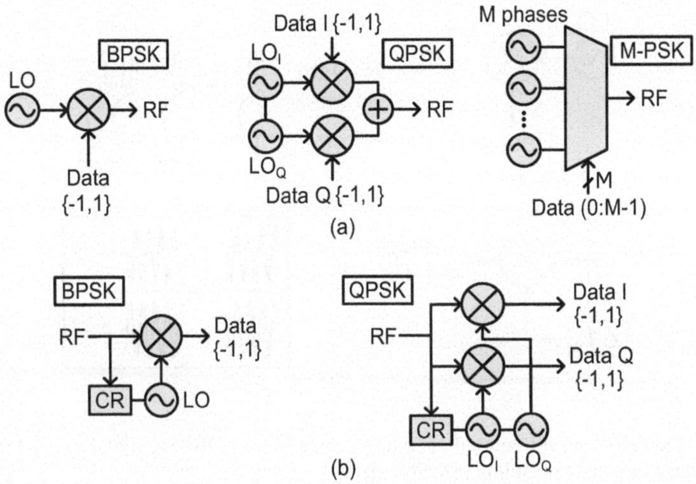

Fig. 2.19 Examples of BPSK, QPSK, and M-PSK modulators (**a**) and BPSK and QPSK demodulator (**b**)

For BPSK, the signal is downconverted by a local oscillator synchronized to the carrier. For QPSK, the signal is typically downconverted by two synchronized LO signals which are 90° out of phase. Higher order M-PSK demodulation uses a similar demodulation architecture with multilevel data outputs to differentiate between more phases. This carrier synchronization or recovery is often performed in the digital domain, based on the downconverted signals, but can also be done in the analog domain, for instance, using a Costas-loop structure [51].

Since coherent detection can be challenging, an alternative PSK modulation exists that can be demodulated incoherently: differential phase-shift keying (DPSK). In DPSK, the information is encoded in the phase transitions as opposed to in the absolute phase. As a result, the data can be incoherently recovered by comparing the DPSK signal with the same signal delayed by one symbol period, also called differential decoding. In essence, the phase reference is provided by a delayed version of the signal instead of separate LO signal that requires carrier synchronization. This differential modulation and incoherent demodulation principle is demonstrated for differential BPSK (DBPSK) in Fig. 2.20.

2.2.4.3 Frequency-Shift Keying

Next to amplitude and phase, frequency is the third parameter that can be modulated. A frequency-shift keying (FSK) signal employs different carrier frequencies depending on the input data. FSK modulation can be split up into two types: discontinuous-phase FSK (DPFSK) and continuous-phase FSK (CPFSK). For DPFSK, the output switches between different, decoupled oscillators resulting in

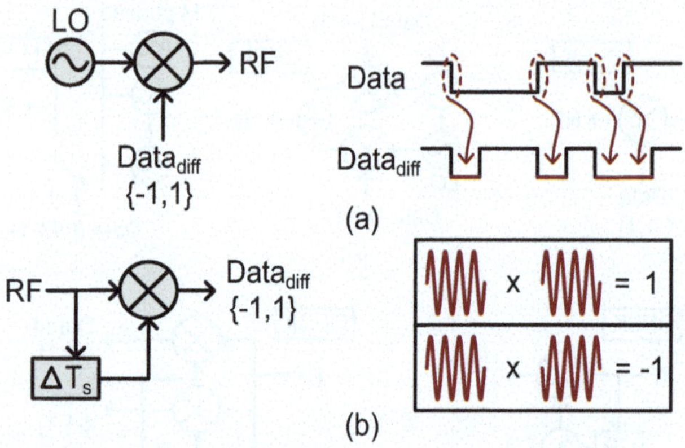

Fig. 2.20 Modulator and differential encoding principle of a DBPSK signal (**a**) and differential DBPSK demodulator and decoding principle (**b**)

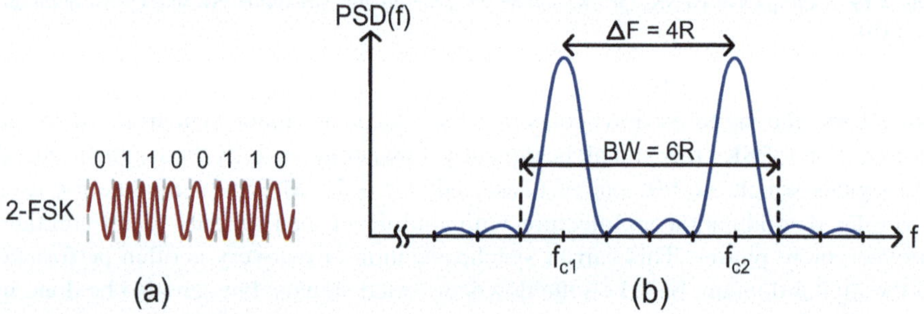

Fig. 2.21 Waveform (**a**) and PSD (**b**) of a 2-FSK signal with $h_{FSK} = 4$

a discontinuity at switching time. For CPFSK a single oscillator is tuned to different frequencies by the data, yielding an output that varies in frequency but whose phase remains continuous, as shown in Fig. 2.21a. Most practical FSK systems employ the second type of signal, CPFSK, which can be described as follows [41]:

$$s(t) = A\cos\left(\omega_c t + D_f \int_{-\infty}^{t} d(\lambda)\, d\lambda\right), \tag{2.50}$$

where D_f is the frequency sensitivity and $d(\lambda)$ the digital data signal.[6] Even though the data can change discontinuously, the phase remains continuous since it is proportional to the integral of the data. When the input data is binary, the resulting

[6] Not to be confused with the differential operator $d\lambda$.

signal is called binary FSK or 2-FSK. A binary input is assumed to discuss the properties of the FSK modulation.

Due to the nonlinear mapping function of the data onto the phase, evaluating the spectra of FSK signals is not trivial. First, the digital modulation index for 2-FSK is defined as [41]

$$h_{\mathrm{FSK}} = \frac{\Delta F}{R}, \tag{2.51}$$

where $R = 1/T_b$ is the bit rate and ΔF the peak-to-peak frequency shift. While the PSD varies significantly with modulation index, an approximate expression for the bandwidth of a 2-FSK signal is given by Carson's rule [41] and can be written as

$$\mathrm{BW} = \Delta F + 2R, \tag{2.52}$$

where a rectangular pulse shape is assumed with a null-to-null bandwidth of $D = R$ in case of binary switching. The PSD of a 2-FSK signal is shown in Fig. 2.21b for $h_{\mathrm{FSK}} = 4$, yielding an approximate bandwidth of $6R$. Of course, a higher number of frequencies can be used to encode information, resulting in higher order M-FSK modulations with greater spectral efficiency.

A special case of 2-FSK is obtained when $h = 0.5$ called minimum-shift keying (MSK). With this condition the phase change across a single symbol time will be exactly $\pi/2$, which can be useful in coherent receiver architectures.

As shown in Fig. 2.22a, a 2-CPFSK signal is most commonly generated by tuning the frequency of a voltage-controlled oscillator (VCO) with a binary data signal, while a 2-DPFSK signal is made by selecting from two distinct VCOs. FSK demodulation can be performed coherently by downconverting the signal with all

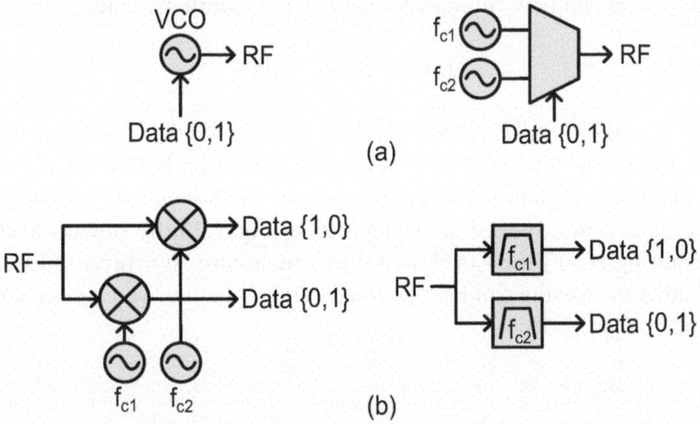

Fig. 2.22 Examples of 2-FSK modulators (**a**) and a coherent and incoherent 2-FSK demodulator (**b**)

possible oscillator frequencies or incoherently by processing the input by various filter circuits tuned at the different frequencies, both visualized in Fig. 2.22b for 2-FSK signaling. Another possible demodulation method is used in [52], where a frequency-dependent phase shift is introduced to convert the frequency offset of the input into a phase difference at the output.

2.2.5 Complex Modulations: QAM and APSK

The simple modulations from the previous section can be extended to modulate more than one parameter simultaneously. The most popular combination is the combination of amplitude and phase modulation, yielding a two-dimensional set of signal states. In its most general form, these amplitude and phase modulated signals can be described under the label of amplitude and phase-shift keying (APSK). A very popular subset of APSK is quadrature amplitude modulation (QAM), which is typically produced from two orthogonal signals.

2.2.5.1 Quadrature Amplitude Modulation

The name quadrature amplitude modulation refers to the fact that a QAM signal can be represented as the signed amplitude modulation of two carriers in quadrature or 90° out of phase:

$$s(t) = x(t)\,\cos(\omega_c t) - y(t)\,\sin(\omega_c t)\,, \tag{2.53}$$

where $x(t)$ and $y(t)$ are the two signed, multilevel data inputs. Using this form, the complex envelope can be written as a vector in the complex plane:

$$g(t) = x(t) + jy(t)\,. \tag{2.54}$$

This is known as the Cartesian representation, referring to the coordinate system that bears the same name. Thanks to the quadrature property, both carriers are orthogonal[7] and can thus be modulated and demodulated independently (ideally) without influencing each other. Indeed, as visualized in Fig. 2.23, by downconverting and low-pass filtering (LPF) the QAM signal by quadrature, synchronized LO signals, both original data streams can be recovered. This operation can be mathematically expressed as follows 2.46:

[7] This orthogonality can be mathematically described as $\int_{t_0}^{t_0+T} \cos(\omega_c t)\sin(\omega_c t)dt = 0$.

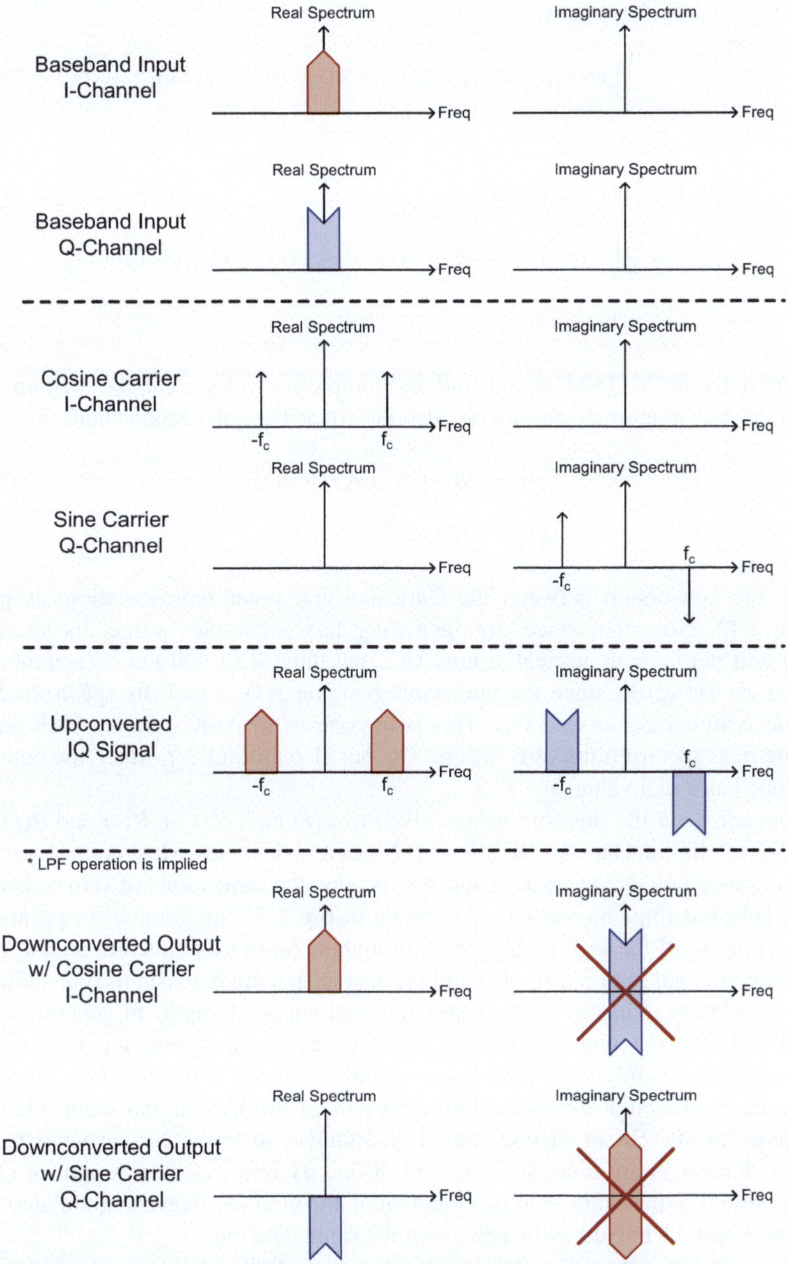

Fig. 2.23 Frequency-domain visualization of the IQ demodulation principle

$$
\begin{aligned}
I(t) &= s(t) \cdot \cos(\omega_c t) \\
&= \frac{1}{2}\big[x(t) + x(t)\cos(2\omega_c) - y(t) \cdot 0 - y(t)\sin(2\omega_c)\big] \\
&\overset{\text{LPF}}{=} x(t)
\end{aligned} \tag{2.55}
$$

$$
\begin{aligned}
Q(t) &= s(t) \cdot \sin(\omega_c t) \\
&= \frac{1}{2}\big[x(t)\sin(2\omega_c t) - x(t) \cdot 0 - y(t) + y(t)\cos(2\omega_c t)\big] \\
&\overset{\text{LPF}}{=} -y(t).
\end{aligned} \tag{2.56}
$$

Alternatively, a QAM signal and its complex envelope can be written as a function of its magnitude and phase, also known as the polar representation:

$$s(t) = R(t)\cos(\omega_c t + \theta(t)) \tag{2.57}$$

$$g(t) = R(t)e^{j\theta(t)}, \tag{2.58}$$

where the conversion between the Cartesian and polar representations is given by Eq. 2.42. Note that since $g(t)$ generally has a complex value, its spectrum $G(f)$ will not be symmetrical around DC, and thus $S(f)$ will not be symmetrical around f_c. However, since the upconverted signal $s(t)$ is real, its spectrum $S(f)$ will be symmetrical around DC. This is in contrast to ASK signals which have a symmetrical spectrum not only around DC but also around $\pm f_c$ since the complex envelope has a real value.

Depending on the possible values given to $x(t)$ and $y(t)$ or $R(t)$ and $\theta(t)$, the QAM constellation can take on different shapes and densities. Most common are the square constellations, where $x(t)$ and $y(t)$ occupy the same range of values, leading to the simplest implementations. As shown in Fig. 2.24, the number of points can range from 4, all the way to 256, or even higher. As more points are added, more bits are contained within a single symbol, but the spacing between the constellation points is clearly reduced for the same maximal signal strength. In general, an M-QAM constellation with M distinct signal states can represent $\log_2(M)$ bits of information per symbol and leads to a spectral efficiency of $0.5 \log_2(M)$ bits/s/Hz. However, even though the spectral efficiency of an M-QAM signal seems identical to that of an M-ASK or M-PSK signal, the number of levels per carrier is halved since two carriers are modulated instead of one. Herein lies the strength of QAM modulation: For the same occupied bandwidth, the number of levels modulated onto the carrier can be halved while achieving the same data rate.

As can be seen from Fig. 2.24, within the magnitude circles of maximal signal amplitude, a lot of signal space is lost because of the square shape. For this reason, nonsquare constellation shapes are sometimes used to more efficiently fill up the circle, where the number of points is an odd power of 2, such as 32-QAM and 128-QAM, as illustrated in Fig. 2.25.

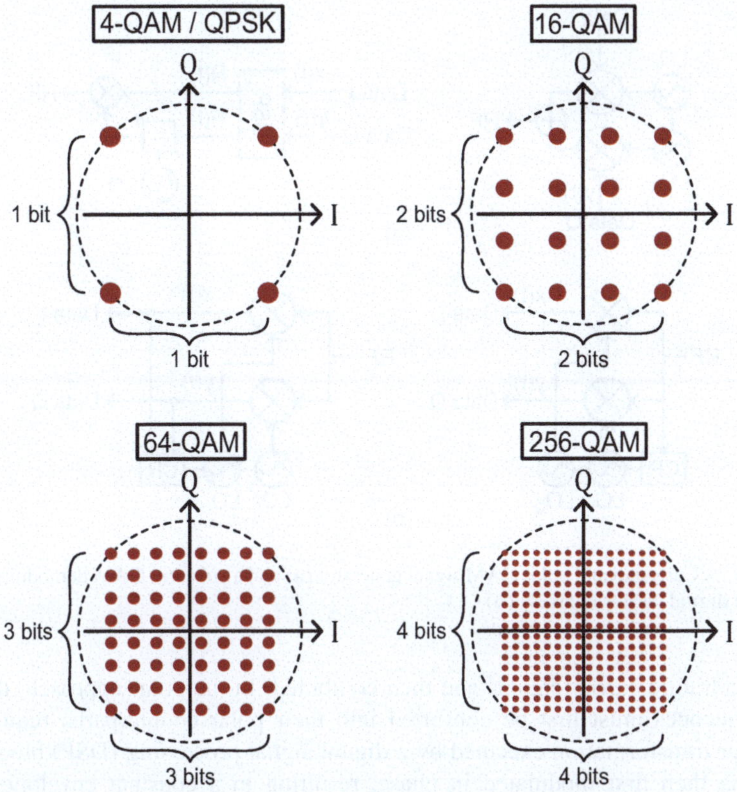

Fig. 2.24 Signal constellations of various square QAM signals: 4-QAM or QPSK, 16-QAM, 64-QAM, and 256-QAM

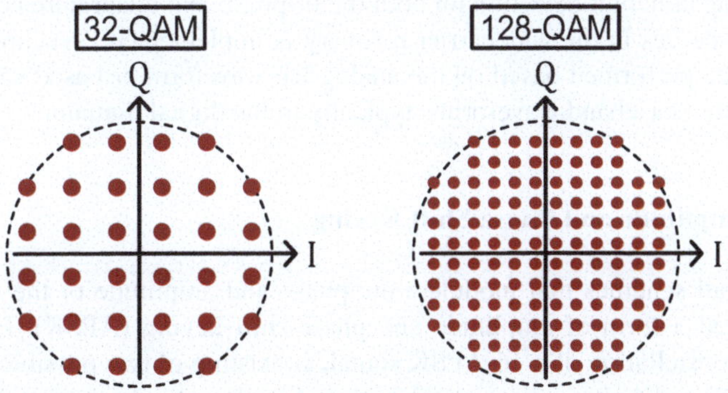

Fig. 2.25 Signal constellations of different nonsquare QAM signals: 32-QAM and 128-QAM

Generating M-QAM signals can be achieved via a Cartesian or polar approach, as shown in Fig. 2.26a. In the Cartesian approach, the two data streams are upconverted

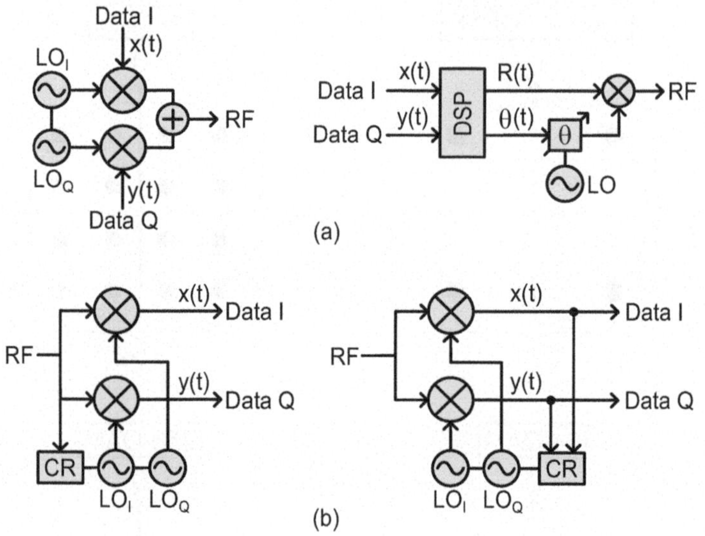

Fig. 2.26 A Cartesian and polar QAM signal generator (**a**) and a coherent QAM demodulator with analog or digital carrier recovery (**b**)

onto quadrature carrier signals and then combined. In the polar approach, the two data sequences must first be converted into their polar counterparts, requiring a nonlinear transformation executed by a digital signal processing (DSP) block. The carrier is then first modulated in phase, resulting in a constant envelope phase modulated signal. This phase modulated signal is then modulated in amplitude to construct the desired M-QAM signal.

Figure 2.26b shows two possible M-QAM demodulation architectures, both implementing coherent detection on both the in-phase and quadrature components. The difference lies in the way carrier recovery is implemented. Synchronizing the carrier can be performed based on the analog RF waveform or based solely on the downconverted baseband waveforms, typically in the digital domain.

2.2.5.2 Amplitude and Phase-Shift Keying

In general, all schemes that modulate the phase and amplitude of the carrier can be considered a form of amplitude and phase-shift keying (APSK). Figure 2.27 shows the constellation of a 16-APSK signal, consisting of two possible amplitude levels with 12 and 4 phase states on the high and low amplitude, respectively. APSK signals generally have a more circular constellation, compared to the more square constellation of QAM signals, which are a special case of APSK. An interesting consequence is that, for the same number of points, the signal takes on a lower number of amplitude levels compared to the equivalent QAM constellation (e.g., 16-APSK vs. 16-APSK). This shape is therefore beneficial for linearity specifications

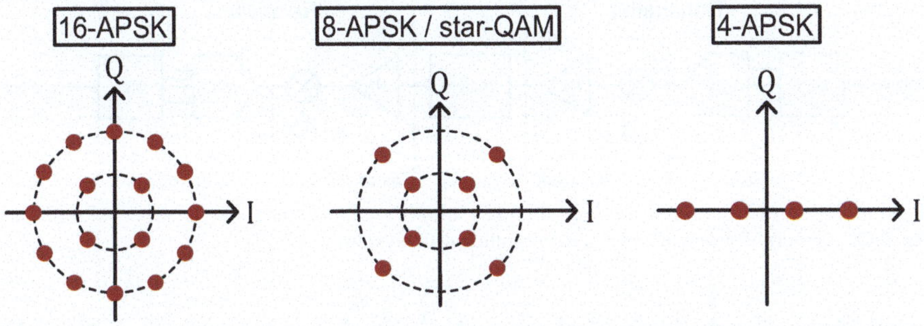

Fig. 2.27 Signal constellations of different APSK signals: 4-APSK, star-QAM, and 16-APSK

and can easily be generated with a polar modulator architecture (Fig. 2.26a). Traditional IQ demodulation is more challenging, however, since the constellation points typically do not map nicely onto the in-phase and quadrature axes.

Figure 2.27 also depicts an 8-APSK constellation that is often referred to as star-QAM. This signal can be viewed as a QPSK signal with an additional amplitude level. Adding an additional amplitude bit improves the spectral efficiency compared to QPSK while still only occupying four distinct phase states, which makes modulation and demodulation more straightforward.

A special case of APSK arises only when one of the carriers is modulated by a signed data input, which can be considered as an intermediate step toward QAM, as two quadrature APSK signals can be combined to form a QAM signal. Figure 2.27 shows an example of a 4-APSK signal modulated onto a single carrier, which forms a 16-QAM signal when combined on the in-phase and quadrature components.

2.3 Performance of Telecommunication Systems

With the information from the previous section, we can now analyze the performance of telecommunication systems and compare the different modulation techniques. Before discussing the most important performance degrading effects, the general framework of a communication system is first explained.

2.3.1 General Framework and Components

Figure 2.28 shows the general framework of a telecommunication system with its basic components. To transform digital input data into digital output data a certain distance away, the system includes a transmitter (TX) and a receiver (RX), both

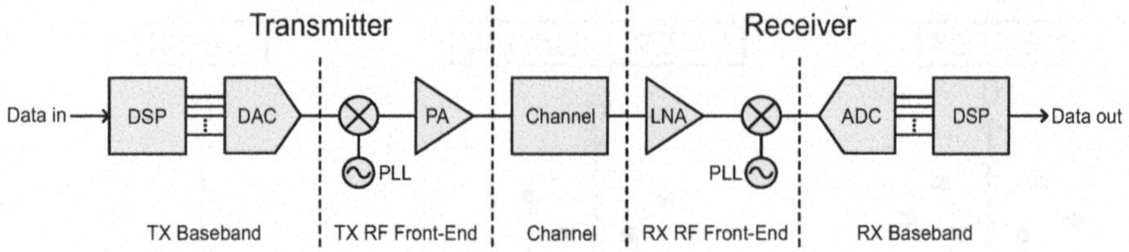

Fig. 2.28 General framework of a telecommunication system

containing a digital processing back end, a mixed-signal converter, and a radio frequency (RF) front-end, with a channel in between.

The raw binary input data is first processed by a digital signal processing (DSP) block and converted into an analog modulated signal by a digital-to-analog converter (DAC). The RF transmitter front-end then upconverts the modulated signal onto the carrier and transmits it over the channel. The RF receiver front-end conditions the received signal and downconverts it to a lower frequency. After digitizing the downconverted signal by an analog-to-digital converter (ADC), it is processed by another DSP block and converted into a binary output stream.

While many different architectures exist (e.g., incoherent receiver systems, direct-digital systems, etc.) and will be discussed later on, this basic framework provides a suitable reference to explain the basic components and performance metrics of a general, coherent communication system.

2.3.1.1 TX-Side DSP

On the transmitter side, the raw input data is processed by a digital processing block to improve the performance of the whole system. For instance, forward error correction is often employed to increase the reliability of the system against bit errors, and hence the data needs to be encoded before transmission. Other popular signal operations executed in the TX's DSP block include digital predistortion, pulse shaping, and equalization, all of which will be elaborated on later. Optionally, a deserializer is required to convert the incoming serial data into a parallel bit stream which eases digital processing and can be used to control the DAC.

2.3.1.2 Digital-to-Analog Converter

The DAC converts the processed digital data into a modulated analog signal. The DAC can generate a baseband modulated signal, or a modulated passband signal at an intermediate frequency (IF), depending on the specifications of the system. The required resolution and the sample rate of the converter depend on the characteristics of the desired output signal, such as bandwidth and amplitude information. For instance, generating a modulated IF signal typically demands a much higher sample rate and a number of bits compared to baseband signal. Furthermore, certain DSP algorithms can also make the signal more complex and thus make the converter specifications more stringent.

2.3.1.3 TX RF Front-End

The RF front-end of the transmitter is responsible for upconverting the modulated DAC output onto the RF carrier and transmitting the signal over the channel. The upconversion is usually performed by a mixer driven by a local oscillator (LO) signal. The LO is typically generated using a phase-locked loop (PLL) which uses an external frequency reference at a lower frequency (such as a crystal oscillator) to produce a stable carrier signal at the desired frequency. The final stage of the transmitter is a power amplifier (PA) that is specifically designed to generate a large output power while maintaining adequate signal quality.

2.3.1.4 Channel

The channel is the physical medium through which the modulated signal propagates. For radio signals, the medium can be free space or a dielectric waveguide, both needing a different interface to the RF circuitry. For free space propagation, antennas are required to radiate the radio signal, while a coupler is necessary to launch the radio signal into the waveguide. Both medium attenuate the signal and can cause specific distortions, both influencing the overall performance of the system.

2.3.1.5 RX RF Front-End

The receiver RF front-end receives the modulated signal and performs the downconversion to a lower frequency for demodulation. The initial amplification is carried out by a low-noise amplifier (LNA) designed to add minimal noise and thus preserve the signal quality as best as possible. The downconversion is carried out by a mixer controlled by a synchronized LO signal. The LO can once again be generated by a PLL but must be synchronized to the carrier of the modulated signal to ensure optimal coherent detection performance. The carrier frequency and phase recovery can be achieved in the analog domain (e.g., using a Costas-loop) or in the digital domain based on the digitized downconverted signals.

2.3.1.6 Analog-to-Digital Converter

The ADC digitizes the downconverted modulated signal to be processed and demodulated in the digital domain. Similar to the DAC, the required sample rate and resolution is determined by the nature of the signal (baseband or passband)

and the complexity of the algorithms in the digital baseband. After all, advanced algorithms may need more time or amplitude resolution to work optimally.

2.3.1.7 RX-Side DSP

The DSP block in the receiver processes the digitized signal and tries to recover the transmitted data. Possible digital operations include equalization, pulse shape filtering, carrier frequency and phase recovery, clock[8] and data recovery, demodulation, and decoding the error correcting code. In case the input represents a modulated passband signal, demodulation may include a downconversion in the digital domain. Finally, the output data may be serialized depending on what block or system follows.

2.3.1.8 Evaluating Performance

In the end, the main goal of a communication system is to reliably reproduce the data modulated by the transmitter again at the receiver. Hence, the most important performance metric is the bit-error rate in the system, which is determined by the quality of the signal when the bit or symbol decision is made.[9] Depending on the application, the latency might be an important specification as well but will not be the focus of this book.

In ideal conditions, it is easy to determine what symbol was originally transmitted. However, in reality the signal quality is degraded by noise, nonlinearities, and undesired filtering, which makes the symbol decision more uncertain leading to sporadic errors. In the rest of the chapter, the most important signal degrading effects are investigated, and their influence on the bit-error rate performance is assessed.

2.3.2 Noise

2.3.2.1 Origin and Properties of Noise

Noise is an inherent phenomenon of our physical world, generated by the random motion of charge carriers (e.g., electrons) in a conductor with electrical resistance. Noise fundamentally puts a limit on the smallest signal that can be reliably detected by a receiver before it is indistinguishable from the noise. Various types of noise exist with different origins and characteristics. The most common type of noise is

[8] This is the clock used to sample the digital data, not the carrier used for modulation.

[9] Concretely, this is done when the signal is quantized by a series of comparators comparing the signal to a set of references.

thermal noise, caused by the random thermal motion of charged particles. Other types include shot noise or flicker noise, arising from different physical effects in semiconductors.

Since noise is a random process, it is impossible to exactly predict its instantaneous value. However, noise can be modeled as a statistical process allowing us to make predictions on average quantities. Thermal noise, for example, is a type of white noise, meaning that the power spectral density (PSD) is constant over frequency. In case of thermal noise, the PSD increases with temperature and can be expressed as[10]

$$\mathrm{PSD}(f) = N_0 = kT \ , \tag{2.59}$$

where $k = 1.38 \cdot 10^{-23}\, J/K$ is the Boltzmann constant, a physical constant relating the average thermal energy of particles with its temperature. At a room temperature of 300 K, this corresponds to a thermal noise power density of −174 dBm/Hz. The integrated noise power over a specified bandwidth Δf can then be calculated as follows:

$$P_n = \int_{f_0}^{f_0+\Delta f} kT\, df = kT\, \Delta f \ . \tag{2.60}$$

Furthermore, the probability density function (PDF) of the thermal noise amplitude $p(V_n)$ follows a Gaussian distribution with zero mean:

$$p(V_n) = \frac{1}{\sqrt{2\pi}\,\sigma_n} e^{-V_n^2/2\sigma_n^2} \ , \tag{2.61}$$

where the standard deviation σ_n corresponds to the root mean square (RMS) noise voltage $V_{n,rms}$. Figure 2.29 shows a time-domain waveform, the PSD, and PDF of thermal noise. How circuits generate and influence noise will be discussed in a later chapter.

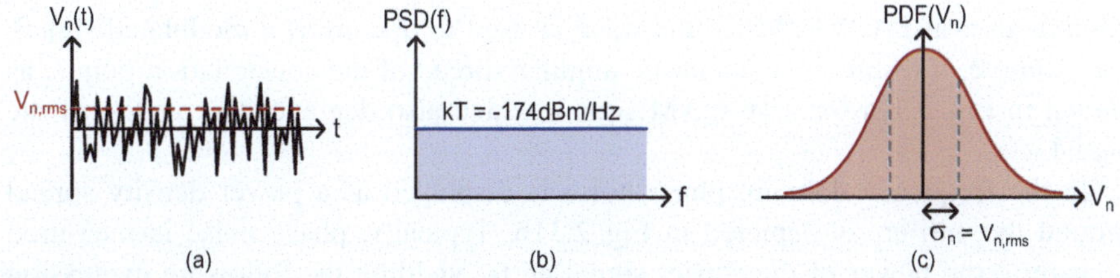

Fig. 2.29 Time-domain waveform (**a**), power spectral density (**b**), and probability density function (**c**) of thermal noise

[10] Defined over positive frequencies only, otherwise, a factor of 1/2 should be added.

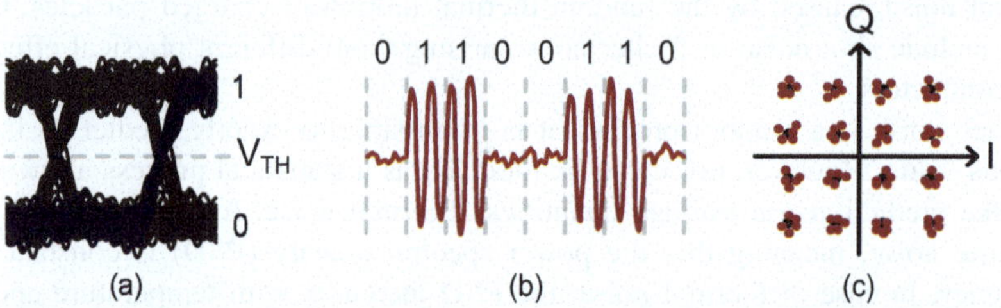

Fig. 2.30 Illustration of how noise affects a binary eye diagram (**a**), an OOK signal (**b**), and a 16-QAM constellation (**c**)

2.3.2.2 Additive White Gaussian Noise

When this white, Gaussian noise is simply added on top of the signal, the model is called additive white Gaussian noise (AWGN). The AWGN model is used very often for theoretical analyses of communication systems as this type of noise is the most prevalent. Figure 2.30 shows how a binary eye diagram, an OOK signal, and a 16-QAM constellation are affected by AWGN. It is clear that noise influences the quality of the signal, and this is usually characterized by the signal-to-noise ratio (SNR), which can be expressed as follows assuming that AWGN is the main source of noise:

$$\mathrm{SNR}_{\mathrm{AWGN}} = \frac{P_s}{P_n} = \frac{P_s}{ktB}, \tag{2.62}$$

where the noise is integrated over the occupied signal bandwidth B.

2.3.2.3 Phase Noise

Another important manifestation of noise is phase noise (PN). This type of noise, present in oscillators, is expressed as a random fluctuation of the phase of a continuous wave (CW). When this wave is used to upconvert a modulated signal, for example, it results in a stochastic angular spread of the constellation points, as shown in Fig. 2.31a for a 16-QAM signal, which also degrades the quality of the signal.

In the frequency domain, phase noise is displayed as a power density spread around the carrier, as depicted in Fig. 2.31b. Typically, phase noise is measured relative to the power of the carrier signal at f_0, yielding the following expression of the RMS phase error:

$$\Phi_{\mathrm{rms}}^2 = \int_{-B}^{B} PN(f)\,df, \tag{2.63}$$

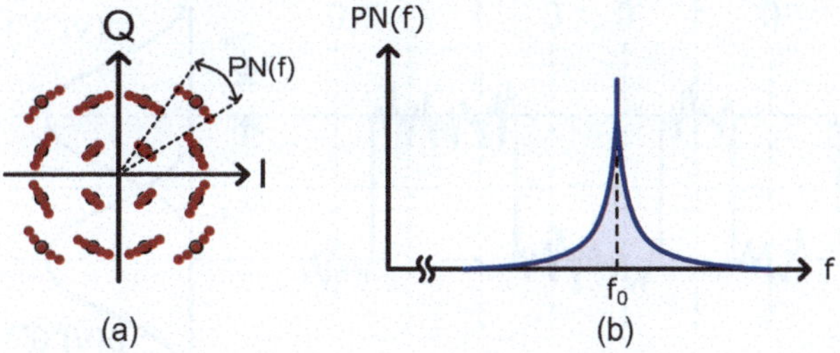

Fig. 2.31 Effect of phase noise on a 16-QAM constellation (**a**) and typical frequency profile of phase noise (**b**)

where the $[-B, B]$ is the occupied RF signal bandwidth. The SNR degradation caused by phase noise can be calculated based on the RMS phase error as follows:

$$\mathrm{SNR}_{PN} = \frac{1}{\Phi^2_{\mathrm{rms}}}, \tag{2.64}$$

since the signal (the carrier in this case) is normalized to one and noise power is equal to the square of the RMS phase error.

2.3.2.4 BER for Baseband Signaling

To quantify the influence of noise on the performance of the system, one must find the relationship between SNR and the bit-error rate (BER). First, the expression for BER in case of binary signaling is derived in an intuitive way.

In the absence of noise, an ideal binary signal can take on two discrete values: V_H for a binary "1" and V_L for a binary "0," resulting in a trivial bit decision. However, when white Gaussian noise is added to this signal (AWGN), a continuous range of values is possible which are characterized by a Gaussian PDF centered around the expected values V_H and V_L in case a binary "1" or "0" was transmitted, respectively. These PDFs are denoted as $p(V|''1'')$ and $p(V|''0'')$ and can be written as

$$p(V|''0'') = \frac{1}{\sqrt{2\pi}\sigma_n} e^{-(V-V_L)^2/2\sigma_n^2} \tag{2.65}$$

$$p(V|''1'') = \frac{1}{\sqrt{2\pi}\sigma_n} e^{-(V-V_H)^2/2\sigma_n^2}, \tag{2.66}$$

where σ_n is again equal to the RMS noise voltage. Figure 2.32 visualizes the Gaussian noise distributions along with the signal waveform.

To make the bit decision, the signal value is compared to a threshold voltage that is placed in the middle of the expected values: $V_{TH} = (V_H + V_L)/2$. When the

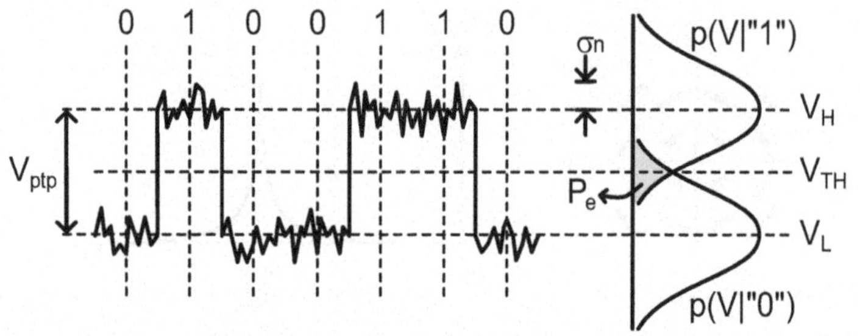

Fig. 2.32 The effect of AWGN on the signal and bit-error rate

signal is greater than this threshold, a binary "1" is assumed, while a binary "0" is assumed when it is smaller. A bit error thus occurs whenever the PDF of either level crosses this threshold value, leading to the mathematical expression of the bit error probability P_e of

$$P_{e,0} = P(V > V_{TH}|''0'') = \int_{V_{TH}}^{\infty} p(V|''0'') \tag{2.67}$$

$$P_{e,1} = P(V < V_{TH}|''1'') = \int_{-\infty}^{V_{TH}} p(V|''1'')\,. \tag{2.68}$$

Assuming equal probabilities of transmitting a "1" or "0," the total BER (P_e) can be written as [41]

$$P_e = Q\left(\frac{V_{ptp}}{2V_{n,rms}}\right) = Q\left(\sqrt{\frac{2E_b}{N_0}}\right), \tag{2.69}$$

where V_{ptp} is the peak-to-peak signal amplitude ($V_H - V_L$). As shown in the last part of the expression, the bit error rate is also often expressed as a function of the energy per bit E_b over the noise density N_0, also called the SNR per bit. $Q(x)$ is a cumulative distribution function for the Gaussian distribution, also called the Q-function, and is defined as

$$Q(x) = \frac{1}{\sqrt{2\pi}} \int_{x}^{\infty} e^{-u^2/2}\,du\,. \tag{2.70}$$

For multilevel signals, the expression can be extended taking into account that V_{ptp} needs to be divided between all possible levels. For the same V_{ptp} and noise power $V_{n,rms}$, the margin between the different signal levels is decreased, leading to a higher bit-error rate for the same noise power.

2.3.2.5 BER for Passband Signaling

For passband signals, a more formal approach is taken to derive BER expressions for all of the discussed modulations, as described in [53]. Only the most important expressions and concepts are given here, while a more elaborate description and explanation is outlined in Appendix A.1.

First of all, since we want BER expressions as a function of SNR, the SNR can be written as a function of the received power P_r, the noise power density N_0, and the energy per symbol E_s.

$$\mathrm{SNR} = \frac{P_r}{N_0 B} = \frac{E_s}{N_0 B T_s} = \frac{E_s}{N_0}, \tag{2.71}$$

which shows that E_s/N_0 is equivalent to SNR in case $B = 1/T_s$. Furthermore, E_b/N_0 corresponds to SNR as well if binary signaling is concerned.

As explained in Appendix A.1, a general expression for BER can be derived by reasoning how the total signal energy is divided among the different symbols and by taking into account the number of neighbors of each symbol in the constellation [53]. Setting L equal to the number of levels per dimension and M to the total number of constellation points, the bit-error rate can be expressed as [54]

$$P_b = \frac{2(1-1/L)}{\log_2(L)} Q\left(\sqrt{\frac{\log_2(L)}{\log_2(M)} \frac{6\,\mathrm{SNR}}{L^2-1}}\right). \tag{2.72}$$

For M-PAM modulations, $L = M$, while for rectangular M-QAM modulations, $L = \sqrt{M}$. For instance, for 16-QAM, $L = 4$ and $M = 16$. In these expressions, it is assumed that the received is a coherent system with perfect carrier synchronization. For incoherent detectors, the error rates can be derived as well [41], typically needing more SNR for the same BER. The lower sensitivity of incoherent receivers does come with the advantage of significantly lower implementation complexity and power consumption.

Figure 2.33 shows the bit-error rate versus SNR, based on Eq. 2.72, for various modulation schemes. It is clear that as the modulation becomes more dense, meaning more constellation points are used for the same signal power, the required SNR increases for the same bit-error rate. This is of course as expected, as the same amount of noise will give a higher chance of overlapping with neighboring symbols as the distance among them becomes smaller. Table 2.2 gives an overview of the needed SNR values for a certain desired BER, again for various modulations.

It must be emphasized that this equation is derived assuming the system is AWGN-limited. In case other effects are also degrading the performance, like phase noise, nonlinearities, etc., the impact of these factors should be studied on a case-by-case basis. However, since finding these relationships is usually quite challenging, the given expression for BER is often used with an equivalent SNR that quantifies the signal quality including other effects as well.

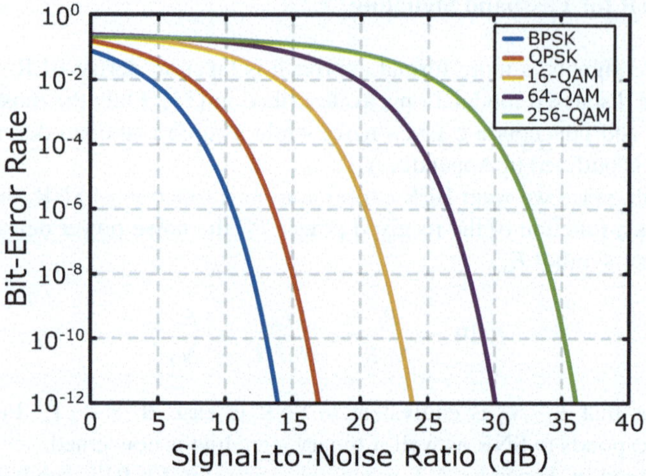

Fig. 2.33 Bit-error rate versus SNR for various modulation schemes in AWGN-limited systems

Table 2.2 Required SNR values for various BERs and modulations

SNR @ BER =	10^{-3}	10^{-6}	10^{-12}
BPSK	6.8 dB	10.5 dB	13.9 dB
QPSK	9.8 dB	13.6 dB	17.0 dB
16-QAM	16.6 dB	20.4 dB	23.9 dB
64-QAM	22.6 dB	26.6 dB	30.1 dB
256-QAM	28.4 dB	32.6 dB	36.1 dB

2.3.2.6 SNR and EVM

Another important metric to quantify the quality of a modulated signal is the error vector magnitude (EVM), which evaluates how far a measured constellation point is from its ideal location. The EVM is defined as the normalized RMS difference between a collection of measured symbols and ideal symbols. For each measured constellation point, the difference can be calculated by taking the magnitude of the error vector going from the measured point to the ideal reference, as demonstrated in Fig. 2.34. Most commonly, the RMS error is normalized to the RMS magnitude of the constellation points, yielding the following expression:

$$\mathrm{EVM_{RMS}} = \frac{\sqrt{\frac{1}{N}\sum_{i=1}^{N}|S_{\mathrm{meas},i} - S_{\mathrm{ref},i}|^2}}{\sqrt{\frac{1}{M}\sum_{i=1}^{M}|S_{\mathrm{ref},i}|^2}} = \frac{\sqrt{\frac{1}{N}\sum_{i=1}^{N}|S_{\mathrm{error},i}|^2}}{\sqrt{\frac{1}{M}\sum_{i=1}^{M}|S_{\mathrm{ref},i}|^2}}. \tag{2.73}$$

Another EVM definition uses the peak magnitude of the constellation for normalization, leading to smaller EVM values compared to $\mathrm{EVM_{RMS}}$. For the rest of this book, the RMS definition will be used, and EVM thus always refers to $\mathrm{EVM_{RMS}}$.

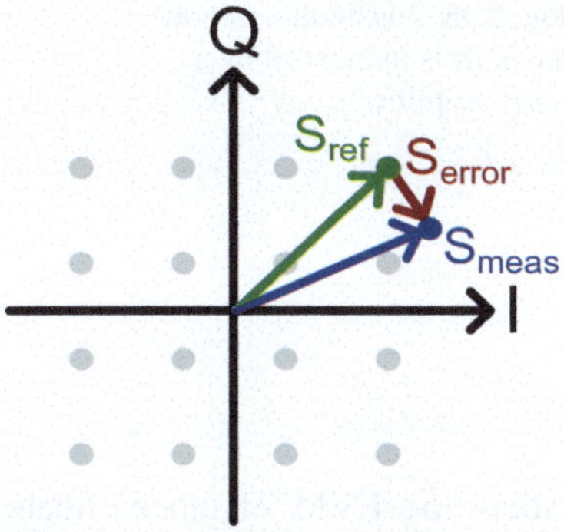

Fig. 2.34 Visualization of the error vector for the calculation of the EVM

The EVM is able to more easily describe signal degradation caused by factors other than noise, such as nonlinearities, and is therefore often used to characterize modulated signals where noise is not the bottleneck, for example, in the transmitter.

Calculating the error vector requires the knowledge of what symbol was transmitted to be used as the reference for the error. When the signal quality is high, the closest reference point to the measured point is usually the correct reference. However, in practical systems with poor signal quality, this is not always. As a result, inaccurate EVM values are obtained when the transmitted symbols are unknown, and the nearest reference is simply taken as a reference. Hence, EVM can be a good metric of the system performance but should be taken with a grain of salt, especially when the reported SNR is poor. It is always a good idea to visually inspect the signal constellation diagram and assess the signal quality based on the plot.

It is clear that the SNR and EVM are closely related. In fact, in AWGN-limited systems, the error vector is fully determined by the Gaussian noise, and there is a clear relationship between SNR and EVM [54]:

$$\mathrm{SNR}_{\mathrm{AWGN}} \approx \frac{1}{\mathrm{EVM}^2_{\mathrm{AWGN}}} . \tag{2.74}$$

As stated before, to evaluate the performance in the presence of multiple degrading effects, the EVM and SNR of the whole system can be calculated by summing the squares of all EVM sources:[11]

$$\mathrm{EVM}_{tot} = \sqrt{\frac{1}{\mathrm{SNR}_{tot}}} = \sqrt{\mathrm{EVM}^2_{\mathrm{AWGN}} + \mathrm{EVM}^2_{\mathrm{PN}} + \dots} , \tag{2.75}$$

[11] This summation is only valid when the error sources can be modeled as independent statistical processes [55].

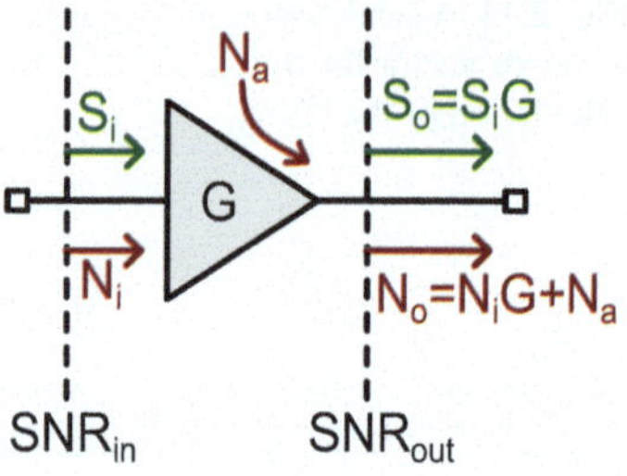

Fig. 2.35 Illustration of how the SNR is influenced by a noisy amplifier

where the EVM of other effects can be added as well, for instance, the EVM quantifying nonlinearities, IQ imbalances, etc. The bit-error rate of the whole system can then be approximated by filling out expression 2.72 with SNR_{tot}.

2.3.2.7 Noise Figure

Knowing what noise is and does to the signal quality, we must be able to describe what the influence is of a circuit on the noise, or better, signal-to-noise ratio. An ideal amplifying block will amplify the signal and the noise equally and thus not alter the SNR. However, any realistic block will add some noise to the signal and therefore decrease the SNR, which is expressed by the noise factor of the block F:

$$F = \frac{\text{SNR}_{\text{in}}}{\text{SNR}_{\text{out}}} . \tag{2.76}$$

This quantity is often expressed in decibel and called the noise figure $NF = 10 \log_{10}(F)$. In the rest of this book, both the linear and decibel quantity will often be referred to as noise figure. As illustrated in Fig. 2.35, the expression of the NF can be further expanded as follows:

$$F = \frac{\frac{S_i}{N_i}}{\frac{S_0}{N_0}} = \frac{\frac{S_i}{N_i}}{\frac{S_i\,G}{N_i\,G+N_a}} = 1 + \frac{N_a}{N_i\,G} , \tag{2.77}$$

showing that the NF of an ideal, noiseless (added noise $N_a = 0$) block is equal to 1 or 0 dB.

The NF is a useful metric to describe how SNR degrades as the signal travels through a circuit. To evaluate the achievable SNR in the whole receiver, all of the different noise sources can be referred back to the input of the receiver and combined into a single NF value. The total NF, as the result of a cascade of noisy stages, can be calculated as [56]

$$F_{tot} = F_1 + \frac{F_2 - 1}{G_1} + \frac{F_3 - 1}{G_1 G_2} + \dots , \tag{2.78}$$

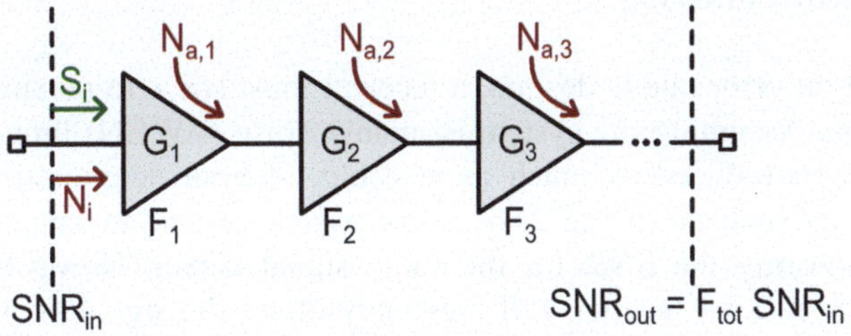

Fig. 2.36 Noise figure calculation of a cascade of noisy stages

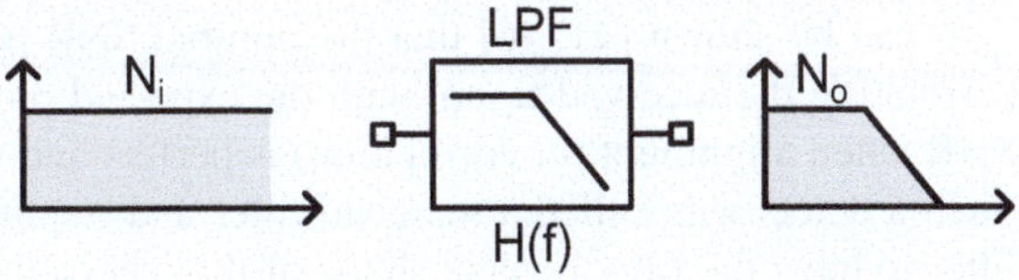

Fig. 2.37 Effect of a low-pass filter on white noise

where F_n and G_n are the noise figure and power gain (both in linear scale) of the nth stage, as depicted in Fig. 2.36. This equation shows that noise added further along the chain is attenuated by the gain delivered by earlier stages. In other words, the first stage(s) in the receiver chain is the most critical for the noise performance as their noise contributes the most. For this reason, most receiver architectures consist of an amplifier chain optimized for low-noise performance, also called a low-noise amplifier (LNA), in the beginning of the chain. The later stages can then be optimized for other specifications as their noise has negligible contribution to the total noise figure of the receiver, as long as the signal maintains a reasonable power level.

The general expressions for noise figure are also valid when the noise figures or gain characteristics vary over frequency. Indeed, a general transfer function $H(f)$, describing a circuit or a channel, has the following effect on the incoming noise power density N_i:

$$N_O = N_i\, |H(f)|^2\,, \tag{2.79}$$

where the output noise power density is not only amplified or attenuated but also shaped by the frequency response of $H(f)$. The filtering or shaping of noise is illustrated in Fig. 2.37 in case of a low-pass filter. Since $H(f)$ is generally a transfer function characterized for signals (voltage or current), it should be squared to determine its effect on a noise power density such as noise. The gain functions G_n in Eq. 2.78 can thus be replaced by any frequency-dependent power transfer function.

2.3.2.8 Matched Filtering

As a minimal bit-error rate is desired, a receiver must strive to maximize the SNR upon detection. Assuming the system is mainly noise (AWGN) limited, the noise power should be reduced as much as possible, without deteriorating the signal of interest. As visualized in Fig. 2.37, a low-pass filter can attenuate the received noise, thus boosting the SNR for the same signal power. However, as will be discussed in Sect. 2.3.4, a filter will inevitably affect the signal as well, possibly causing signal distortions that negatively impact the link performance. Designing the optimal receiving filter thus entails a trade-off between noise suppression and signal distortion.

It can be shown [41, 46] that the optimal SNR performance can be obtained by correlating the received signal with the expected pulse shape, leading to maximum SNR when alignment (or correlation) is perfect and suboptimal SNR when it is not. Such a detector is called a matched filter and requires the impulse response of the filter to have the time-reverse shape of the received pulse. Assuming that the pulse shape is fully determined by a filter on the transmit side, with negligible influence of the front-ends, then the matched filter can be designed identical to this filter if it has a symmetrical impulse response. However, in high-speed systems the pulse shape is affected by many blocks in the transceiver and hence challenging to predict and replicate in a single filter. Therefore, instead of trying to construct an exact pulse shape correlator, the desired noise filtering is obtained using a low-pass filter with optimized response and bandwidth. This way, close to optimal SNR and thus detection performance is achieved with a more realistic and robust implementation. More details on matched filtering can be found in [41, 46].

2.3.2.9 Forward Error Correction

In practical telecommunication systems, forward error correction (FEC) or channel coding is a crucial technique employed to mitigate errors in data transmission over noisy and unreliable channels [57]. The concept involves encoding the information in a redundant manner, often utilizing error correcting codes (ECCs). This redundancy enables the receiver to not only detect errors throughout the message but also correct a limited number of errors. However, the inclusion of redundancy increases the necessary bandwidth for the same effective bit rate, as some transmitted bits do not carry new information. In the 1940s, Richard Hamming was a pioneer in this field and introduced the first error correcting code, known as the Hamming (7,4) code, in 1950 [58].

A simple example of an ECC is the (3,1) repetition code, where each bit is transmitted three times. In the event of an error in one of the bits, the redundancy allows for error recovery using the other two correctly received redundant bits. However, this type of code is highly inefficient, as it requires three times the bandwidth to transmit the same information rate. Alternatively, for the same bandwidth, the effective bit rate is reduced by a factor of 3. Moreover, if two out of three bits are decoded incorrectly, error recovery becomes impossible. The maximum number of correctable errors is determined by the ECC design, making different ECCs suitable for different scenarios.

Designing ECCs involves a trade-off between reliability or error-rate improvement and effective data rate. Increasing redundancy or employing more complex codes can reduce the bit-error rate after ECC decoding, known as post-FEC BER, but typically leads to a decrease in effective data rate. Furthermore, encoding and decoding complex codes at high data speeds requires sophisticated algorithms, introducing latency in the link and consuming substantial power in the digital processor. Overall, applying FEC with varying degrees of complexity can significantly enhance system performance, as achieving the same BER improvement without FEC would be more challenging and less efficient, e.g., by improving the link SNR through other means. In practice, some type of FEC is usually assumed when determining the SNR specification of the link.

There are two main types of ECCs: block codes and convolutional codes, each with its own advantages and applications. Decoding can be implemented using hard-decision or soft-decision algorithms. An example of an advanced code is the Continuously Interleaving version of traditional Bose–Chaudhuri–Hocquenghem (CI-BCH) cyclic error correcting codes [59]. This code can convert a raw BER below 4.4×10^{-3} to a post-FEC BER $<10^{-15}$ with a 7 % overhead in effective bit rate. This is why many published communication links in literature strive for a raw BER around 10^{-3} even though an error-free system is desired in practical situations. Despite the implementation complexity and data rate overhead of codes like CI-BCH, using such advanced FEC will enhance the efficiency of most links compared to a link that has to obtain a raw BER better than $<10^{-15}$ without any FEC, which would require an immense SNR.

FEC is a vast and extensive topic that deserves in-depth exploration beyond the scope of this book. However, the reader is encouraged to delve further into this fascinating subject [53].

2.3.3 Nonlinearity

Apart from noise corrupting the useful signal, the signal can also be distorted by the circuits. While circuits are often approximated by a linear model, which is valid for small signals, large signal amplitudes can lead to nonlinear behavior that may significantly distort the signal. Assuming a memoryless system, the input $x(t)$–output $y(t)$ characteristic of a nonlinear system can be approximated by Razavi [60]

$$y(t) = a_1 x(t) + a_2 x(t)^2 + a_3 x(t)^3 , \tag{2.80}$$

where higher order terms are negligible in most systems. This equation describes systems with so-called soft-linearity, where the nonlinear behavior follows a reasonably continuous function. "Hard-nonlinearity," which occurs in systems with a sharp cut-off for example, requires more tedious mathematical models, such as Volterra series, which will not be discussed in this book. Also, the physical effects that are responsible for the nonlinear behavior will be examined in a later chapter and will be disregarded for now.

2.3.3.1 Harmonic Distortion

As we are mainly interested in RF systems, it is instructive to examine the consequences of this nonlinear behavior on a sinusoidal input. If $x(t) = A \cos \omega t$, then [60]

$$\begin{aligned} y(t) &= a_1 A \cos \omega t + a_2 A^2 \cos^2 \omega t + a_3 A^3 \cos^3 \omega t \\ &= a_1 A \cos \omega t + \frac{a_2 A^2}{2}(1 + \cos 2\omega t) + \frac{a_3 A^3}{4}(3 \cos \omega t + \cos 3\omega t) \\ &= \frac{a_2 A^2}{2} + \left(a_1 A + \frac{3 a_3 A^3}{4}\right) \cos \omega t + \frac{a_2 A^2}{2} \cos 2\omega t + \frac{a_3 A^3}{4} \cos 3\omega t \,. \end{aligned} \tag{2.81}$$

Clearly, the output contains frequency components at the input frequency, called the fundamental, and components at integer multiples of the input frequency, called the harmonics. The amplitude of each nth harmonic grows in proportion to A^n, indicating that the harmonic distortion becomes more important as the input amplitude increases. Furthermore, a DC shift is obtained caused by the second-order nonlinearity or even-order nonlinearities in general.

As we will see in subsequent chapters, many RF and analog circuits employ a fully differential topology, resulting in an odd symmetrical characteristic which eliminates even-order harmonics from the output. Even though random mismatch can corrupt this symmetry, even-order harmonics can be neglected in many systems.

While the effects of nonlinearity are clear on a single-tone input, communication systems work with modulated signals carrying information over a certain bandwidth. To investigate the influence of nonlinearity on the modulation, assume a generic, complex APSK signal with a modulated amplitude and phase as input to a nonlinear system, only coping with the Nth order or nonlinearity:

$$y(t) = \left[A(t) \cdot e^{\omega t + \phi(t)}\right]^N = A(t)^N \cdot e^{N\omega t + N\phi(t)} \,. \tag{2.82}$$

From this equation it follows that the amplitude information experiences the same nonlinearity, while the phase is multiplied by a factor of the order of the nonlinearity. Clearly, a nonlinear characteristic can not only degrade the quality or EVM of some constellations but also leads to an expansion of the signal bandwidth as the effective modulation speed is increased. While the first consequence is only of importance when a certain harmonic is the desired output (e.g., the tripler-last transmitter in Chap. 6), the second one increases the bandwidth of undesired spurs that may overlap with the desired signal. The remaining analyses will mostly be performed with single-tone inputs but can be extended to modulated signals bearing in mind the bandwidth expanding property caused by nonlinearity.

2.3.3.2 Gain Compression

Many systems exhibit a compressive behavior, meaning a decreasing gain as the input amplitude increases. When even-order nonlinearities are neglected and all harmonics are filtered out, the signal gain A experienced by input $x(t) = A\cos\omega t$ is equal to

$$A = \frac{y(t)}{x(t)} = a_1 + 3\,a_3\frac{A^2}{4}, \tag{2.83}$$

where $a_1 \cdot a_3 < 0$, causing a decrease of gain as the input amplitude A grows. This effect is often quantified by the 1-dB compression point, defined as the input level A_{in} that causes a gain drop of 1 dB (or 10%) and thus an output level 1 dB below the ideal value, as illustrated in Fig. 2.38. 1-dB compression is also often described using the output level at which it occurs. Where $A_{\mathrm{in,1\,dB}}$ and $A_{\mathrm{out,1\,dB}}$ are amplitude (voltage or current) quantities, in RF circuits these metrics are usually specified in terms of power quantities, yielding the input- and output-referred 1-dB compression points: P1dB and OP1dB.

By equating the compressed gain (Eq. 2.83) to 1 dB less than the ideal gain (a_1), the input 1-dB compression points can be calculated as

$$A_{\mathrm{in,1\,dB}} = \sqrt{0.145\left|\frac{a_1}{a_3}\right|}. \tag{2.84}$$

While gain compression is no problem for CW signals, it can distort a modulated signal that carries information in its amplitude. As illustrated in Fig. 2.39, when the

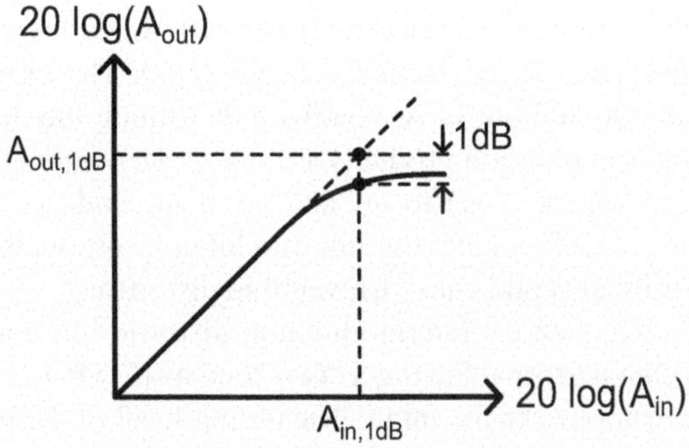

Fig. 2.38 Definition of 1-dB compression point in a compressive system

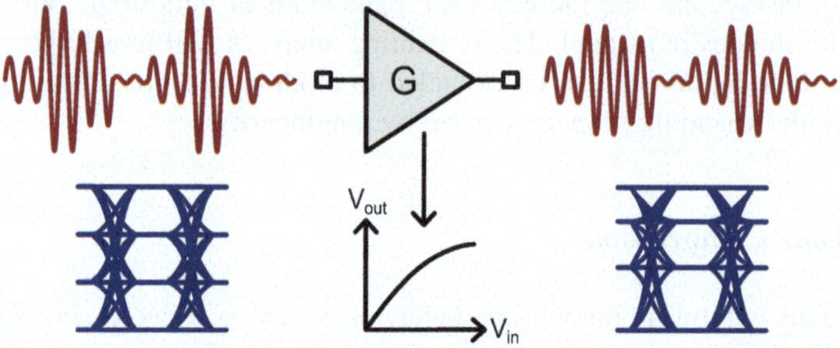

Fig. 2.39 Effect of gain compression on a 4-ASK signal and downconverted 4-PAM signal

four amplitude levels of a 4-ASK signal experience a different gain, the spacing among the levels becomes unequal. More specifically, the margin between the highest signal levels decreases, which reduces the noise margin of SNR during detection, as can be clearly seen from the downconverted 4-PAM eye diagram in the figure.

Gain compression can also be problematic if another undesired signal, called an interferer, is received together with the desired signal. In particular, if the interferer has a large amplitude, it can cause gain compression (or saturation) in the system even when the signal of interest has a small amplitude. This phenomenon is called desensitization because the decrease in gain lowers the SNR of the signal and therefore the sensitivity of the receiver.

2.3.3.3 Intermodulation Distortion

Next to harmonic distortion and desensitization, a nonlinear system presented with two signals comparable in amplitude can output components that are not harmonics of these input frequencies. This effect is called intermodulation (IM) and arises from the mixing of two components as their sum is raised to a power rather than the individual signals. Assuming an input $x(t) = A_1 \cos \omega_1 t + A_2 \cos \omega_2 t$, the output will not only contain components at the fundamental frequencies, ω_1 and ω_2, and their harmonics but also at the difference between all fundamental and harmonic frequencies, as visualized in Fig. 2.40. The exact amplitudes of these components are worked out in Appendix A.2. Among the different components, the third order intermodulation products are particularly interesting since $2\omega_1 - \omega_2$ and $2\omega_2 - \omega_1$ can appear close to ω_1 and ω_2 if ω_1 and ω_2 have similar values. Especially for wideband signals, the intermodulation products can easily overlap with the signal of interest and cause irreversible distortion.

To quantify intermodulation distortion in nonlinear systems, characterization is often performed using a "two-tone test," where two sinusoids with equal amplitudes are applied to the input. The output level of the intermodulation products (Eq. A.14) is then normalized to that of the fundamental ($A_1 = A_2 = A$) to arrive at the relative third order intermodulation strength IM_3:

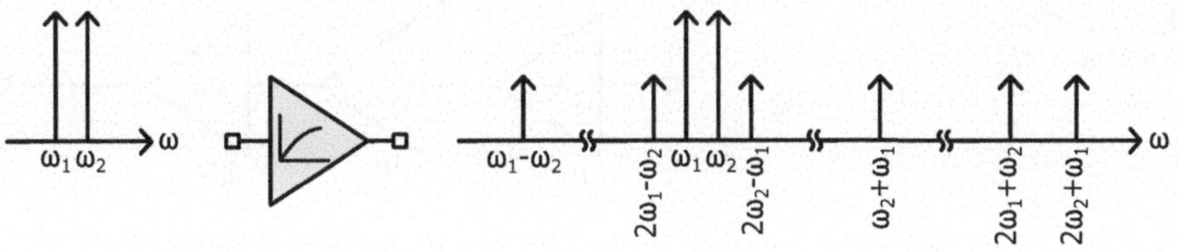

Fig. 2.40 Generation of intermodulation products in a two-tone test of a nonlinear system

Fig. 2.41 Definition of IP_3

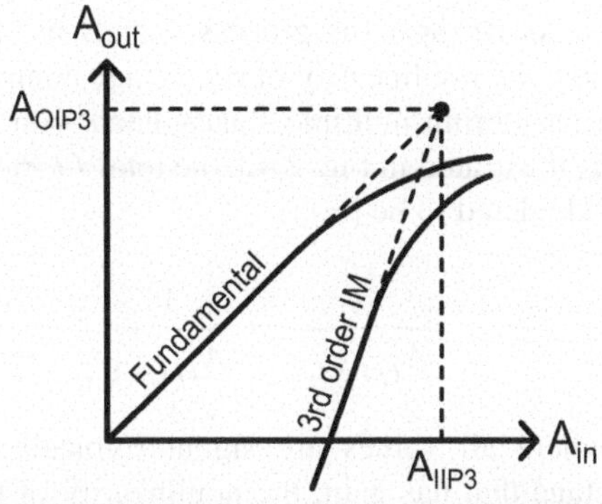

$$IM_3 = 20 \log \left(\frac{3}{4} \frac{a_3}{a_1} A^2 \right) , \tag{2.85}$$

which is expressed in units dBc to denote the normalization with respect to the carrier (i.e., the fundamental amplitude in this case). As the amplitude of each tone is increased by 6 dB (a factor of 2), both intermodulation products grow by 18 dB ($\propto A^3$) and thus the relative IM by 12 dB ($\propto A^2$). To quantify the intermodulation behavior independent of input amplitude, a measure called "third intercept point" (IP_3) is used which is defined as the amplitude where the intermodulation product becomes equal in amplitude to the fundamental tones. As illustrated in Fig. 2.41, the input and output levels where this occurs are called the input- and output-referred third order intercept points (IIP3 and OIP3), respectively.

By equating the amplitudes of the intermodulation product and fundamental component, the IIP_3 can be found:

$$A_{IIP3} = \sqrt{\frac{4}{3} \left| \frac{a_a}{a_3} \right|} \approx A_{1\,\mathrm{dB}} + 9.6\,\mathrm{dB} , \tag{2.86}$$

where the last part of the equation relates the third intercept point of the 1-dB compression point. Since A_{IIP3} can be quite high, the gain will in reality be heavily compressed at this point, rendering the above characteristic inaccurate as higher order harmonics come into play. In practice, the strength of the fundamental and

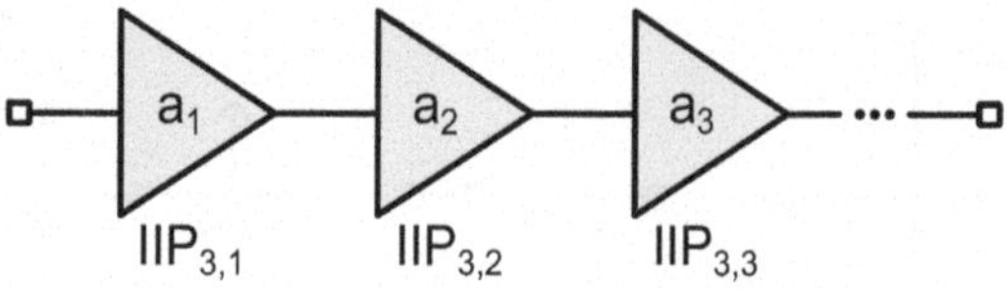

Fig. 2.42 Cascade of nonlinear stages

third order IM products will be measured at lower input levels and extrapolated to obtain the IIP_3, as demonstrated by Fig. 2.41.

As RF systems process signals by multiple stages, it is instructive to analyze how the nonlinearity of each stage contributes to the overall linearity of the system. Considering multiple stages, each with specified linear gain a_n and IIP_3 (A_{IIP3}) as illustrated in Fig. 2.42, the total IIP_3 of the cascade (referred to the input) can be calculated to be [60]

$$\frac{1}{A^2_{IIP3,tot}} = \frac{1}{A^2_{IIP3,1}} + \frac{a_1^2}{A^2_{IIP3,2}} + \frac{a_1^2 a_2^2}{A^2_{IIP3,3}} + \ldots, \tag{2.87}$$

where all values are signal (typically voltage) quantities. Therefore, for each stage that has gain, the nonlinearity of the latter stage becomes increasingly more important as the IIP_3 is scaled down by the gain in the preceding stages. A similar conclusion can be drawn for the 1-dB compression point or linearity in general of a cascade of stages. For this reason, the last stage in the transmitter is the most critical in terms of linearity performance as the signal swing is the highest at this point. Typically the last stage is implemented as a power amplifier (PA), which is designed specifically to deliver a large output power with the least amount of nonlinearity.

2.3.3.4 Phase Distortion

Thus far we have only studied the effects of amplitude nonlinearity. However, in some systems, amplitude variation can result in a different phase response if the input level is large enough. This dependence of the phase shift on the signal amplitude can be expressed as

$$V_{\text{out}}(t) = A_{\text{out}} \cos[\omega t + \phi(A_{\text{in}})]\,. \tag{2.88}$$

Such an amplitude-dependent phase shift is undesirable for modulated signals with multiple phases and amplitudes. As illustrated in Fig. 2.43 for a 16-APSK signal, a relative phase shift is introduced in the outer constellation points compared to the inner points as a result of a different phase response for the larger signal amplitude. Such phase distortion can decrease the distance between some constellation points, making the detection more challenging for the receiver.

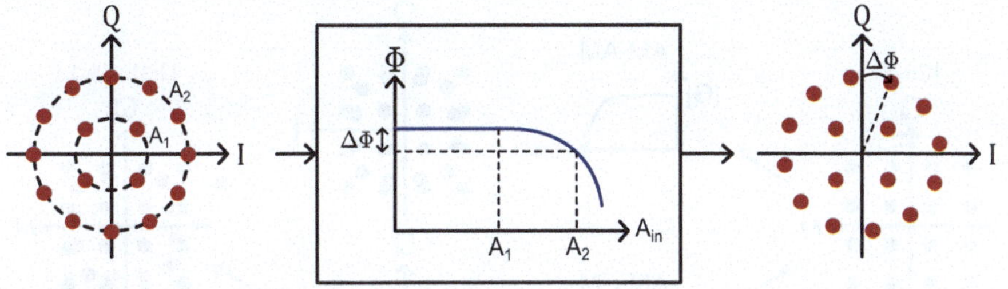

Fig. 2.43 Illustration of the effects of phase distortion on a 16-APSK constellation

2.3.3.5 Linearity Requirements for Passband Modulations

Understanding the effects of nonlinearity, it is time to investigate the linearity requirements for various passband modulations. Due to the directive nature and large propagation losses of telecommunication at very high frequencies, large interferer signals are quite uncommon and will not be considered for this assessment. Furthermore, as discussed, the most critical block in the system that mostly determines the linearity performance is the PA, since it processes the largest signals. Additionally, the PA is also the most power-hungry block in the system. The following considerations are therefore very important when determining the system specifications, such as total power consumption or transmitter output power (and thus communication range).

The most important types of distortion in the PA (or RF system in general) can be subdivided into a variation of either the magnitude response or the phase response as a function of input amplitude, both of which have been discussed before. The variation of the gain or magnitude response versus input amplitude is generally called AM–AM distortion. This mostly corresponds to the gain compressing effect discussed earlier, although other effects, such as gain expansion, can occur as well. The variation of the phase response versus input amplitude is referred to as AM–PM distortion, as amplitude modulation (variation) leads to phase modulation (variation).

The amount of distortion obviously depends on the variation of the amplitude or envelope of the modulated signal, often quantified by the peak-to-average power ratio (PAPR). The PAPR is defined as the ratio of the peak power to the average power of a signal, often expressed in decibels, and can be calculated as follows:

$$\mathrm{PAPR} = \frac{|S_{\mathrm{peak}}|^2}{|S_{\mathrm{RMS}}|^2} = \frac{\max\{|S_i|^2\}}{\frac{1}{M}\sum_{i=1}^{M}|S_i|^2}, \tag{2.89}$$

where S_{peak} corresponds to most outer of all possible constellation points S_i and the squaring operation explicitly transforms the signal magnitudes to power quantities. A signal like QPSK, which has only one possible amplitude state, has a PAPR of

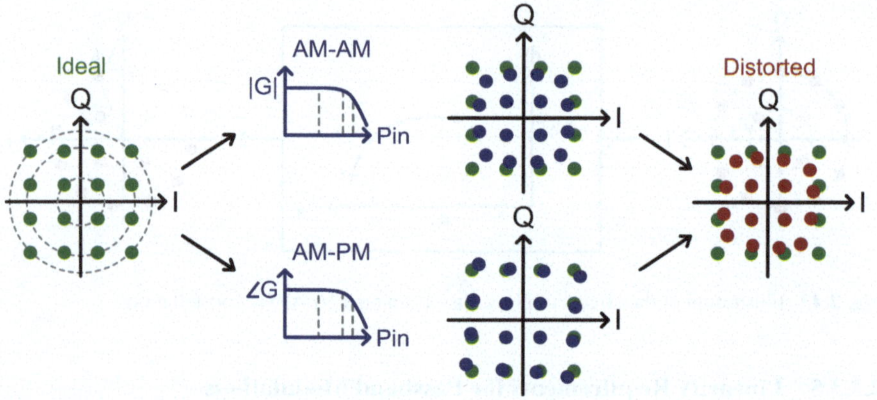

Fig. 2.44 Illustration of the effects of AM–AM and AM–PM distortion on a 16-QAM constellation

Table 2.3 Overview of the PAPR values for different ideal constellations

Modulation	QPSK	16-QAM	64-QAM	256-QAM
PAPR	0 dB	2.6 dB	3.7 dB	4.2 dB

1 or 0 dB. Such a signal is called a constant envelope signal as it experiences no envelope variations and is therefore insensitive to AM–AM or AM–PM distortions. A 16-QAM signal on the other hand experiences three distinct envelope levels and therefore has a higher PAPR of 2.6 dB. As illustrated in Fig. 2.44, a 16-QAM constellation is seriously distorted when experiencing nonlinear behavior. AM–AM distortion will cause the outer constellation points to come inwards, while AM–PM distortion gives the outer points a relative phase shift with respect to the inner points, both degrading the EVM.

As the modulation gets more dense, the PAPR inevitably decreases, as summarized in Table 2.3 for various schemes.[12] Furthermore, increasing the modulation density means that the distance between the constellation points reduces for the same peak envelope, making the signal more sensitive to distortions. In other words, the same amount of distortion (amplitude or phase) will cause a larger degradation in EVM (or SNR) compared to lower density modulations. For this reason, in order to limit the amount of distortion and thus EVM degradation in dense modulations, the PA must operate at a lower power level than the PA is able to deliver, called a power back-off (PBO). This should ensure that the signal swing is low enough not to experience excessive nonlinear behavior. How much nonlinearity can be tolerated depends on the EVM specification of the system.

[12] These PAPR values assume ideal symbol transitions. As will be discussed later, applying pulse shaping can alter the transitions and possibly increase the PAPR as a result.

The required power back-off, more formally defined as the difference between the actual peak output power (P_{peak}) and the saturated output power (P_{sat}) of the PA, generally increases as the adopted modulation scheme becomes more dense. The exact value depends on the definition of P_{sat} and the nonlinear characteristic (e.g., the sharpness of the response) of the PA.

These discussed effects have important consequences for the performance of RF transmitters and therefore the choice of the modulation scheme.

First, for the same PA specifications (with a certain $Psat$) the maximum attainable average signal power will drop as the modulation scheme becomes more dense, limiting the attainable system SNR or communication range. This is because (1) a high PBO is required to reduce nonlinearity and hence lower the peak output power relative to the saturate output power and (2) the PAPR increases therefore the average signal power decreases relative to the peak output power. Combining both effects, the average signal output power can be calculated as follows (in decibel units):

$$P_{sig} = P_{sat} - \mathrm{PBO} - \mathrm{PAPR}\,. \tag{2.90}$$

Second, also the power efficiency of the PA, and therefore also the efficiency of the system, will degrade as the order of the modulation is increased. This is because the PA typically has the highest efficiency at saturation and drops sharply in efficiency as it operates further away from saturation.

To conclude, it is clear that the increased spectral efficiency of higher order modulation schemes comes at a price of lower system SNR and power efficiency.

2.3.3.6 Predistortion

To mitigate the degradation in PA output power and efficiency, predistortion techniques can be applied to partially compensate the effects of nonlinearity. Gain compression can be compensated by applying some gain expansion, while phase distortion can be offset by inserting the opposite amplitude-dependent phase shift.

Predistortion techniques can be implemented directly in the RF domain using various circuit techniques. For example, as will be discussed, gain expansion can be achieved by biasing an amplifier in class B, and the AM–PM distortion of a PA can be compensated by adding a complementary amplifier. However, all of these techniques have a rather limited predistortion range and come with inevitable signal loss to implement, limiting the effectiveness of RF/analog predistortion.

More commonly, predistortion is applied in the baseband chain. Using complex digital signal processing, the amplitude and phase of the constellation points can be predistorted to account for the PA nonlinearity to come. However, as the exact distortion characteristic depends on temperature and can vary over time, the predistortion algorithms should be able to detect and adapt to those changes. Furthermore, the data converters have to be able to generate the predistorted waveforms (generally with an increase in resolution) over a broad bandwidth. Not only does digital predistortion make the design of the required DSP algorithms and data converters significantly more complex, but it can also lead to a serious increase of power consumption. Hence, predistortion will not be able to completely compensate the aforementioned consequences of nonlinearity, but it has the potential to provide a reasonable improvement when applied judiciously.

2.3.4 Bandlimiting Filtering

While noise is omnipresent and inevitable nonlinearity in circuits causes signal distortion at large signal swings, another effect that can possibly corrupt the signal is filtering. Filtering occurs whenever the frequency characteristic (magnitude or propagation delay) of a system (circuits or channel) is not flat over frequency and limits the usable bandwidth for the signal. After all, if the magnitude response changes within the bandwidth of the signal, some frequency components of the signal will be attenuated more than others, causing signal distortion. Furthermore, if different parts of the signal experience another propagation delay, it can also result in pulse distortion. Ideally, the frequency response is as flat as possible within the signal bandwidth to process the signal with minimal distortion.

In this section, filtering caused by a nonflat magnitude response (low-pass and high-pass) and by a varying group delay (called dispersion) will be discussed. In general, some type of filtering is unavoidable as it is inherent to most physical structures, especially low-pass filtering. However, this part focuses on the manifestations of filtering without mentioning physical implementations of filters. It should also be noted that filtering can be intentional, for instance, to limit the signal bandwidth in a controlled way or to control the pulse shape, but is mostly an undesired effect.

Lastly, filtering can be analyzed on a baseband or a passband signal. Similar to how a baseband spectrum is converted into a passband spectrum by upconverting to a certain carrier (Fig. 2.11), a filter response can be characterized at baseband or around a nonzero frequency. The effect of a particular filtering operation depends on what frequency components of the signal are being attenuated. This principle of equivalent representations at baseband or passband is useful for analyzing filtering effects and is illustrated by Fig. 2.45. When the high frequency information of a baseband signal is attenuated while the low frequency information is unfiltered, called a low-pass filter (LPF), this is equivalent to filtering out the information far away from the center frequency of a passband signal and leaving the information close to the center unaffected, called a band-pass filter (BPF). The inverse also holds true, making a high-pass filter (HPF) at baseband equivalent to a band-stop filter (BSF) at passband around the center frequency. The following analyses discuss filtering from a baseband perspective but will be applied to passband systems as well in later chapters.

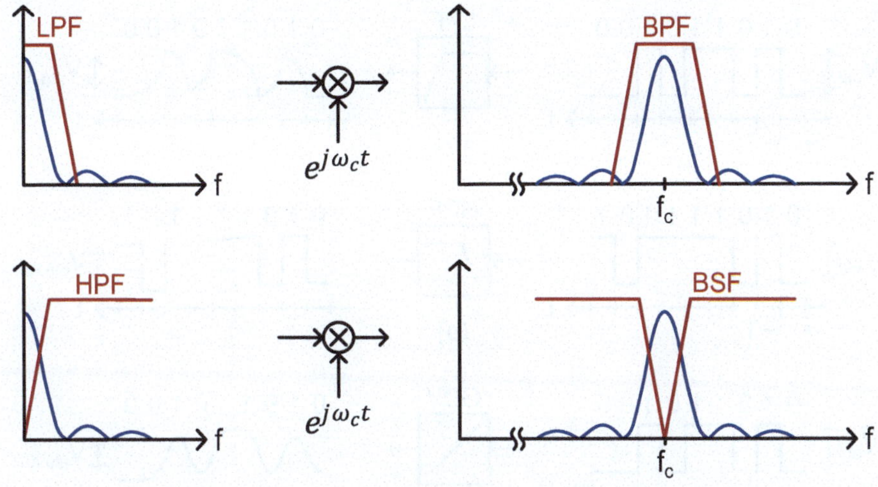

Fig. 2.45 Equivalence of filtering representations as a baseband or passband response

2.3.4.1 Types of Filtering

Most filtering effects can be distinguished into three types: low-pass filtering, high-pass filtering, and dispersion.

Low-Pass Filtering

Low-pass filtering occurs when low frequencies remain undisturbed while higher frequencies are attenuated. As visualized in Fig. 2.46a, if a significant portion of the high-frequency information of the signal is filtered out, the symbol transitions become slower and need time to settle to its final value. Whenever the necessary settling time is longer than a single symbol period, the signal will not be able to fully settle and reach its intended value.

Moreover, the exact value that is obtained at the end of a symbol period depends on the symbols that precede the current one. The phenomenon where a symbol spills over into succeeding symbols, causing interference and therefore influencing the signal level of these symbols, is called intersymbol interference (ISI). ISI can cause signal distortion and reduce the effective amplitude of the signal. ISI caused by low-pass filtering is the most excessive when many symbol transitions occur quickly after one another.

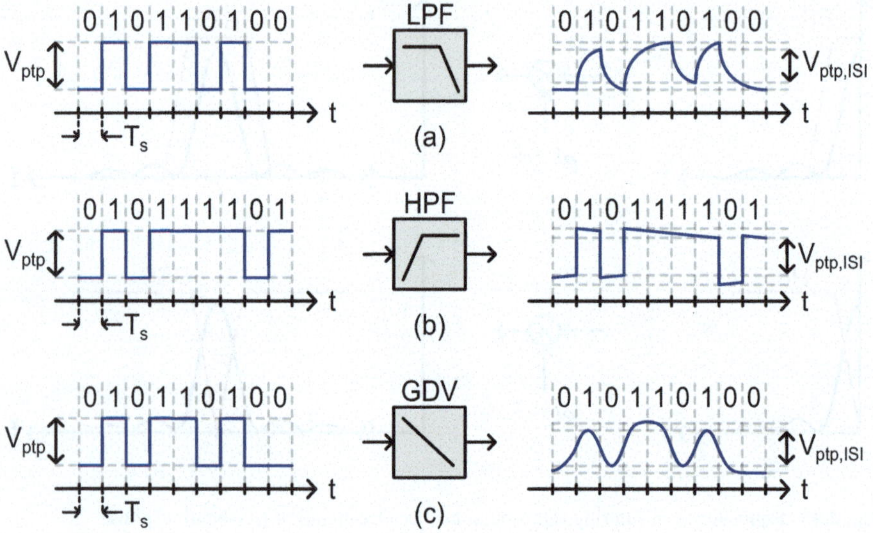

Fig. 2.46 The effects of low-pass filtering (**a**), high-pass filtering (**b**), and dispersion (**c**) on baseband data

High-Pass Filtering

High-pass filtering refers to the attenuation of low-frequency signal information while passing the high-frequency components. As illustrated in Fig. 2.46b, if signal power at low frequencies and DC is suppressed, symbol transitions appear quickly, but the output level starts to drift when the symbol stays constant for a while, which is also called baseline wander. This effect is most visible for long consecutive runs of the same symbol where a large temporary shift in the DC value is induced that reduces the effective amplitude of the signal. Since the signal level of each symbol depends on the preceding symbol pattern, this effect can also be viewed as ISI.

Dispersion

Unlike the two previous types of filtering, this effect is not caused by the magnitude response, but by the phase response of the filter. More specifically, if the signal experiences a variation of propagation delay, also called group delay, over its bandwidth, the various frequency components of the signal arrive at different times at the output, causing the signal shape to distort. This effect is called dispersion and usually results in the spreading of the pulse shape over time. As shown in Fig. 2.46c, the spreading of the pulse can cause overlap between different symbols, thus also resulting in ISI.

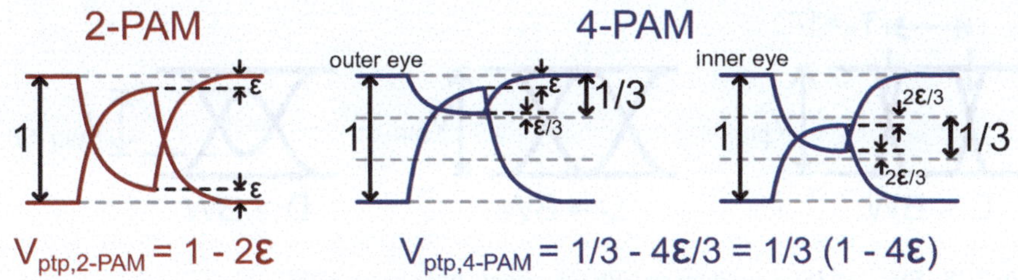

Fig. 2.47 Effect of finite settling on a 2-PAM and 4-PAM signal

2.3.4.2 BW, ISI, and Power Penalty

While the exact filtering mechanisms are different, they all cause signal distortion and interference between neighboring symbols. In the end, the main problem of ISI is that the usable signal range to make the symbol decision reduces, leading to a lower effective SNR and hence higher BER for the same noise power. The degradation in effective SNR can be viewed as a power penalty on the signal as a result of filtering. Assuming a peak-to-peak signal amplitude of V_{ptp} and $V_{ptp,ISI}$ before and after filtering, respectively, the power penalty (PP) caused by ISI can be expressed as (in decibel units)

$$PP_{ISI} = 20 \log \left(\frac{V_{ptp}}{V_{ptp,ISI}} \right). \tag{2.91}$$

Without taking the analysis too far, it should be mentioned that a certain amount of ISI, for instance, settling error caused by an LPF, will cause a more detrimental degradation depending on the density of the modulation. As illustrated in Fig. 2.47, the same amount of settling will result in a relative signal level degradation twice as big in a 4-PAM signal compared to a 2-PAM signal.

Furthermore, the exact amount of signal distortion, and thus reduction in amplitude, depends on the characteristics of the filter (type of filter, steepness, and bandwidth) and properties of the signal (distribution of power over frequency). While this is a vast topic, some rudimentary guidelines will be outlined for each filtering type.

Low-Pass Filtering

The effects of low-pass filtering, or equivalently band-pass filtering in RF circuits, are by far the most common and thus important to analyze. To predict the amount of ISI, the bandwidth of the filter (typically, 3-dB bandwidth) is often compared to the symbol rate of the signal (as the baseband null-to-null bandwidth is equal to the symbol rate for rectangular pulses). Figure 2.48 shows the eye diagram of an NRZ signal which is filtered by an LPF with various bandwidths. As the filter bandwidth

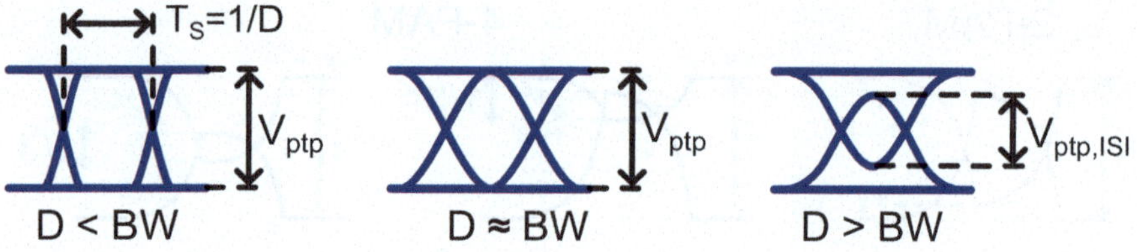

Fig. 2.48 Effect of ISI caused by an LPF for various filter bandwidths

becomes much smaller than the symbol rate, a significant power penalty will occur due to ISI.

It must be noted that a bit of filtering is usually desired as it also filters noise, thus increasing the SNR as long as the ISI-induced power penalty is not too large. Clearly, an optimal filter bandwidth can be found that provides as much noise filtering as possible without introducing too much ISI. As a general rule of thumb, as long as the 3-dB filter bandwidth is larger than half the symbol rate, the amount of ISI should be manageable. For example, to transmit and receive a 10 GBaud OOK signal, a system RF bandwidth of at least 10 GHz is required, as the OOK signal occupies a null-to-null RF bandwidth of 20 GHz. But again, this is highly dependent on the filter characteristics, modulation density, and other system impairments that degrade the signal quality.

Finally, one must remark that the mentioned bandwidth specification refers to the equivalent bandwidth of the total system. In reality, the system will consist of various analog and RF blocks, each with a limited bandwidth. Assuming each block has a first order response with a 3-dB bandwidth BW, then cascading N identical stages yields the following total bandwidth BW_{tot}:

$$BW_{tot} = BW\sqrt{2^{1/N} - 1}\,, \tag{2.92}$$

showing that the bandwidth of the cascaded system quickly reduces as the number of stages increases. In other words, for a specific system bandwidth specification, the bandwidth of a single stage must be increasingly larger in a multistage configuration. Adding many stages (e.g., to achieve higher gain) must therefore be done mindfully, as achieving a large bandwidth becomes progressively more challenging.

A more mathematical analysis of low-pass filtered systems will be given later on in Chap. 5.

High-Pass Filtering

High-pass filtering only really occurs in baseband blocks, for example, in circuits where the DC signal is blocked for biasing purposes. To limit the effects of baseline wander, the filter ideally only suppresses power at frequencies below the lowest frequency component of the signal. How low that frequency is depends on the symbol rate and the type of data. For random data, the signal contains power all the way down to DC. However, as practical data is often approximated as a pseudorandom binary data sequence (PRBS), the lowest frequency is determined by the periodicity of the sequence [41]. As the PRBS sequence increases in length, the data will appear more random, but any high-pass filtering circuits must present a lower cut-off to avoid ISI.

Dispersion

Dispersion can arise in circuits but is the most problematic in long dielectric waveguide channels, as will be discussed in the next chapter. Dispersion causes pulse spreading which renders neighboring symbols indistinguishable as the amount of spreading exceeds the symbol period. A common rule of thumb for dispersion is therefore that the total delay variation across the signal bandwidth must be lower than the symbol period to avoid detrimental distortion [61]. A more formal derivation of dispersion specifications will be given in Chap. 5 as well.

2.3.4.3 Pulse Shaping

In order to lower the amount of ISI in a system, a viable approach might be to try to reduce the bandwidth of the modulated signal. After all, rectangular pulses with steep edges have a sinc spectrum with an unnecessarily wide bandwidth. By changing the baseband pulse shape, the occupied signal bandwidth can be reduced. This operation is called pulse shaping. Lowering the signal bandwidth is not only beneficial to decrease ISI but might even be necessary to reduce out-of-band emission and comply with the regulated spectral mask, as discussed in Sect. 1.3. Intuitively, one could simply reduce the steepness of the edges, which would reduce the power at higher frequencies in the spectrum. However, filtering generally introduces ISI to the signal, so a more methodical approach is required.

The main question is how can we minimize the signal bandwidth without introducing ISI? Nyquist suggested that in order to eliminate ISI the signal pulse must have a nonzero value at its sampling time, but a zero value at the sampling instants of all over symbols, effectively avoiding interference between symbols even though they may overlap. Signals that obey this criterion are called Nyquist signals and can be implemented in various ways. The Nyquist signal with the tightest spectrum is the sinc pulse, which corresponds to a rectangular or "brickwall" spectrum with a baseband null-to-null bandwidth of only D/2. This is half the bandwidth of a rectangular pulse which has a sinc spectrum.[13] Since sinc pulses are difficult to generate in practice, a very popular Nyquist signal is the raised-cosine

[13] Note the equivalence between the time- and frequency-domain responses of the sinc and rectangular functions.

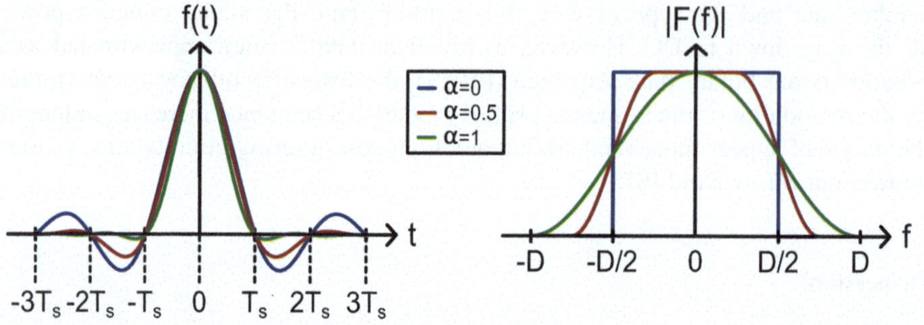

Fig. 2.49 Time-domain waveform and corresponding frequency spectrum of a raised-cosine pulse for different roll-off factors

(RC) pulse, which approximates the shape of the sinc function [60]:

$$f(t) = \frac{\sin(\pi t/T_s)}{\pi t/T_s} \frac{\cos(\pi \alpha t/T_s)}{1 - 4\alpha^2 t^2/T_s^2}, \tag{2.93}$$

where α is called the roll-off factor that determines how close the raised-cosine pulse is to a sinc shape. Figure 2.49 shows the raised-cosine pulse in the time domain with its corresponding spectrum, for various roll-off factors. By observing that each pulse, regardless of the roll-off factor, indeed has a zero value at $t = kT_s$ for all nonzero integers k. If $\alpha = 1$, the shape reduces to a sinc shape. As α decreases, the sidelobes in the time domain become smaller, but the occupied bandwidth increases. In general, the baseband bandwidth of an RC pulse with roll-off factor α is equal to [41]

$$BW_{\mathrm{RC}} = \frac{D}{1+\alpha}. \tag{2.94}$$

Typical values used in practical systems range from 0.3 to 0.5, striking a balance between bandwidth and pulse generation complexity.

Unfortunately, the desirable benefits of pulse shaping come with several challenges. As mentioned, RC pulses with a higher roll-off factor show larger sidelobes and are thus more sensitive to timing variations in the sampling clock. Furthermore, the rather complicated pulse shape puts a heavier burden on the digital-to-analog converter that has to generate the signal, as a higher resolution and significant oversampling is required to accurately reproduce the raised-cosine shape. And lastly, since the effectiveness of the bandwidth reduction relies on the specific signal shape, it is more sensitive to nonlinear effects from the RF front-ends (especially, the PA). Furthermore, the pulse shaping operation causes more envelope variations in the RF signal, resulting in a larger PAPR and therefore aggravating the linearity challenges.

The raised-cosine characteristic is sometimes split between the transmitter and the receiver, yielding a root raised-cosine (RRC) filter on both sides. As a result, the RX-side RRC filter functions as a matched filter (see Sect. 2.3.2) when the incoming pulse is mostly determined by the TX-side RRC filter (assuming the channel and other blocks have negligible influence on the signal shape). While the signal shaped by the TX is not ISI-free, as the RRC filter itself is not a Nyquist filter, all ISI ideally disappears after the RX-side RRC filter since the combined filter response (RC) is a Nyquist filter [46].

2.3.4.4 Equalization

Another way to compensate ISI in the system is to use equalization. Equalization aims to purposefully remove the interference resulting from pulse spreading across multiple symbol intervals. As discussed, ISI is caused by nonflat frequency response and can also be characterized in the time domain by means of a single pulse response. Within a single spread out pulse, different points can be identified at the various sampling instants of the preceding, main, and succeeding symbols. As shown in Fig. 2.50a, the overlapping points at preceding and succeeding symbol intervals are called pre-cursors (C_{-1}) and post-cursors (C_1, C_2), respectively, while the value at the sampling instant of the pulse itself is referred to as the main cursor (C_0). In general, two equalization strategies can be employed, working in the frequency or time domain. One possibility is to compensate the response in the frequency domain, making the overall system response more flat thereby reducing the ISI. Alternatively, the pre- and post-cursors caused by pulse spreading can be directly compensated in the time domain. The most popular equalization techniques will now be briefly discussed.

Continuous-Time Linear Equalizer

A continuous-time linear equalizer (CTLE) is an analog equalizer designed to compensate the low-pass response of circuits or channels. As demonstrated in Fig. 2.50c, this is typically achieved by introducing high-frequency peaking in the frequency response, thereby mimicking a high-pass behavior. If the low-pass response is adequately compensated over a wide enough bandwidth, ISI in the time domain will be canceled. While the CTLE shines in simplicity and effectiveness, its gain characteristic also amplifies high-frequency noise, leading to ISI improvement at the cost of SNR. Furthermore, while a CTLE is most commonly used to compensate the roll-off in the magnitude response, a similar analog equalizer can be designed to compensate group delay variation in the system, which is then referred to as a group-delay equalizer (GDE).

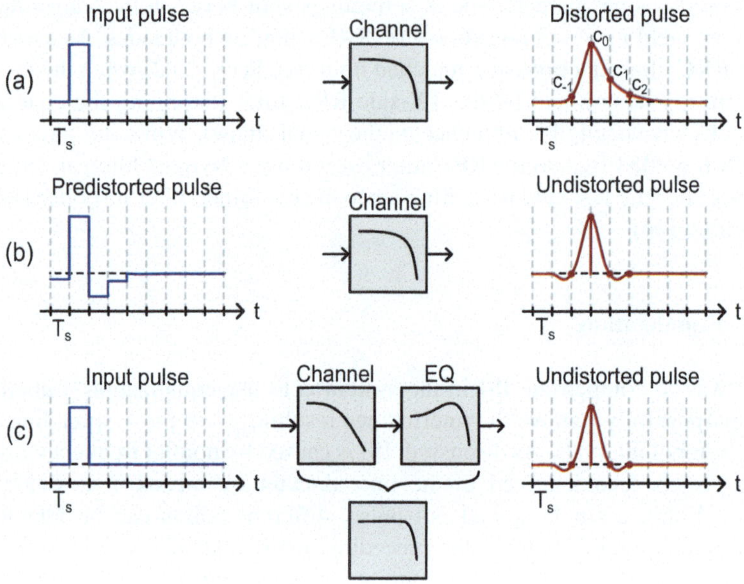

Fig. 2.50 Visualization of (**a**) a pulse being distorted by ISI, (**b**) predistortion of the pulse to compensate ISI, and (**c**) frequency peaking to compensate ISI

Feed-Forward Equalizer

A feed-forward equalizer (FFE) is a finite impulse response (FIR) filter, meaning the filter contains no feedback path, that can alter the pulse response directly in the time domain. On the transmitter side, an FFE can predistort the transmitted pulse to compensate for the distortion of the following blocks, as illustrated in Fig. 2.50b. Since more signal swing is needed for the predistorted shape, the effective signal swing will be lower for the same peak amplitude. Furthermore, the pulse is more sensitive to nonlinearity, for similar reasons as explained for pulse shaping. An FFE can also be implemented on the receiver side, where it can cancel ISI (pre- and post-cursors) that has already occurred. Compared to a TX-side FFE, an FFE on the RX-side has less signal swing degradation but amplifies high-frequency noise similar to a CTLE. In fact, an FFE has a similar characteristic in the frequency domain as the CTLE but is implemented in the time rather than the frequency domain. As shown in Fig. 2.51a, an FFE can be implemented in the analog domain by generating multiple time-delayed versions of the signal (typically multiples or fractions of the symbol period) that are multiplied by the filter coefficients and finally added together. An FFE can also be realized in the digital domain, where physical delays are replaced by discrete memory elements.

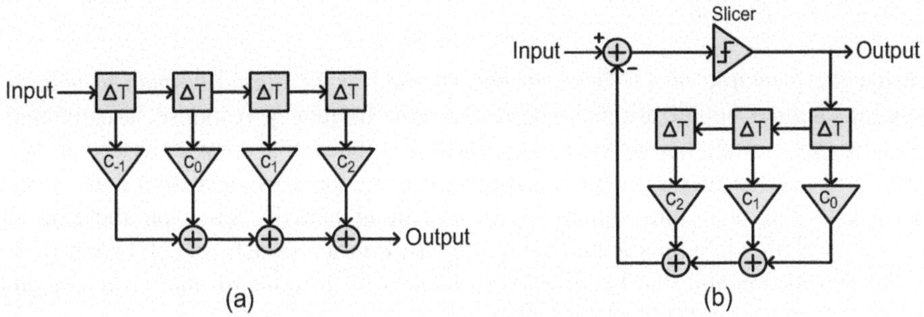

Fig. 2.51 Block diagrams of an FFE equalizer (**a**) and DFE equalizer (**b**)

Decision Feedback Equalizer

A decision feedback equalizer (DFE) is a nonlinear equalizer, as opposed to linear equalizers such as the CTLE or FFE. As Fig. 2.51b shows, a data slicer makes the symbol decision (effectively quantizing the signal) which is then used to subtract the ISI from the signal via a feedback FIR filter. Thanks to its nonlinear behavior, the DFE can boost high-frequency content without noise amplification. However, it cannot cancel precursor ISI and can be challenging to implement at high speeds due to the critical timing constraints in the feedback path. Furthermore, a DFE can give rise to error propagation, since a symbol error in the slicer can lead to more errors due to the feedback. A DFE can once again be implemented in the analog or digital domain and is always performed at the receiver side since it relies on the symbol decision.

Crosstalk Equalization

When multiple communication channels are closely combined (via two polarizations, multiple physical channels, or quadrature carriers (think QAM)), there is often coupling between the channels resulting to crosstalk that degrades the signal quality in each individual channel. In general this crosstalk is frequency-dependent, meaning that not only will the main symbol couple through but also pre- and/or post-cursors might appear on the other channels. To compensate this type of crosstalk, the aforementioned FFE and DFE techniques can be applied with not only the usual blocks but also duplicates controlled by so-called cross-coefficients with inputs from the other channels. These duplicate specifically tuned to compensate undesired ISI arising from other channels. In the absence of ISI, crosstalk can still be compensated, only needing cross-coefficients to account for the coupling of the main cursors.

Channel Equalization

Lastly, the principle of a CTLE can be extended to the digital domain to achieve general channel equalization. First, the system frequency response is estimated, often by transmitting known symbols (called pilot tones) that are used as a reference. Next, the inverse of the estimated response is realized using a digital filter and applied to the incoming signals. This technique heavily relies on the Fourier transform, which is implemented using the Fast Fourier Transform (FFT) algorithm in the digital domain, that becomes very expensive in terms of hardware area and energy consumption for high data rates.

General Considerations

Equalization can be extremely effective at combating the performance degrading effects of ISI but, as many good things in life, comes with its set of challenges. First of all, as the properties of the circuits and the channel vary over time (due to temperature variation, hardware degradation, or physical changes in the channel), the characteristics of the equalizer must be changeable during use. This is usually accomplished by adaptively tuning the different equalizer coefficients, which can be challenging in a high-speed system.

Furthermore, the designer must choose when to implement equalization in the analog or digital domain. In general, analog implementations prove to be more efficient at high speeds and low-to-moderate resolutions but suffer from the usual challenges of analog design, such as a high sensitivity to process and temperature variations. Digital blocks, on the other hand, provide a more robust and repeatable operation and are more efficient to implement for a very large number of filter taps. Nevertheless, the power consumption of these digital blocks grows rapidly as the data speed increases and also places a heavy burden on the data converters (DACs and ADCs), as a higher resolution and often oversampling operation is required for the equalization algorithms to function effectively. In the end, a modest amount of equalization can significantly simplify the specifications of the analog and RF front-end circuits for only a moderate price in area and power.

2.3.5 Propagation and Link Budget

All of the previously described effects can be used to analyze the performance of a full communication link. The only missing link is how the signal propagates from transmitter to receiver, which will be briefly discussed next. RF communication systems use two different types of propagation: wireless propagation and guided propagation through a dielectric waveguide (DWG).

2.3.5.1 Wireless Propagation

Given a certain amount of transmitted output power P_{TX}, the amount of received power P_{RX} over a wireless distance d can be calculated using Friis' transmission equation [62]:

$$P_{RX} = P_{TX} \, G_{TX} \, G_{RX} \left(\frac{\lambda}{4\pi d} \right)^2 , \tag{2.95}$$

where G_{TX} and G_{RX} are the gains of the transmit and receive antennas (with respect to an omnidirectional antenna) and λ the wavelength of the carrier. The factor $(4\pi d/\lambda)^2$ is called the free-space path loss (FSPL) and was discussed in Sect. 1.3.2.1. It is important to note that this expression is only valid when the received wave can be approximated as a planar wave, which is when the distance is large enough such that both antennas are in each other's far field.

In case both antennas are in line of sight (LoS), meaning the wave can travel in a direct path without obstacles and the operating band is chosen to avoid the absorption peaks in the atmosphere, the propagating loss can be accurately predicted by the above expression, and the response of the channel can be considered flat over frequency. However, in the situation where the direct path is obstructed (by buildings, trees, hills, etc.) and only non-line-of-sight (NLoS) propagation is feasible, additional losses have to be considered depending on the nature of the object. Furthermore, even in the case of LoS propagation, multiple reflections of the transmitted wave can arrive indirectly at the receiver's antenna via different paths, all with an arbitrary signal strength and relative delay to the direct wave. Especially for indoor applications, this multipath propagation can be an important impairment, sometimes requiring specialized equalization schemes to recover the original signal. At very high frequencies, however, the FSPL is so large that all practical systems employ highly directive LoS propagation where reflections are mostly nonexistent due to the large propagation loss. Therefore, for the purpose of this book, Friis' equation will be used assuming reflectionless, LoS operation.

2.3.5.2 Guided DWG Propagation

Also discussed in Sect. 1.3.2.1, guided signal propagation through a DWG can overcome the large free-space path loss at high frequencies. The attenuation through a DWG increases exponentially with distance (waveguide length), as opposed to quadratically, leading to a lower propagation loss than wireless links up until tens of meters. In contrast to a wireless channel, the frequency response of a DWG generally is not flat. Not only does the magnitude response change over frequency, but the group delay can also vary significantly. The magnitude response and the dispersion of the waveguide will thus introduce ISI in the signal that degrades the quality of the link.

2.3.5.3 Link Budget

A link budget is used to make a first assessment of the available SNR in the link and thereby estimating the BER performance. Alternatively, for a desired BER performance, the required specifications of the link can be determined. A link budget is constructed by accounting for all the different "gains" and "losses" with respect to SNR within the system, commonly expressed as follows in decibel units:

$$P_{TX} + G_{TX} + G_{RX} = ktB + NF_{RX} + SNR_{\min} + PL + OL + LM\ , \qquad (2.96)$$

where P_{TX} and kTB have absolute power units, and all other terms have relative units (in dB or dBi).

On the gains side, P_{TX} represents the transmitter output power and G_{TX},G_{RX} the gains of the transmit and receive antennas, expressed relative to an isotropic radiator with units dBi. The output power is quantified as the average signal power (including the PAPR of the modulation) and takes into account the nonlinearities of the TX (thus including the PBO if necessary). The product (or logarithmic sum) of TX output power and gain of TX antenna is often called the equivalent isotropic radiated power (EIRP) as it equals the output power that a transmitter with an isotropic antenna (gain of 1 or 0 dB) would have in order to have the same radiated power density. In case of a DWG link, the antenna gains are replaced by the losses of the transmit-side and receiver-side waveguide couplers.

On the side of the losses, kTB equals the noise floor, NF_{RX} the noise figure of the receiver, $\text{SNR}_{\min}$ the minimum SNR needed to achieve the desired BER, PL the path loss, OL other losses, and LM the link margin. The total input-referred noise floor of the receiver is equal to $kTB + NF_{RX}$, which represents the integrated noise at the input together with all noise added by the receiver. The receiver sensitivity $P_{sens,RX}$ is defined as the minimal receiver input power needed to obtain the wanted BER and is calculated by adding the noise floor and the minimum needed SNR determined by the BER. The path loss can be represented by the FSPL in case of a wireless link or by the waveguide loss in case of a guided DWG link. Finally, other losses may include the ISI-induced power penalty (caused by filtering of the circuits or channel), interference, or other remaining nonidealities in the system. Sometimes a link margin is added to the budget to ensure the specified BER performance is attained even in the presence of unexpected losses.

Figure 2.52 visualizes the link budget for a wireless link. A more concrete example can be found in Chap. 7, Fig. 7.11.

2.3.6 Performance Comparison of Modulation Techniques

Finally, all of the considerations discussed so far can be combined to evaluate all modulation schemes. The most promising ASK, PSK, and QAM modulations for high-frequency communication systems will be compared on several metrics:

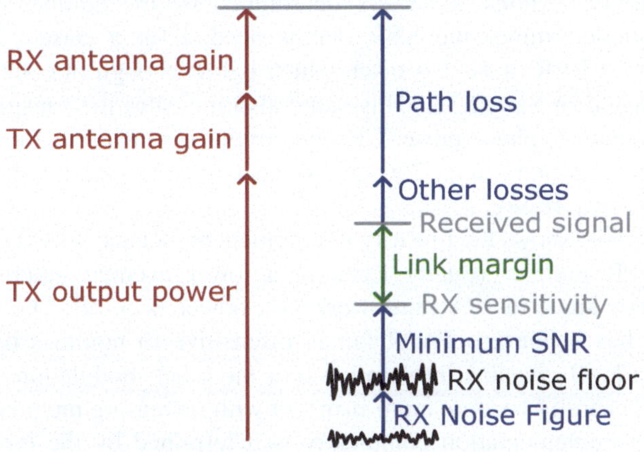

Fig. 2.52 Link budget of a wireless link

Table 2.4 Comparison of different modulation techniques

		Spectr. Efficiency		Noise	Linearity	Implementation	
	Modulation	Bits/Sym	D = BW/2	SNR*	PAPR	Detection	IQ LO
ASK	OOK	1	R	9.8dB	3dB	Incoh.	No
	4-ASK	2	R/2	12.6dB	4.1dB	Incoh.	No
PSK	BPSK	1	R	6.8dB	0dB	Coh.	No
	QPSK	2	R/2	9.8dB	0dB	Coh.	Yes
	8-PSK	3	R/3	11.7dB	0dB	Coh.	Yes
QAM	16-QAM	4	R/4	16.6dB	2.6dB	Coh.	Yes
	64-QAM	6	R/6	22.6dB	3.7dB	Coh.	Yes
	256-QAM	8	R/8	28.4dB	4.2dB	Coh.	Yes

* for BER < 1e-3

spectral efficiency, noise performance, linearity requirements, and implementation complexity. This analysis assumes the same desired data rate R for all modulations, and the effects of pulse shaping or predistortion will not be considered. Table 2.4 provides an overview for different modulation schemes, where the color indicates the degree to which the particular property is beneficial to the system.

The spectral efficiency is quantified by the modulation density (in bits per symbol) and directly determined the necessary bandwidth for the wanted data rate. In this comparison, the (null-to-null RF) bandwidth is set by twice the symbol rate. Binary schemes (e.g., OOK and BPSK) have very poor spectral efficiency and thus have large bandwidth requirements compared to dense QAM modulations (e.g., 64-QAM and 256-QAM).

The sensitivity to noise is mostly determined by the number of levels per dimension and determines the SNR that is needed for a certain BER. In the table, a reference BER of 1e-3 is taken which is low enough in combination with FEC, as discussed in Sect. 2.3.2. This value also indicates the sensitivity to other impairments, such as phase noise. Clearly, simple modulations are quite robust, while dense modulations (e.g., 256-QAM and 64-QAM) are highly sensitive to noise and other distortions.

The PAPR value shows the linearity requirement associated with the modulation. A higher PAPR number usually leads to a lower average output power and worse efficiency since the PA must work at a power back-off. One exception is OOK, which has a nonzero PAPR but is insensitive to nonlinearity since only one amplitude level contains information. For the other modulations the linearity requirements generally become more stringent with increasing modulation density.

Finally, the implementation complexity is determined by the type of circuits necessary for modulation and demodulation. For instance, requiring coherent detection and/or quadrature LO signals adds significant complexity and power consumption to the realized circuitry. Incoherent schemes, such as ASK, have a clear advantage on this front compared to QAM or PSK modulations.

2.4 Conclusion

This chapter introduced fundamental concepts in the field of digital telecommunication and signal modulation. Based on these concepts, the performance of a communication system was analyzed from different perspectives: noise, nonlinearity, and bandlimiting filtering. Finally, all of these concepts and system considerations were combined to compare various modulation techniques.

From observing Table 2.4, a spectrum of modulation techniques are available, all with their respective advantages and disadvantages. A number of schemes, namely QPSK, 8-PSK, and 16-QAM, jump out due to their moderate-to-high spectral efficiency and decent SNR and linearity requirements. These modulations will prove very promising for communication systems above 100 GHz but will require efficient ways to realize coherent, quadrature detection at such high frequencies. The practical considerations of implementing circuits above 100 GHz will be the focus in the next two chapters.

Chapter 3
Basic Electronics and Components

Starting from Maxwell's fundamental electromagnetic equations, circuit theory can be derived as an approximation in specific situations. Transmission line theory, the foundation of high-frequency RF circuit design, can then be seen as the middle ground between hardcore electromagnetism and simplified circuit theory.

Integrated circuits are implemented using both active and passive components. Active devices (mostly transistors) can be used to provide signal amplification, while the passive devices (inductors, transformers, capacitors, and transmission lines) take care of interconnects and ensure optimal operation at high frequencies.

In this chapter, the fundamental physics and circuit theory required for high-frequency circuit design are discussed. Then, the primary active and passive components are introduced and their behavior is described. As explained in Chap. 1, the discussion is mainly focused on the devices in standard CMOS technologies, as CMOS offers great economic and integration benefits.

3.1 Electromagnetism and Circuit Theory

3.1.1 Maxwell and Electromagnetic Waves

Electromagnetism describes the interaction between electrically charged particles via electromagnetic fields. It is one of the four fundamental forces of nature and can be seen as a combination between the theories of electricity and magnetism. We will start the discussion with the fundamental laws that model electromagnetism, Maxwell's equations, and how they predict the existence of electromagnetic (EM) waves.

C. D'heer, P. Reynaert, *THz and Sub-THz CMOS Electronics for High-Speed Telecommunication*, Analog Circuits and Signal Processing,
https://doi.org/10.1007/978-3-031-64439-9_3

3.1.1.1 Maxwell's Equations

In 1873, James Clerk Maxwell[1] published a set of equations that unified previous insights from Gauss, Faraday, and Ampère on electricity and magnetism [63]. These equations can be expressed as follows in their differential forms:

$$\nabla \cdot \mathbf{D} = \rho \tag{3.1}$$

$$\nabla \cdot \mathbf{B} = 0 \tag{3.2}$$

$$\nabla \times \mathbf{E} = -\frac{\partial \mathbf{B}}{\partial t} \tag{3.3}$$

$$\nabla \times \mathbf{H} = \mathbf{J} + \frac{\partial \mathbf{D}}{\partial t}. \tag{3.4}$$

These equations describe four different vector fields containing a vector (with a magnitude and direction) attached to each point in 3D space $(x,\ y,\ z)$. These fields are defined as follows: **D** as the electric displacement or induction field (C/m^2), **B** as the magnetic induction field (Wb/m^2), **E** as the electric field (V/m), and **H** as the magnetic field (A/m).

The sources of these fields are the charge density ρ (C/m^3), a scalar quantity, and the current density **J** (A/m^2), a vector with a scalar value for each point in 3D space. Since electric current is really the flow of charge, these quantities are closely related. Note that all of these scalar or vector values can, in general, be time-varying in nature.

Furthermore, the electric and magnetic field intensities and flux densities are closely related as follows:

$$\mathbf{D} = \epsilon \mathbf{E} = \epsilon_0 \epsilon_r \mathbf{E} \tag{3.5}$$

$$\mathbf{B} = \mu \mathbf{H} = \mu_0 \mu_r \mathbf{H}\ , \tag{3.6}$$

where ϵ and μ represent the permittivity and permeability of the medium. These equations are called the constitutive relations.

In vacuum, these properties are important universal constants: the permittivity of free space $\epsilon_0 \approx 8.854 \cdot 10^{-12}$ F/m and the permeability of free space $\mu_0 = 4\pi \cdot 10^{-7}$ H/m. These constants relate to the speed of light in vacuum as follows:

$$c = \frac{1}{\sqrt{\epsilon_0 \mu_0}}\ , \tag{3.7}$$

[1] This unification was first observed by Faraday, then extended by Maxwell, and partially reformulated by Heaviside and Hertz. Credit should thus be given to all contributors.

which is equal to about $3 \cdot 10^8$ m/s. In an ideal (linear, homogeneous, and isotropic) medium characterized by a specific relative permittivity ϵ_r and permeability μ_r, Maxwell's equations are still valid when combined with the constitutive relations. A higher permittivity results in a larger electric displacement field or polarization within the material in the presence of an electric field. The relative permittivity can vary from unity (in vacuum) to more than a thousand for specialized materials. The permeability is a measure of the amount of magnetization or magnetic induction produced in a material in response to a magnetic field. Paramagnetic materials, for instance, exhibit a relative permeability greater than 1, leading to an attraction toward magnetic fields. In more complicated media (i.e., dispersive, nonhomogeneous, or anisotropic), a more complicated formulation is required using tensors [64], which will not be discussed here.

Maxwell's equations can be viewed as a combination of different laws. The first equation (Eq. 3.1) quantifies Gauss' law which describes the relationship between an electric charge and the static electric field around it. It states that the net outflow of electric field (i.e., the divergence of the electric field vector) is proportional to the enclosed charge.

The second equation (Eq. 3.2) is Gauss' law for magnetism, stating that the net outflow of the magnetic field through a closed surface is always zero and hence that magnetic monopoles do not exist.

The third equation (Eq. 3.3) is Faraday's law of induction describing how a time-varying magnetic field relates to the curl of an electric field.

Finally, the last equation (Eq. 3.4) consists of Ampère's law with an extra term added by Maxwell. It states how a magnetic field arises from an electric current and from a changing electric field, also called displacement current.

These coupled partial differential equations can be solved for each physical situation with particular boundary conditions and field sources (ρ and $\mathbf{J}$). In general, no analytic solution exists, and one must rely on numerical methods to find a solution. However, in very specific cases, an analytic, elegant solution can be found. For instance, by solving Maxwell's equation in free space, the solution turns out to be a propagating planar wave.

Working with so many partial derivatives in space and time can be quite tedious. However, assuming monochromatic fields with sinusoidal time dependence, each field component can be written using a phasor notation, for example,

$$\mathbf{E}(t) = \mathbf{E}\, e^{j\omega t}\,, \tag{3.8}$$

where $\mathbf{E}$ is a complex number quantifying the magnitude and phase of the electric field. With this notation, each time derivative can be replaced with $j\omega$, yielding the following Maxwell equations in phasor form:

$$\nabla \cdot \mathbf{D} = \rho \tag{3.9}$$

$$\nabla \cdot \mathbf{B} = 0 \tag{3.10}$$

$$\nabla \times \mathbf{E} = -j\omega\mathbf{B} \tag{3.11}$$

$$\nabla \times \mathbf{H} = \mathbf{J} + j\omega\mathbf{D}\,, \tag{3.12}$$

which can be solved much more easily. Using the Fourier transform, a solution to these equations for an arbitrary frequency ω can be converted into a solution for arbitrary time dependence.

Lastly, for the sake of completeness, a fully descriptive formulation of a physical system should include how the free current relates to the electric field, also known as Ohm's law:

$$\mathbf{J} = \sigma\mathbf{E}\,, \tag{3.13}$$

where σ is the conductivity of the material. This relationship is crucial when relating physical currents to the arising fields.

3.1.1.2 Electromagnetic Wave Propagation

In a perfectly linear, isotropic, homogeneous medium without any sources ($\rho = 0$ and $\mathbf{J} = 0$), Maxwell's curl equations in phasor form can be expressed as [65]

$$\nabla \times \mathbf{E} = -j\omega\mu\mathbf{H} \tag{3.14}$$

$$\nabla \times \mathbf{H} = j\omega\epsilon\mathbf{E}\,, \tag{3.15}$$

which are two equations with two unknowns, **E** and **H**, and hence can be solved. Using vector calculus, these equations can be transformed into the following symmetrical set of ordinary differential equations for **E** and **H**:

$$\nabla^2\mathbf{E} + \omega^2\mu\epsilon\mathbf{E} = 0 \tag{3.16}$$

$$\nabla^2\mathbf{H} + \omega^2\mu\epsilon\mathbf{H} = 0\,, \tag{3.17}$$

which are called the wave equations or Helmholtz equations for **E** and **H**. The constant $k = \omega\sqrt{\mu\epsilon}$ is defined as the propagation constant or wave number and characterizes the propagation characteristics of the waves following from these equations.

Assuming a lossless medium, i.e., ϵ, μ, and k are real numbers, a basic plane wave solution can be found. Assuming the electric field only has a component along the x-axis that only varies along the direction of propagation (the z-axis), the Helmholtz equation for **E** becomes

$$\frac{\partial^2 E_x}{\partial z^2} + k^2 E_x = 0\,, \tag{3.18}$$

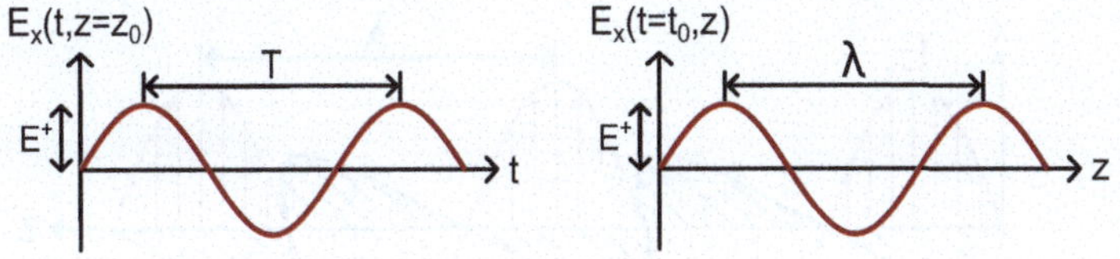

Fig. 3.1 The periodicity of a wave in time and space

which has a general solution of the form:

$$E_x(z) = E^+e^{-jkz} + E^-e^{jkz} , \tag{3.19}$$

where E^+ and E^- are two arbitrary constants. The solutions thus consist of a forward traveling wave (in the $+z$ direction) and a backward traveling wave (in the $-z$ direction). Since this solution holds for a time harmonic signal at frequency ω, the time-domain solution can be expressed as

$$E_x(z,t) = E^+ \cos(\omega t - kz) + E^- \cos(\omega t + kz) . \tag{3.20}$$

The phase velocity of the wave, being the velocity at which a fixed phase point on the wave travels, is given by

$$v_p = \frac{dz}{dt} = \frac{\omega}{k} = \frac{1}{\sqrt{\mu\epsilon}} , \tag{3.21}$$

which has a value of $v_p = 1/\sqrt{\mu_0\epsilon_0} = 2.998 \cdot 10^8$ m/s in vacuum, which is the speed of light.

As illustrated in Fig. 3.1, the wave is periodic in space and time, where the periodicity is determined by k and ω, respectively. In time, the period defines the time difference between two identical points on the wave at a fixed point in space and relates to the wave frequency f or angular frequency ω as

$$T = \frac{1}{f} = \frac{2\pi}{\omega} . \tag{3.22}$$

In space, the distance between two successive maxima (or any points having a multiple of 2π phase difference) at a fixed moment in time is defined as the wavelength λ:

$$\lambda = \frac{2\pi}{k} = \frac{v_p}{f} . \tag{3.23}$$

Using Maxwell's curl equation, the expression for the magnetic field can be derived from the electric field, yielding only a component in the y direction [65]:

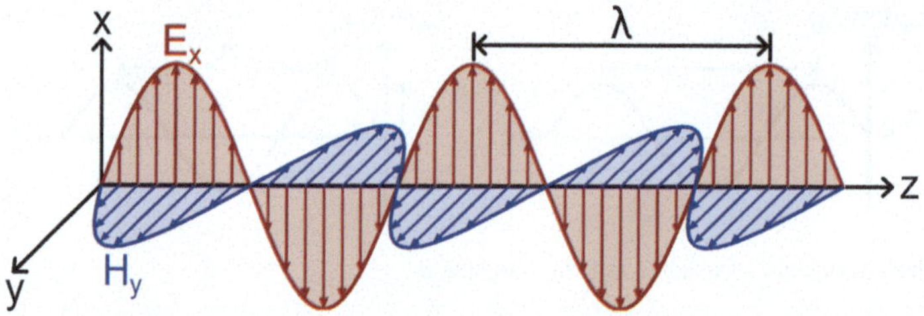

Fig. 3.2 Visualization of a propagating TEM wave

$$H_y = \frac{j}{\omega\mu}\frac{\partial E_x}{\partial z} = \frac{1}{\eta}(E^+ e^{-jkz} - E^- e^{jkz}) , \tag{3.24}$$

where the ratio between the electric and magnetic field, η, is known as the intrinsic impedance of the medium or wave impedance in the case of plane waves:

$$\eta = \frac{\mu}{\epsilon} . \tag{3.25}$$

In vacuum, this wave impedance is equal to $\eta_0 = \mu_0/\epsilon_0 \approx 377\,\Omega$.

3.1.1.3 Basic Properties of TEM Waves

As visualized in Fig. 3.2, the described EM wave has electric and magnetic fields that are perpendicular to each other and to the direction of propagation ($+z$ or $-z$). Furthermore, both fields exhibit the described periodic behavior in the z direction and are in phase with each other. Such a wave is called a transverse electromagnetic (TEM) wave since electric (E) and magnetic (M) only have components in the transverse (T) plane, i.e., perpendicular to the propagation direction. Maxwell's equations explain how such a wave can physically propagate through space. The time-varying magnetic field creates a changing electric field through Faraday's law, while the changing electric field in turn generates a varying magnetic field through the extended formulation of Ampère's law. This perpetual cycle enables an EM wave to propagate through space at velocity v_p.

Furthermore, if the properties of the wave (amplitude, phase, and polarization) are independent of the transversal coordinates, i.e., are constant through any plane that is perpendicular to the propagation direction, it is called a plane or a planar wave. The planar wave approximation is valid in many situations, for instance, far away from a radiation source, and significantly simplifies calculations.

Another property ascribed to EM waves is the polarization, denoting the orientation of the transverse waves. If the fields in the transverse plane are oriented along a single axis, the wave is linearly polarized. Alternatively, if the axis of orientation continuously rotates within the transverse plane while propagating, the wave is said to be circularly or elliptically polarized, depending on whether or not the magnitude of the fields changes while rotating. As an example, the EM wave in Fig. 3.2 is linearly polarized along the x-axis.

In general, an EM wave can be superposition of different frequencies and polarizations, which still satisfies Maxwell's equations.

3.1.1.4 Power Flow and Poynting Vector

Knowing how the electric and magnetic fields propagate through space, the flow of electromagnetic power can be quantified using the Poynting vector which represents the directional flux of energy (energy transfer per unit area per unit time) or power density (power per unit area). The Poynting vector **S** is defined as the following cross product [64]:

$$\mathbf{S} = \mathbf{E} \times \mathbf{H} \,, \tag{3.26}$$

where **E** and **H** represent the electric field and magnetizing field vector, as used previously. The Poynting vector measures the amount of power transfer and its direction. For a planar TEM wave, the direction of power transfer coincides with the direction of propagation, unexpectedly, which is perpendicular to both electric and magnetic fields. The instantaneous power transfer is given by the magnitude of the Poynting vector:

$$|\mathbf{S}| = \frac{|\mathbf{E}|^2}{\eta} \,, \tag{3.27}$$

which rapidly oscillates in magnitude like the EM fields. To find the average power contained in an EM wave, this expression must be averaged over the wave period, yielding

$$\langle|\mathbf{S}|\rangle = \frac{E_{\mathrm{rms}}^2}{\eta} = \frac{E_{\mathrm{peak}}^2}{2\eta} \,, \tag{3.28}$$

where E_{rms} is the root mean square of the electric field amplitude, which is a factor $\sqrt{2}$ smaller than the peak amplitude in case of a sinusoidal varying wave.

3.1.1.5 Planar Waves in a Lossy Medium

In a general lossy medium with conductivity σ, Maxwell's curl equations can be rewritten as

$$\nabla \times \mathbf{E} = -j\omega\mu\mathbf{H} \tag{3.29}$$

$$\nabla \times \mathbf{H} = j\omega\epsilon\mathbf{E} + \sigma\mathbf{E} \, , \tag{3.30}$$

where σ can be viewed as the imaginary part of the permittivity of the medium: $\epsilon' = \epsilon - j\sigma/\omega$. The wave solutions can then be found from those in a lossless medium by replacing the wave number by a complex propagation constant γ equal to

$$\gamma = \alpha + j\beta = j\omega\sqrt{\mu\epsilon}\sqrt{1 - j\frac{\sigma}{\omega\epsilon}} \, , \tag{3.31}$$

where α is the attenuation constant and β the phase constant. The positive traveling wave can then be expressed as follows:

$$E_x(z,t) = E^+e^{-\gamma z} = E^+e^{-\alpha z}e^{-j\beta z} \, , \tag{3.32}$$

or in the time domain:

$$E_x(z,t) = E^+e^{-\alpha z}\cos(\omega t - \beta z) \, . \tag{3.33}$$

The wave thus travels with a phase velocity $v_p = \omega/\beta$ and a wavelength $\lambda = 2\pi/\beta$. However, due to the loss of the medium, the strength of the field decays exponentially with distance as given by the attenuation constant α. Furthermore, the intrinsic impedance of the medium is now complex:

$$\eta = \frac{j\omega\mu}{\gamma} \, . \tag{3.34}$$

It must be noted that by setting $\sigma = 0$, all expressions reduce to those derived in the lossless case.

3.1.2 Electromagnetic Waves in Dielectric Waveguides

Next to wireless communication, this book also discusses communication through dielectric waveguides (DWGs). Instead of transmitting an EM wave through free space, the wave can be confined within a waveguide leading to different propagation characteristics, as discussed in Chap. 1. These waves can also be described using Maxwell's equations, albeit with different boundary conditions. In this subsection, the most important propagation characteristics of EM waves through DWGs are outlined.

3.1.2.1 Propagation Principles

To derive the propagation characteristics through a DWG, the wave equation derived from Maxwell's equations must be solved within the dielectric medium with the appropriate boundary conditions. In this section, we will make abstraction of the rigorous physics and focus on the most insightful results that are relevant for this book. More mathematical details can be found in [64].

In general, for guided modes propagating along a straight path in the z direction, all components of the EM wave (i.e., **E** and **H**) can be represented in the following form:

$$f(x, y)\, e^{-\gamma z}\, e^{-j\omega t} . \tag{3.35}$$

In this form, $f(x, y)$ represents the transversal (i.e., perpendicular to the propagation direction) shape of the electric and magnetic field that does not change during propagation. It corresponds to a standing wave pattern that arises due to reflections at the waveguide interface and is the biggest difference compared to the TEM fields of EM waves in a homogeneous medium.

The evolution of the propagating fields over distance is characterized with the factor $e^{-\gamma z}$, where γ represents the propagation constant, similar to planar waves, that can be further split into a real and an imaginary component:

$$\gamma = \alpha + j\beta , \tag{3.36}$$

where α is called the attenuation constant and β the phase constant, similar to the propagation of planar waves in a lossy medium. The attenuation constant quantifies the degradation of the field strength as the wave propagates along the DWG, while the phase constant measures the phase change of the traveling wave.

Lastly, the factor $e^{-j\omega t}$ illustrates the periodic wave-like behavior in the time domain. In most analyses, this time-domain behavior is implicitly assumed and excluded from the expressions using the phasor notation.

Waveguides typically have a rectangular or circular cross-section and can have a solid or hollow (i.e., air) core. The exact field distributions $f(x, y)$ and associated propagation constants γ, corresponding to a specific waveguide mode, depend on the physical shape and dimensions of the DWG and the used materials.

A mode can be defined as a transverse field pattern whose profile (amplitude and polarization) remains constant along the longitudinal or propagation dimension of the waveguide. The different modes can be classified based on which field components it contains relative to the propagation dimension. Transverse electric (TE) modes only contain an electric field component in the transverse (x, y) plane (perpendicular to the propagation direction), while the magnetic field has a transverse (x, y) and longitudinal (z) component (along the propagation direction). Transverse magnetic (TM) modes contain a transverse and longitudinal electric field component, but only a transverse magnetic field component. In TEM modes, both the electric and magnetic fields are perpendicular to the direction of propagation.

This is the operating mode for planar EM waves in free space, as discussed before, but TEM waves cannot propagate within a waveguide. Another special case is a hybrid electromagnetic (HEM) mode, in which both the electric and magnetic fields have a component in the longitudinal direction. The HE_{11} mode, for instance, is a hybrid mode where the TE component dominates and is supported in symmetrical circular waveguides. Of course, in addition to mode type, the waves can also be linearly, circularly, or elliptically polarized, as discussed for planar waves before.

Each mode, with its corresponding phase constant β, is obtained as a solution to the transcendental wave equation that arises from satisfying the boundary conditions of the waveguide. For a given frequency, this equation yields a finite number of solutions for β corresponding to the supported modes of the waveguide. These solutions can be elegantly visualized via a dispersion diagram, plotting the β of each mode versus frequency ω. While the exact solution depends on the dimensions and material of the DWG, using normalized parameters enables the calculation of general solutions for a specific DWG geometry, independent of the material or dimensions. For instance, for a circular waveguide with permittivity ϵ_1 and ϵ_2 inside and outside the core, respectively, and core radius R, the normalized frequency V and normalized propagation constant b are defined as follows:

$$V = k_0 R \sqrt{n_1^2 - n_2^2} = 2\pi \frac{R}{\lambda} \sqrt{n_1^2 - n_2^2} \tag{3.37}$$

$$b = \frac{\beta / k_0 - n_2}{n_1 - n_2} = \frac{n - n_2}{n_1 - n_2}, \tag{3.38}$$

where the optical index relates to the permittivity as $n_{1,2} = \sqrt{\epsilon_{1,2}/\epsilon_0}$. The frequency is typically normalized relative to the size of the waveguide (R/λ), while the propagation constant is normalized to the phase constant in free space (β/k_0).

For each normalized frequency V, the number of supported modes (meaning $b > 0$) can be calculated from these equations. For instance, it can be shown that only one mode is supported for $V < 2.4$ (the fundamental mode HE_{11}), and this value thus quantifies the bandwidth of a waveguide in single-mode operation. Increasing the normalized frequency (by increasing the frequency or the core radius) will increase the number of supported modes.

Furthermore, while most of the field propagates within the core of the waveguide, a part of the field exists outside of the core. This field, called the evanescent field, decays exponentially in strength away from the core. To avoid the user from interacting with this field when the waveguide is touched, a cladding is typically added around the core of the waveguide, as shown in Fig. 3.3a. The difference between the dielectric constants of the core and cladding materials, ϵ_1 and ϵ_2, respectively, determines the confinement of the fields within the core and thus the strength of the evanescent field. In case of a large difference, most of the fields will propagate within the core with a small evanescent field in the cladding. If the difference is small, the wave is only loosely coupled to the DWG. In the extreme case where $\epsilon_1 = \epsilon_2$, the wave propagates like a TEM wave through

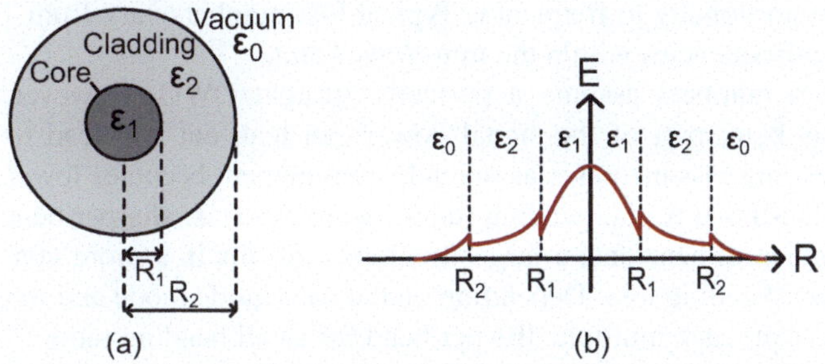

Fig. 3.3 Cross-section of a circular DWG with a core and cladding (**a**) and electric field intensity distribution for the HE_{11} mode across the radial distance of the cross-section (**b**)

a continuous medium, assuming the cladding radius is large enough to contain most of the field. Figure 3.3b shows the electric field intensity distribution of the dominant HE_{11} mode over radial distance from the center. Within the core, the field intensity decreases approximately quadratically with distance (according to the Bessel function to be exact), while the field decays exponentially outside the core (evanescent field). Also notice the discontinuity in electric field at the material interfaces, which is a necessary consequence from the ensuring the $\mathbf{D} = \epsilon\, \mathbf{E}$ field is constant across the boundary.

3.1.2.2 Attenuation

Most of the signal loss through a DWG arises from the dissipation of EM energy within the dielectric materials as heat, quantified by the loss tangent (tan δ) of the material. The attenuation constant of a DWG thus depends not only on the materials used for the core and cladding but also on the field distribution in both regions and generally varies over frequency. For circular DWGs, the attenuation constant α can be calculated by Yeh and Shimabukuro [64]

$$\alpha(\omega) = \frac{\pi}{\lambda_0}(\sqrt{\epsilon_{r,1}}\, R_1 \tan\delta_1 + \sqrt{\epsilon_{r,2}}\, R_2 \tan\delta_2)\ , \tag{3.39}$$

where $\epsilon_{r,1}$ and $\epsilon_{r,1}$ are the relative permittivities and tan δ_1 and tan δ_2 the material loss factors of the core and cladding, respectively. R_1 and R_2 are the geometric loss factors which can be calculated based on the ratio between the power loss and average power flow in the core and cladding [64]. For example, when most of the fields are confined within the core, $R_1 \approx 1$ and $R_2 \approx 0$, and the loss will be largely determined by the core material. Hollow fibers, which have a core material made out of air, will have lower losses since only little power propagates within lossy dielectric material (of the cladding). One can also observe that waveguide loss is roughly inversely dependent on the free-space wavelength, λ_0, and therefore increases proportionally to frequency. Typical loss numbers vary from 1 to 5 dB/m for operation frequencies within the mm-Wave band.

These loss numbers assume a perfectly straight DWG. However, when the waveguide is bent, part of the signal power can leak out and lead to additional loss. The bending loss increases as the field confinement becomes lower (i.e., more evanescent field) and as the bending radius decreases (i.e., sharper bends). For the same waveguide dimensions, a larger frequency results in a more confined mode and thus lower bending loss. Depending on the waveguide mode and frequency, the bending loss can reach multiple dBs per bend for small bending radii.

3.1.2.3 Dispersion

Ideally, the phase constant of the waveguide, $\beta(\omega)$, changes linearly over frequency, which would result in identical group (propagation) delays for all spectral components of the signal, thus conserving the pulse shape. However, for practical DWGs, the group delay varies over frequency, causing pulse distortion, as discussed in Sect. 2.3.4.

Dispersion can have different origins. The first type of dispersion is material dispersion which is caused by a variable dielectric constant of the material over frequency. The polymer materials that the DWGs in this book are made up from vary negligibly over frequency [66], and as a result, material dispersion is neglected.

The second type of dispersion is modal dispersion. In the case where multiple different modes propagate through the waveguide, each mode has a distinct propagation delay and thus leads to distortion of the overall pulse. This type of dispersion can be avoided by ensuring that the DWG can only support a single mode at the frequency of operation (e.g., by ensuring a normalized frequency $V < 2.4$ in circular DWGs).

A third type of dispersion is polarization dispersion, where the different polarizations propagate at different speeds causing distortion of the signal. When using only a single polarization in the waveguide, this type of dispersion is not relevant.

Lastly, the most important type of dispersion is chromatic dispersion. Due to a change in the field distribution in the DWG as the frequency varies, the effective dielectric constant becomes frequency-dependent causing the group delay to vary as well. This can be expressed as follows:

$$\beta(\omega) = \omega\sqrt{\mu_0 \epsilon_{eff}(\omega)} = \frac{\omega}{c}\sqrt{\epsilon_{r,eff}(\omega)} , \tag{3.40}$$

where c is the speed of light in vacuum. Due to a change in the field distribution in the DWG as the frequency varies, the effective dielectric constant, ϵ_{eff}, becomes frequency-dependent causing the group delay to vary as well. For instance, a well-confined wave will have a dielectric constant close to that of the core, while a loosely confined wave will have one closer to that of the cladding. Typically, as the frequency increases, the EM energy becomes more confined to the core, resulting in

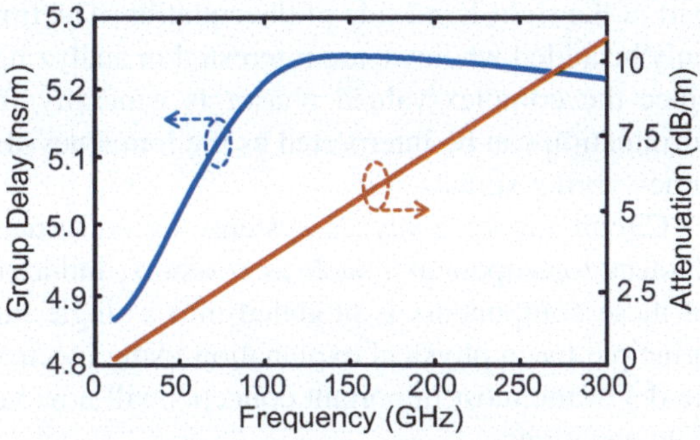

Fig. 3.4 Simulated attenuation and group delay of a 1 m DWG

an increase in effective dielectric constant and hence an increase in group delay. This phenomenon is the main cause of dispersion in single-mode, single-polarization fibers and is one of the most important bottlenecks in high-speed DWG links.

Figure 3.4 shows a simulation plot of the attenuation (in dB/m) and group delay (in ns/m) of a typical optimized DWG. The attenuation increases linearly with frequency, while the group delay varies significantly but is the most flat around 140 GHz by design. As a reference, light would take about 3.3 ns to propagate over 1 m in vacuum. Within the DWG, the waves travel at a lower speed, a factor of $\sqrt{\epsilon_{r,eff}}$ lower to be exact, depending on the exact frequency and mode of the wave.

3.1.3 Basic Circuit Theory

While rigorous field analysis is necessary to fully describe the behavior of EM fields, it is not effective to analyze electronic circuits or networks. Rather, the analysis of circuits can be simplified significantly by making abstraction of the EM fields and making some valid approximations. Indeed, when the circuit dimensions are much smaller than the wavelength of the EM fields, signal variation over the physical dimension of the elements can be discarded, and the behavior can be described using lumped components. Furthermore, all 3D vectors (**E** and **H**) are transformed into scalars (V and I), and the partial differential equations from Maxwell are converted into ordinary differential equations, which makes calculations much easier.

Similar to electromagnetism, circuit theory mostly works with voltages and currents that vary sinusoidally in time. These signals can thus be more easily represented by complex numbers, using the exponential notation introduced before. A time-varying voltage V(t) will, for instance, be written as follows:

$$V(t) = Ve^{j\omega t} , \tag{3.41}$$

where V represents a complex number, independent of time, with a real and an imaginary part. The actual physical time-varying signal is of course given by the real part of the right-hand side of the equation. The time-dependent factor will generally only be added when we are interested in analyzing the time-domain behavior. Also, since the complex-valued phasor is generally frequency-dependent, this type of formulation can be interpreted as the frequency-domain equivalent of working with time-varying signals.

Circuit theory analyzes systems by constructing networks from a set of fundamental components, such as resistors, inductors, and capacitors. The behavior of these components is modeled into a single lumped element, but their working principle has a physical explanation using electromagnetism. The lumped-element model of the most important concepts will now be derived starting from Maxwell's field equations.

3.1.3.1 Voltage and Current

Circuit theory works with voltages (V) and currents (I), as opposed to electric fields (**E**) and magnetic fields (**H**). To see their relationship, the first approximation is to assume all fields are static and therefore do not change through time. This approximation is not only exact at DC (i.e., $\omega = 0$) but also accurate for very low frequencies.

Faraday's law of induction (Eq. 3.3 from Maxwell's equations) then becomes

$$\nabla \times \mathbf{E} = -\frac{\partial B}{\partial t} = 0 \,. \tag{3.42}$$

From vector calculus, we know that a field whose rotor is equal to zero can be written as the gradient of a scalar function. The electric field can thus be written as

$$\mathbf{E} = -\nabla \phi \,, \tag{3.43}$$

where ϕ represents the potential of the field. By integrating both sides of the equation, the voltage (difference), defined as the electric potential difference, can be calculated by integrating the electric field along a path:

$$\Delta V = \phi(2) - \phi(1) = -\int_1^2 \mathbf{E} \cdot d\mathbf{l} \,, \tag{3.44}$$

where $d\mathbf{l}$ is the differential vector displacement along the line path. Physically, the electric potential can be interpreted as the amount of work required to move a unit of electric charge along the path in an electric field. While this expression is exact at DC, it is also valid for sufficiently low frequencies.

Electric current denotes the flow of charged particles moving through a conducting space and was already related to a change in magnetic field in Ampère's law:

$$I = \oint \mathbf{H} \cdot d\mathbf{l} \,. \tag{3.45}$$

On a microscopic scale, it is also defined as the rate of flow of electrical charge, Q, through a surface:

$$I = \frac{dQ}{dt} \,. \tag{3.46}$$

In electric circuits, electrons are used as charge carriers, which have a negative net charge. As a result, the flow of electric current is opposite to the flow of physical electrons.

3.1.3.2 Resistor

As discussed before, Ohm's law describes how the current density relates to the electric field in a conductive material with conductivity σ on a microscopic level:

$$\mathbf{J} = \sigma \mathbf{E} \,. \tag{3.47}$$

Now assume we have a component with two terminals, made out of a material with conductivity σ, with a certain cross-section and length. The total current through this conductor can be calculated by integrating the current density over its cross-section:

$$I = \int_A \mathbf{J} \cdot d\mathbf{S} = \int_A \sigma \mathbf{E} \cdot d\mathbf{S} \,. \tag{3.48}$$

Furthermore, the voltage difference across the conductor is calculated by taking the line integral of the electric field from one end of the conductor to the other:

$$V = -\int \mathbf{E} \cdot d\mathbf{l} \,. \tag{3.49}$$

With these expressions, it can be deduced that the ratio between the voltage over the conductor and the current through the conductor is constant and called the resistance [67]:

$$R = \frac{V}{I} \,, \tag{3.50}$$

where R is the resistance, defined as the inverse of the conductance G. A component that exhibits this perfectly linear relationship between voltage and current is often called a resistor and is our first foundational circuit element. The resistance of a component can be calculated based on the conductivity of the material and geometry of the component. Assuming a homogeneous conductive material with constant

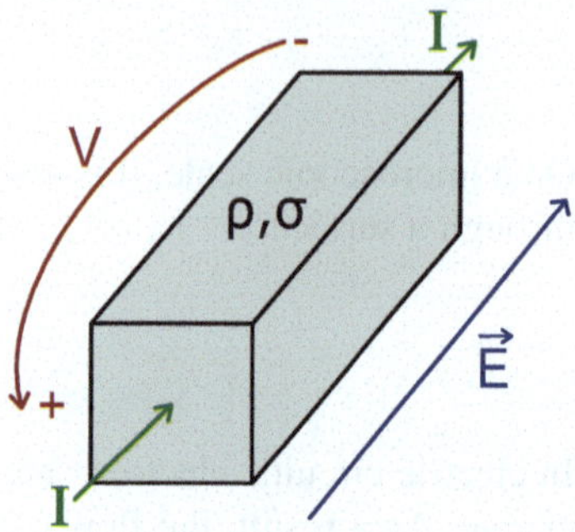

Fig. 3.5 Physical representation of a resistor with corresponding voltage, current, and electric field

cross-section A and length l, then the resistance R can be calculated with the given expression:

$$R = \rho \frac{l}{A}, \tag{3.51}$$

where ρ $(\Omega \cdot m)$ is called the resistivity (not to be confused with the symbol for charge density!), defined as the inverse of conductivity (S/m): $\rho = 1/\sigma$.

Figure 3.5 shows the physical representation of a resistor with corresponding voltage, current, and electric field. Notice how the current (oriented opposite to the movement of the electrons) flows along the orientation of the electric field from a high voltage potential to a lower potential.

3.1.3.3 Inductor

Suppose we have a loop that bounds a certain surface, Faraday's law of induction (the third of Maxwell's equations) then states that a change of magnetic flux Φ_B through that area generates an electromotive force $\mathcal{E}$ or voltage difference V_L across that loop, which can be expressed as [68]

$$V_L = -\mathcal{E} = \frac{d\Phi_B}{dt}, \tag{3.52}$$

where the magnetic flux is calculated by the integrating all magnetic field lines going through the surface:

$$\Phi_B = \int_S \mathbf{B} \cdot d\mathbf{S}. \tag{3.53}$$

This law can be directly derived from Maxwell's equations by transforming the path integral of the voltage difference into a surface integral (using Stoke's theorem):

$$V_L = -\int_C \mathbf{E} \cdot d\mathbf{l} = -\int_S (\nabla \times \mathbf{E}) \cdot d\mathbf{S} = \int_S \frac{d}{dt}\mathbf{B} \cdot d\mathbf{S} = \frac{d\Phi_B}{dt} . \tag{3.54}$$

Furthermore, since the magnetic field is generated by the current flowing through the loop, the magnetic flux is proportional to the current:

$$\Phi_B = L\, I , \tag{3.55}$$

where the proportionality constant is called the inductance of the loop. This yields the following relation between the voltage across the loop V and current through the loop I:

$$V = L \frac{dI}{dt} , \tag{3.56}$$

which is the governing equation of an inductor component. In the phasor notation, where $dI/dT = j\omega I$, the ratio between voltage and current can be written as

$$Z_L = \frac{V}{I} = j\omega L , \tag{3.57}$$

where Z is called the impedance, the more generic complex-valued ratio between voltage and current compared to Ohm's law.

In general, any loop where current flows behaves somewhat like an inductor and thus has an associated inductance property. This current loop can be an explicit conducting wire or more abstractly within a conducting material. The exact value of the inductance mostly depends on the geometry of the loop and will be discussed later on. While any current loop exhibits inductive properties, an inductor component is a two-terminal device that is specifically designed to store maximal energy in its magnetic field that is generated by an electric current flowing through it. Figure 3.6 shows the physical representation of an inductor as a loop of current with generated magnetic field. The electric field is orientated along the wire, like the flow of current, and is generated/maintained by the gradient of surface charges around the wire.

3.1.3.4 Capacitor

A capacitor typically consists of a pair of conducting plates separated by a specific distance. The region between the plates can be filled by any nonconducting material, such as vacuum or a dielectric. From Coulomb's law we know that a charge on one conductor will generate a force on the charge carriers within the other conductor. As a result, the charges on both conducting plates will always be equal and opposite in polarity to reach equilibrium, leading to an electric field and thus potential difference between the plates.

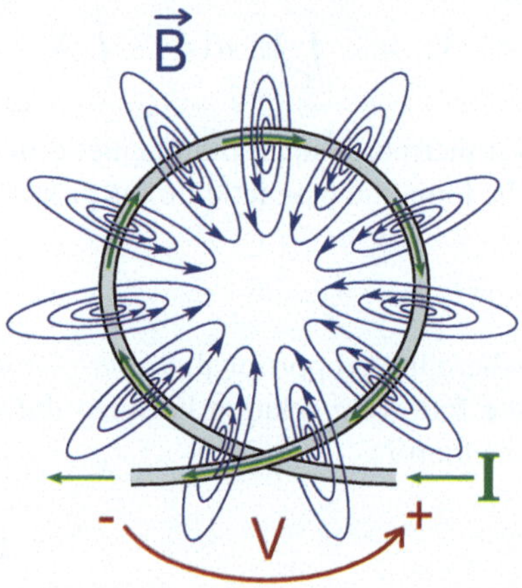

Fig. 3.6 Physical representation of an inductor with corresponding voltage, current, and magnetic field

In an ideal capacitor, the amount of charge on each conducting plate is proportional to the voltage between them:

$$Q = C\,V\,, \tag{3.58}$$

where the proportionality constant C is called the capacitance. Since current quantifies the change of charge per unit time, the relation between voltage over and current through a capacitor is given by

$$I = C\,\frac{dV}{dt}\,. \tag{3.59}$$

Again, using the phasor notation, the impedance of a capacitor can be found:

$$Z_C = \frac{V}{I} = \frac{1}{j\omega C}\,. \tag{3.60}$$

In general, any two conducting plates (however small) with an electrical field behave like a capacitor and thus has a capacitance property associated with it. However, a capacitor component is optimized to store maximal energy in the electric field for a given applied voltage. The exact value of the capacitance depends on the geometry of the conductors, the spacing in between, and the material in between. For example, for a parallel-plate capacitor with plate area A, spacing d, and the dielectric constant of the filling material ϵ, the capacitance can be written as [68]

$$C = \frac{\epsilon\,A}{d}\,. \tag{3.61}$$

To increase the capacitance, the plates should be made as large as possible and places as closely together as possible. Furthermore, special dielectric materials are often used with a high permittivity to further boost the capacitance. Figure 3.7 shows the physical representation of a parallel-plate capacitor with associated current,

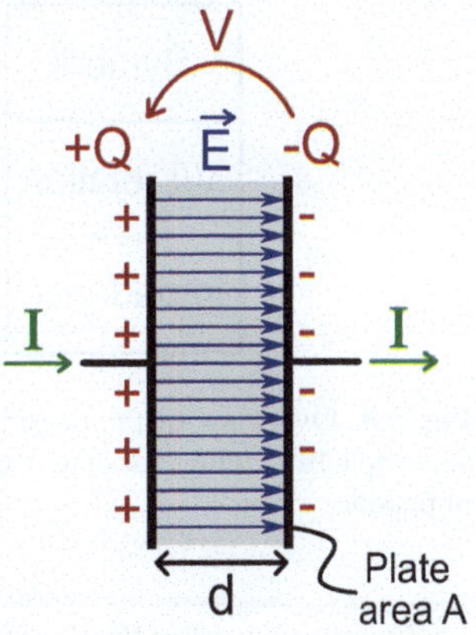

Fig. 3.7 Physical representation of a capacitor with corresponding voltage, current, charges, and electric field

charge distribution, voltage, and electric field lines. The charge buildup generates an electric field and thus a voltage potential across the plates. A change of charge through time thus leads to a current flowing through the capacitor.

3.1.3.5 Rules of Circuit Theory

Gustav Kirchhoff developed two laws that enable the analysis of electrical circuits using the discussed lumped elements in a relatively straightforward way. The first is Kirchhoff's voltage law, which states that the sum of the potential differences around a complete loop in a network is equal to zero:

$$\sum_{i=1}^{N} V_i = 0\,. \tag{3.62}$$

This law can be easily derived from Maxwell's equations assuming that there is no magnetic field in the region outside of the individual lumped circuit elements. In that case, the line integral of the electric field (and thus voltage difference) across all elements in the loop is equal to zero [68]:

$$\sum V_i = \oint \mathbf{E} \cdot d\mathbf{l} = 0\,. \tag{3.63}$$

The second law is Kirchhoff's current law and states that the algebraic sum of the currents which enter any node in the network must be zero:

$$\sum_{i=1}^{N} I_i = 0\,, \tag{3.64}$$

which easily follows from conservation of charge.

Symbol	L	C	R
V/I relation	$V = L\dfrac{dI}{dt}$	$I = C\dfrac{dV}{dt}$	$R = \dfrac{V}{I}$
Impedance	$Z_L = j\omega L$	$Z_C = \dfrac{1}{j\omega C}$	$Z_R = R$

Fig. 3.8 Overview of the three main passive components in circuit theory: the resistor, inductor, and capacitor, with associated expressions of their current–voltage characteristic and complex impedance

Together with Ohm's law generalized for complex impedances, most circuit networks can be analyzed. Sometimes it is easier to describe components by means of their admittance Y, being the reciprocal of the impedance Z, both consisting of a real and an imaginary part expressed as follows:

$$Z = R + jX = \frac{1}{Y} = \frac{1}{G + jB}\,, \tag{3.65}$$

where R (the resistance) and X (the reactance) are the real and imaginary parts of the impedance, while G (the conductance) and B (the susceptance) are the same for the admittance. Multi-port elements (such as transformers) are then often described by their Z-parameters or Y-parameters, quantifying the impedance or admittance between the different ports. For instance, Z_{ij} is defined as the ratio of the voltage at port i over the current into port j, while Y_{ij} is defined as its reciprocal.

Figure 3.8 gives an overview of the three main (passive) components used in circuit theory with their discussed voltage–current characteristics and complex impedance expressions. Next to these passive components, circuits also consist of active components, e.g., transistors, which can generate power gain when supplied with DC power, as will be discussed in the next section.

3.1.4 Transmission Line Theory

Finally, we arrive at the discussion of transmission line theory, the core of all circuit design at high frequencies. The key difference between basic circuit design and transmission line theory is the electrical size of the circuits. When deriving the basic equations and concepts of circuit theory in the previous section, we assumed the voltages and currents stayed constant over the physical dimension of the components. However, when the component size becomes a considerable fraction of the wavelength or even many wavelengths in the case of a transmission line, this assumption is invalid. To accommodate changes in voltage and current across an element, it should be modeled as a distributed parameter network, rather than a lumped element.

Transmission line theory can be viewed as an extension of circuit theory or a specific case of Maxwell's equations. This section discusses the model of a transmission line, from a lumped-element and EM field perspective, and several interesting characteristics.

3.1.4.1 Lumped-Element Model of a Transmission Line

The following derivation is a summarized version of the one described in [65]. Figure 3.9 shows the schematic model of a transmission line, consisting of two conducting wires orientated along the z direction. A transmission line can be modeled as a cascade of infinitesimally short line sections of length Δz. Each infinitesimal piece of line can be modeled as a lumped-element circuit with per-unit-length components: R, L, G, and C, as shown in Fig. 3.9.

The series inductance L (in H/m) represents the self-inductance of the two wires, while the shunt capacitance C (in F/m) arises from the close proximity of the two conductors. The series resistance R (in Ω/m) quantifies the finite conductivity of the material, and the shunt conductance G (in S/m) originates from the dielectric loss in the material between the conductors.

By applying Kirchhoff's voltage law and current law to the circuit from Fig. 3.9 and taking the limit as $\Delta z \to 0$, the following differential equations are obtained:

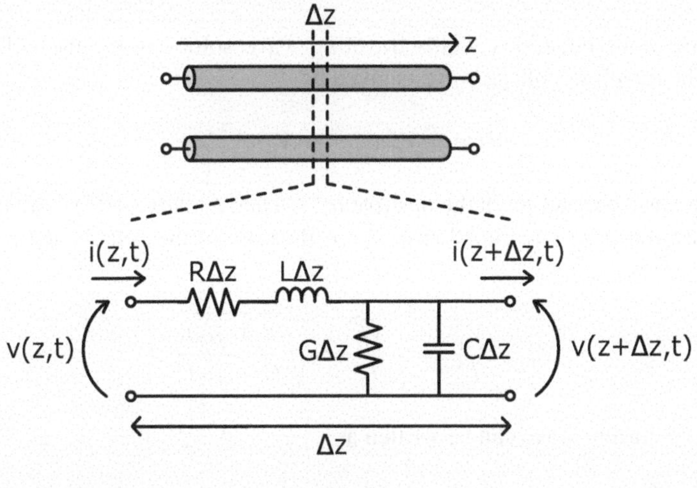

Fig. 3.9 Lumped-element model of a transmission line

$$\frac{\partial v(z,t)}{\partial z} = -R\,i(z,t) - L\,\frac{\partial i(z,t)}{\partial t} \tag{3.66}$$

$$\frac{\partial i(z,t)}{\partial z} = -G\,v(z,t) - C\,\frac{\partial v(z,t)}{\partial t}\,. \tag{3.67}$$

These equations are the time-domain transmission line equations, also known as the telegrapher equations. Similar to Maxwell's curl equations, these time-domain equations can be simplified assuming sinusoidal signaling using the phasor notation:

$$\frac{dV(z)}{dz} = -(R + j\omega L)\,I(z) \tag{3.68}$$

$$\frac{dI(z)}{dz} = -(G + j\omega C)\,V(z)\,. \tag{3.69}$$

These equations can be solved together to yield the wave equations for the voltage $V(z)$ and current $I(z)$ along the transmission line:

$$\frac{d^2V(z)}{dz^2} - \gamma^2 V(z) = 0 \tag{3.70}$$

$$\frac{d^2I(z)}{dz^2} - \gamma^2 I(z) = 0\,, \tag{3.71}$$

where γ is the complex propagation constant given by

$$\gamma = \alpha + j\beta = \sqrt{(R + j\omega L)(G + j\omega C)}\,. \tag{3.72}$$

Solving the wave equations yields traveling wave solutions for the voltage and current. The traveling voltage wave is given by

$$V(z) = V_0^+ e^{-\gamma z} + V_0^- e^{\gamma z}\,, \tag{3.73}$$

containing a wave traveling in the positive ($e^{-\gamma z}$) and negative ($e^{\gamma z}$) z direction. By defining the characteristic impedance, Z_c, as the ratio of the voltage and current on the line:

$$Z_c = \frac{V_0^+}{I_0^+} = -\frac{V_0^-}{I_0^-} = \sqrt{\frac{R + j\omega L}{G + j\omega C}}\,, \tag{3.74}$$

the traveling current wave can be written as

$$I(z) = I_0^+ e^{-\gamma z} + I_0^- e^{\gamma z} = \frac{V_0^+}{Z_c} e^{-\gamma z} - \frac{V_0^-}{Z_c} e^{\gamma z}\,. \tag{3.75}$$

In many cases, the loss of the line is quite small and can be neglected which results in much simpler and more intuitive results. Setting the loss elements $R = G = 0$ gives a zero attenuation constant and hence a purely imaginary propagation constant:

$$\gamma = j\beta = j\omega\sqrt{LC}\,. \tag{3.76}$$

Furthermore, the characteristic impedance of the line simplifies to

$$Z_c = \sqrt{\frac{L}{C}}\,. \tag{3.77}$$

Lastly, the wavelength λ and phase velocity v_p can be written as

$$\lambda = \frac{2\pi}{\beta} = \frac{2\pi}{\omega\sqrt{LC}} \tag{3.78}$$

$$v_p = \frac{\omega}{\beta} = \frac{1}{\sqrt{LC}}\,. \tag{3.79}$$

3.1.4.2 Field Interpretation of a Transmission Line

Even though the traveling voltage and current wave formulation of transmission lines is quite an elegant one, their operation is still fundamentally governed by electric and magnetic waves. Generally, a transmission line is a structure that supports the propagation of TEM waves, meaning the fields are always perpendicular to the direction of propagation (along the z-axis). For this to be true, the conductor must be perfectly lossless ($\sigma = \infty$) since $J_z \neq 0$, but E_z must be zero to satisfy Ohm's law.

Suppose we have a coaxial cable, consisting of an inner conductor and an outer conductor which form a transmission line. As illustrated in Fig. 3.10, the electric field lines run between the conductors can be integrated along that path to find the voltage difference between the inner (1) and outer (2) conductors:

$$V(z) = -\int_1^2 \mathbf{E}(z) \cdot d\mathbf{l}\,. \tag{3.80}$$

Furthermore, the magnetic field lines run around the inner conductor whose strength can be found by using Ampère's law, where the path integral of the magnetic field along a loop around the conductor is determined by the enclosed current:

$$I(z) = \oint \mathbf{H}(z) \cdot d\mathbf{l}\,. \tag{3.81}$$

For TEM propagation, both fields thus have a zero axial component, and the flow of power, i.e., Poynting vector, runs in the propagation direction, as expected. While

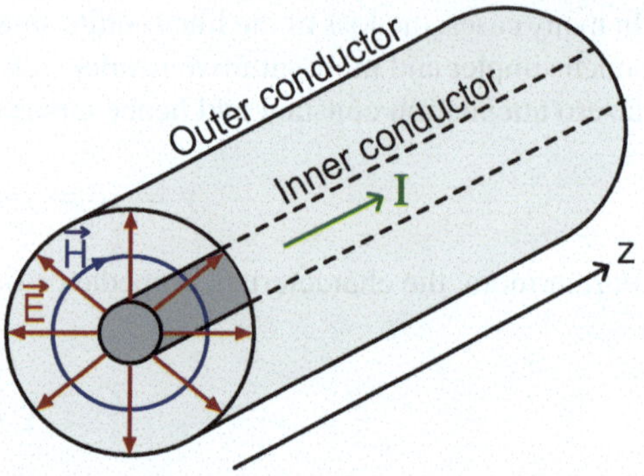

Fig. 3.10 Illustration of the electric and magnetic field lines in a two-conductor coaxial transmission line

this interpretation is only valid for lossless conductors, practical low-loss conductors essentially also behave like TEM guides. It can be proven that TEM waves require two or more conductors to propagate. A single conductor can, however, support (a combination of) TE or TM waves.

It is also remarkable how similar the expressions of the traveling voltage $V(z)$ and current $I(z)$ waves are to the expressions of the electric $\mathbf{E}(z)$ and magnetic $\mathbf{H}(z)$ fields of planar waves. This is of course to be expected since both describe TEM waves. Both are characterized by a propagation constant that has a purely imaginary value in the case of a lossless medium. The speed, wavelength, and impedance of the waves also have similar expressions, albeit as a function of circuit elements (L, C, R, G) in the case of transmission lines instead of medium characteristics (μ, ϵ, σ). Of course, since both formulations are equivalent, the circuit elements can be expressed as a function of the characteristics of the medium given the specific geometry of the transmission line. For example, the equivalence between the lumped-element model and fields-based model is described in [65] in case of a coaxial transmission line.

3.1.4.3 Behavior of Transmission Lines

Knowing the basic characteristics of a transmission line, we can analyze how it behaves when used within an electrical circuit [65]. An important consequence of distributed components is the concept of mismatch and the reflection of waves.

Figure 3.11 shows the schematic of a lossless transmission line terminated by a load impedance Z_L. Assume the transmission line is excited by an incident, forward traveling wave of the form $V_0^+ e^{-j\beta z}$ generated from a source at $z < 0$. While the ratio between the voltage and the current on the line is determined by Z_c, at the load (at $z = 0$) it is determined by Z_L. When there is a mismatch between these impedances, i.e., $Z_L \neq Z_c$, a reflected wave must be excited with a certain amplitude to satisfy this condition. Indeed, it can be shown that the ration between the amplitude of the reflected voltage wave and the incident voltage wave, defined as the reflection coefficient Γ, can be expressed as [65]

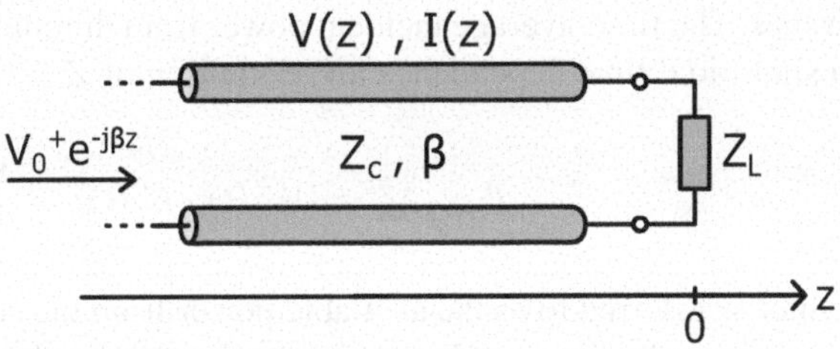

Fig. 3.11 A lossless transmission line with a load Z_L

$$\Gamma = \frac{V_0^-}{V_0^+} = \frac{Z_L - Z_c}{Z_L - Z_c} . \tag{3.82}$$

The total voltage and current waves on the line can then be written as a function of the incident wave amplitude V_0^+ as

$$V(z) = V_0^+ \left(e^{-j\beta z} + \Gamma e^{j\beta z}\right) \tag{3.83}$$

$$I(z) = \frac{V_0^+}{Z_c} \left(e^{-j\beta z} - \Gamma e^{j\beta z}\right) . \tag{3.84}$$

The voltage and current on the line thus consist of a superposition of an incident and a reflected wave that together form a standing wave across the line. To avoid a reflected wave, the load impedance Z_L must be equal (matched) to the characteristic impedance of the line Z_c in order to obtain $\Gamma_L = 0$. In this scenario, the load is said to be matched to the line since there is not reflection at the load.

If the load is matched, the voltage across the line has a constant magnitude equal to the magnitude of the incident source voltage $|V_0^+|$. However, when the load is mismatched, a standing wave pattern appears across the line leading to a nonconstant voltage on the line. The magnitude of the voltage over the line can be written as [65]

$$|V(z)| = |V_0^+| \, |1 + \Gamma e^{2j\beta z}| = |V_0^+| \, |1 + |\Gamma| e^{j(\theta + 2\beta z)}| , \tag{3.85}$$

where $\Gamma = |\Gamma| \, e^{j\theta}$ can be written in its polar form. The voltage magnitude thus oscillates with z position along the line. As $|\Gamma|$ increases, the maximum variation of this voltage increases, which is often quantified using the voltage standing wave ratio (VSWR):

$$VSWR = \frac{V_{\max}}{V_{\min}} = \frac{1 + |\Gamma|}{1 - |\Gamma|} \tag{3.86}$$

and is another popular metric for the amount of mismatch in the system.

In many electronic systems, power is of big concern and is calculated as the product of the voltage and current, equivalent to the Poynting vector ($\mathbf{E} \times \mathbf{H}$) of EM waves. The time-average incident power from the source can be calculated assuming a sinusoidal time dependence and definition of Z_c:

$$P_{\text{avg, in}} = \langle V^+(t) \cdot I^+(t) \rangle = \frac{1}{2} \frac{|V_0^+|^2}{Z_c} , \tag{3.87}$$

which is referred to as the available power from the source. The total average power flow along the line, and hence delivered to the load, is equal to [65]

$$P_{\text{avg}} = \frac{1}{2} \frac{|V_0^+|^2}{Z_c} (1 - |\Gamma|^2) \tag{3.88}$$

and is thus equal to the incident power minus the reflected power. If $\Gamma = 0$, the maximum amount of power is delivered to the load, i.e., the available power from the source. However, when the load is mismatched, some of the power from the source is not delivered to the load and is thus lost. This loss is called return loss (RL) and is defined in decibel units as

$$RL = -20 \log(|\Gamma|) . \tag{3.89}$$

A matched load ($\Gamma = 0$) has a return loss of ∞ dB, while total reflection leads to a return loss of 0 dB. A higher return loss is thus a desirable characteristic, which is somewhat confusing.

The reflection coefficient defined at the load can be generalized to any point along the line as follows [65]:

$$\Gamma(z) = \frac{V^-(z)}{V^+(z)} = \frac{V_0^- \, e^{-j\beta z}}{V_0^+ \, e^{j\beta z}} = \Gamma(0) \, e^{-2j\beta z} , \tag{3.90}$$

where $\Gamma(0)$ is the reflection coefficient at the load ($z = 0$). This expression allows analyzing the effect of load mismatch at different points on the line. Since the power flow along the line is constant, but the voltage magnitude generally is not, the impedance looking into the line depends on the exact value of the reflection coefficient at the load Γ_L and the length of the line. Assuming a lossless line of length ℓ terminated with load Z_L, as shown in Fig. 3.12, the input impedance can be expressed as [65]

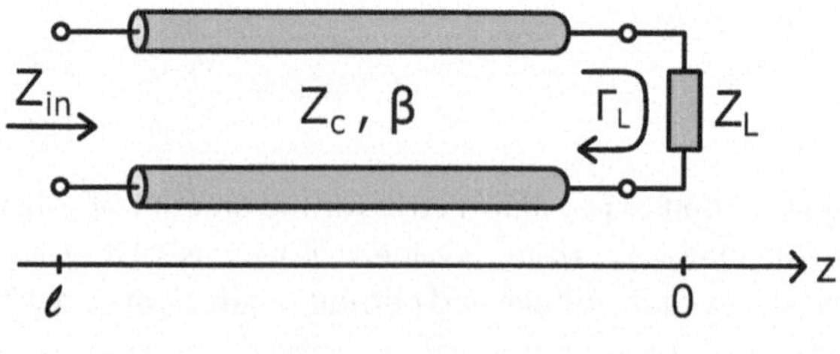

Fig. 3.12 A lossless transmission line of length ℓ terminated with Z_L

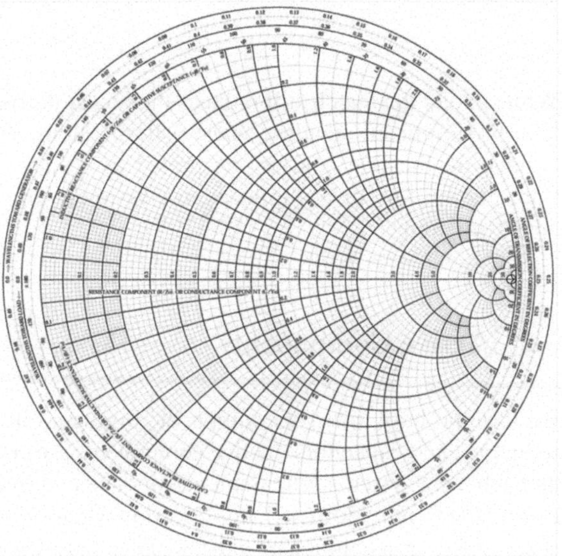

Fig. 3.13 Example of the Smith chart

$$Z_{\text{in}} = \frac{V(-\ell)}{I(-\ell)} = Z_c \, \frac{1 + \Gamma \, e^{-2j\beta\ell}}{1 - \Gamma \, e^{-2j\beta\ell}} = Z_c \, \frac{Z_L + j Z_c \, \tan(\beta\ell)}{Z_c + j Z_L \, \tan(\beta\ell)} \,. \tag{3.91}$$

To avoid working with complicated equations, transmission line problems are often solved using a graphical diagram called a Smith chart. Figure 3.13 shows an example of the Smith chart, which is basically a polar plot of the reflection coefficient $\Gamma = |\Gamma| \, e^{j\theta}$. The magnitude $|\Gamma|$ is plotted as the radius from the center, while the angle θ is measured counterclockwise from the rightmost point on the graph. Any passive ($|\Gamma|$) reflection coefficient can then be visualized within this chart. An open load ($Z_L = \infty$), corresponding to a $\Gamma = 1$, is located at the rightmost point, while a shorted load ($Z_L = 0$), corresponding to a $\Gamma = -1$, is located at the leftmost part of the chart. The center point represents a perfectly matched load ($\Gamma = 0$). The Smith chart can provide a unique insight and method of visualization of transmission lines phenomena, and the reader is encouraged to explore the capabilities of this chart [65].

It must be noted that a transmission line system can also have a mismatch at the source or along the line, in addition to at the load. This will lead to multiple reflections with quite complicated expressions for the waves and impedances along the line. While a practical case of multiple reflections will be discussed later on in Chap. 7, these unnecessary details will be omitted from this foundational chapter but can be found in [65].

Finally, it must be noted that all derived expressions simplify considerably when the physical size of the line is much smaller than the wavelength of the signal, i.e., when

$$\beta \ell = \frac{2\pi}{\lambda} \ell \ll 1 . \tag{3.92}$$

While a load mismatch still reduces the power deliver to the load, reflections do not lead to significant standing wave patterns along the line, but rather the resulting voltage on the line can be interpreted as a classic voltage divider. In that case, transmission line theory reduces to classic circuit theory with its lumped-element approximations.

3.2 Active Components

Besides the basic passive components discussed in the previous section (resistors, inductors, and capacitors), active components are crucial for high-performance THz and sub-THz systems. The main active device in integrated circuits is the transistor and is able to provide signal amplification when supplied with DC power. Numerous semiconductor technologies exist, such as bulk CMOS, Silicon-on-Insulator (SOI) CMOS, SiGe, and III/V technologies (e.g., InP, GaAs, and GaN), all with multiple types of transistors with different properties. As discussed in Sect. 1.4, many applications prefer a CMOS technology for its low production cost at high volumes and its integration potential with a digital back end, despite the inferior performance at high frequencies compared to SiGe and III/V technologies. This discussion will therefore mainly focus on the characteristics of transistors in CMOS technologies.

3.2.1 *MOSFETs in CMOS*

In a complementary metal-oxide-semiconductor or CMOS process, two main types of transistors can be implemented: an N-type MOS (NMOS) and a P-type MOS (PMOS) transistor, both shown in Fig. 3.14 with their respective symbol representations. The name "CMOS" has its origin from the being able to combine both types enabling the fabrication of CMOS circuits, a very efficient type of digital circuits. Since the flow of current is controlled by an electric field, the transistor is called a field-effect transistor or FET. However, the terms MOSFET, MOS transistor, or even just transistor will be used interchangeably throughout this book. While a rigorous description of the physics is outside of the scope of this book, the working principle of the MOSFET will be briefly explained. A more thorough explanation can be found in [43].

As shown in Fig. 3.14, a MOSFET is controlled by four primary terminals: gate (G), drain (D), source (S), and bulk (B). The traditional MOS structure consists of a gate metal on top of an oxide layer deposited onto a semiconductor material called the bulk or substrate. In the case of an NMOS, this substrate is a p-type semiconductor, meaning it contains an excess of holes. On both sides of this MOS

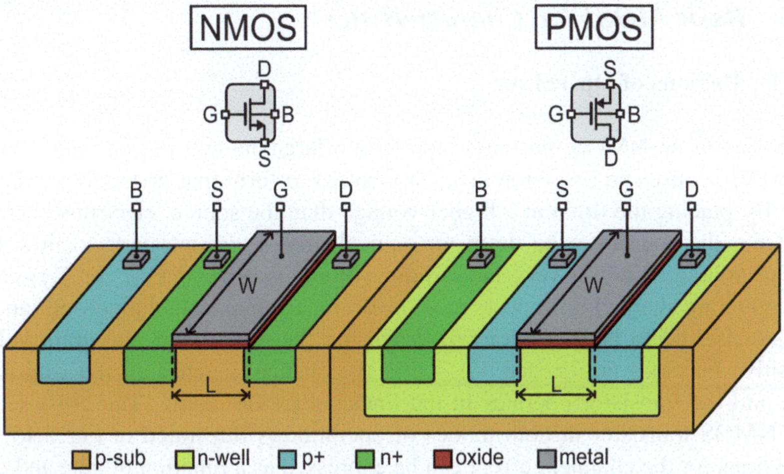

Fig. 3.14 Structure of an NMOS and PMOS transistor in a CMOS process

structure, regions with an n-type semiconductor material are placed with an excess of electrons, called the source and drain regions. By applying a large enough positive voltage at the gate (i.e., larger than the so-called threshold voltage of the transistor V_T), the underlying substrate is depleted from free holes and filled with electrons from the neighboring n-type material, forming a negatively charged inversion layer under the interface. This inversion charge forms an electrically conductive connection between the source and the drain, also called the channel, where charges and thus current can freely flow. The transistor thus functions as a three-terminal device where the flow of current between the source and drain can be controlled via the voltage on the gate.

For a PMOS transistor, the working principle is similar, but all charges are inverted with respect to the NMOS. Also, since all devices in the circuit are integrated onto the same p-type substrate, PMOS devices are placed in a separate n-well as they require an n-type substrate to form its positively charged inversion layer creating a channel for holes to flow. The following discussion will mainly focus on n-type devices as they provide a larger gain than p-type devices due to the larger carrier mobility and are thus used almost exclusively in high-frequency circuits.

Furthermore, while Fig. 3.14 shows the structure of a bulk CMOS process, another CMOS variation exists called silicon-on-insulator (SOI) CMOS. In an SOI CMOS process, a thin insulating layer is added under the channel to isolate it from the substrate. This can have various benefits, such as an improved power efficiency of digital circuits and enhancement of the speed of the device. Nevertheless, in this work only devices and circuits in bulk CMOS are discussed.

It should also be noted that in more advanced processes, the gate metal is replaced by a polysilicon gate (with lower resistance) and the oxide by a high-k dielectric material (to obtain a better control over the channel of the transistor). The operation principle remains the same.

3.2.2 *Basic MOSFET Characteristics*

3.2.2.1 Regions of Operation

In the case of an NMOS transistor, applying a large enough positive gate voltage ($V_G > V_T$) causes an inversion layer or channel to form that connects source and drain. By placing the drain at a higher voltage than the source, electrons begin the flow from the source to the drain, or equivalently, a current starts to flow from drain to source. When a very large drain voltage is applied ($V_D > V_G - V_T$), the conducting channel disappears near the drain terminal, called pinch-off, and the transistor is said to be in saturation. When, on the other hand, the drain voltage is positive but does not exceed $V_G - V_T$, the channel reaches all the way to the drain, and the transistor operates in the linear or triode region. The cross-section of an NMOS transistor in both modes of operation is illustrated in Fig. 3.15. The expressions for the channel current can be expressed as a function of gate and drain voltages in the general case where the source voltage is not zero. It will be assumed that the bulk is placed at an identical voltage as the source, to avoid bulk effects [69], and as such, the bulk terminal will not be explicitly shown on the MOS symbol.

Figure 3.16a plots the drain–source current I_{DS} versus gate–source voltage V_{GS}, showing a steep increase in current as the voltage exceeds the threshold voltage V_T of the NMOS. As a function of drain–source voltage V_{DS} and assuming $V_{GS} > V_T$, the drain–source current I_{DS} is plotted in Fig. 3.16b. First the current increases linearly, like a resistor, and this is called the linear or triode region. For large V_{DS} the current starts to saturate to a nearly constant value. This value is referred to as the saturation region. Both regions are separated by a parabola described by the condition for pinch-off at the drain region: $V_{DS} = V_{GS} - V_T$. The transistor exhibits very different behavior in both regions which will be discussed separately.

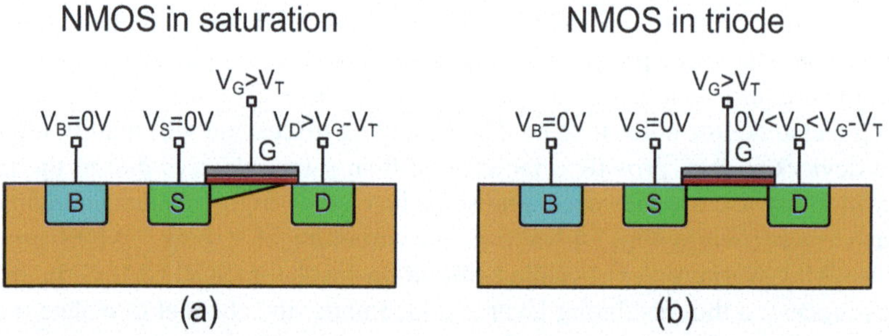

Fig. 3.15 Cross-section of an NMOS transistor biased in saturation (**a**) and in triode (**b**)

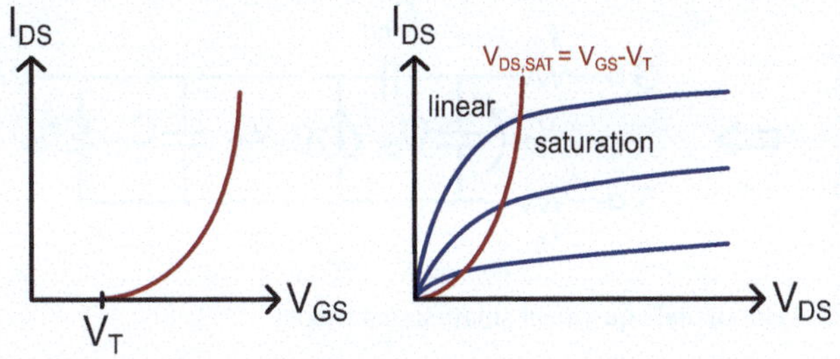

Fig. 3.16 Drain–source current of an NMOS transistor versus gate–source voltage (**a**) and drain–source voltage (**b**)

3.2.2.2 MOSFET as a Transconductance

For a transistor in the saturation region, the drain–source current can be expressed as [43]

$$I_{DS} = \frac{\mu C_{ox}}{2} \frac{W}{L} (V_{GS} - V_T)^2 (1 + \lambda V_{DS}) , \tag{3.93}$$

where μ is the mobility of the charge carriers (i.e., electrons in case of an NMOS) and C_{ox} the gate-oxide capacitance (in F/m^2) calculated via $C_{ox} = \epsilon_{ox}/t_{ox}$ with t_{ox} being the oxide thickness. For a physical derivation of these and other technology parameters, the reader is referred to specialized literature [43, 69]. The width and length of the transistor channel are denoted by W and L, respectively, as illustrated in Fig. 3.14. Equation 3.93 shows a strong dependence of the current on the gate–source voltage, but weak dependence on the drain–source voltage, the latter being quantified by the parameter $\lambda = 1/(V_E L)$, with V_E being a technology parameter called the early voltage. Furthermore, the ratio W/L determines how much current the transistor can deliver for the same $V_{GS} - V_T$ (also called overdrive voltage).

It must be noted that the actual behavior is more complex in reality and that multiple distinct operation regions can be distinguished within the saturation region: weak inversion, strong inversion, and velocity saturation [69]. While this distinction is an important one for low-frequency design, it becomes less relevant when working at higher frequencies and is thus not discussed.

When a transistor is driven by small signals, meaning that the transistor operation does not change appreciably, it can be represented by a small-signal scheme. The small-signal model of a three-terminal ($V_S = V_B$) NMOS transistor is given in Fig. 3.17 and aims to model the small-signal transistor behavior using several linear parameters. The different components refer to the following elements: $r_{g/s/d}$ the gate/source/drain resistance, C_{gs} the gate–source capacitance, C_{gd} the gate–drain capacitance, C_{ds} the drain–source capacitance, r_{ds} the drain–source resistance, and g_m the transistor transconductance.

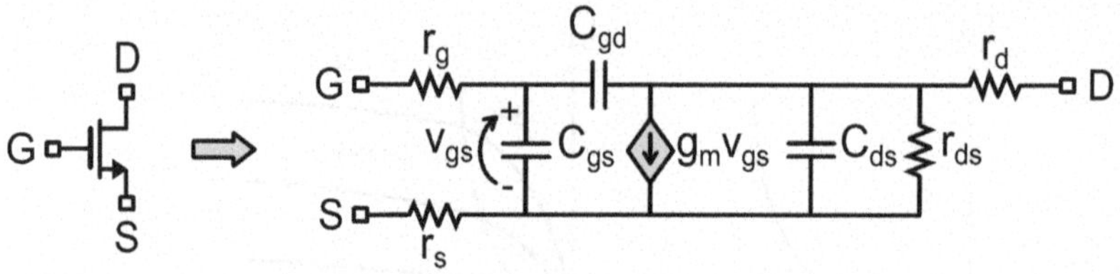

Fig. 3.17 NMOS symbol and equivalent small-signal model

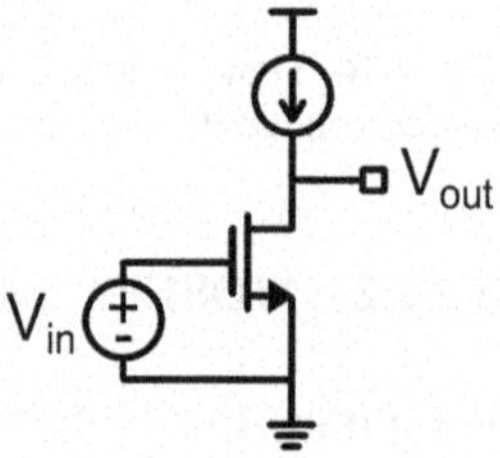

Fig. 3.18 Example of an NMOS transistor used in a common-source amplifier

The most important parameter is the transistor transconductance, defined as the change in small-signal drain–source current i_{ds} when varying the gate–source voltage v_{gs}, and can be expressed as (neglecting λ)

$$g_m = \frac{dI_{DS}}{dV_{GS}} = \mu C_{ox}\frac{W}{L}(V_{GS} - V_T) = \frac{2I_{DS}}{V_{GS} - V_T}\,. \tag{3.94}$$

Clearly, the transistor is able to provide more transconductance gain as W/L is increased.

Next, while the input resistance is approximately infinite, the output resistance is mainly determined by r_{ds}, expressed as

$$r_o = r_{ds} = \frac{dI_{DS}}{dV_{DS}} = \frac{V_E L}{I_{DS}}\,. \tag{3.95}$$

To illustrate the transistor operating as an amplifier, consider the basic common-source amplifier circuit in Fig. 3.18 with an ideal current source (infinite output impedance). This current source supplies the DC current required to bias the transistor into saturation. After extracting the small-signal parameters in the DC operating point, the small-signal voltage gain of this amplifier can be expressed as [69]

$$A_v = g_m r_{ds} = \frac{2V_E L}{V_{GS} - V_T}\,. \tag{3.96}$$

To maximize the voltage gain, the transistor should be biased at a low overdrive voltage (to increase g_m for the same current) and should have a large length (to increase r_{ds} for the same current). While this result might be true at lower frequencies, a different design strategy is required to optimize the gain at very high frequencies, as will be discussed later.

3.2.2.3 MOSFET as a Switch

For a transistor in the triode region, the drain–source current expression can be approximated as follows [43]:

$$I_{DS} = \mu C_{ox} \frac{W}{L} \left[(V_{GS} - V_T) V_{DS} - \frac{V_{DS}^2}{2} \right], \tag{3.97}$$

which shows a nearly linear dependence on the drain–source voltage. Neglecting the quadratic term leads to the following equivalent resistance:

$$R_{DS} = \frac{1}{\mu C_{ox} \frac{W}{L} (V_{GS} - V_T)} . \tag{3.98}$$

A MOS transistor in the linear regions thus behaves like a resistor controlled through its gate voltage. For a large gate voltage ($V_{GS} > V_T$), the switch is on with an equivalent on-resistance R_{DS} determined by the size of the switch (W/L) and its overdrive voltage. For a small gate voltage ($V_{GS} < V_T$), the transistor is turned off, resulting in a very small I_{DS} and hence large R_{DS}. Note that a PMOS switch (composed of a PMOS transistor in the triode region) works similarly but is turned on and off by a small and large gate voltage, respectively.

Voltage-controlled switches are useful in many situations, but most notably form the building blocks of logic gates for digital processors. Figure 3.19 shows the schematic, symbol, and transfer curve of the CMOS inverter, the most fundamental logic building block. When the input voltage is high, the PMOS transistor M_P is turned off and the NMOS transistor M_N is turned on, resulting in a low resistance path from the output to ground, and hence the output is low. Alternatively, when the input is low, the NMOS is turned off, while the PMOS is turned on and pulls the output high toward the positive supply rail V_{DD}. The inverter thus, unexpectedly,

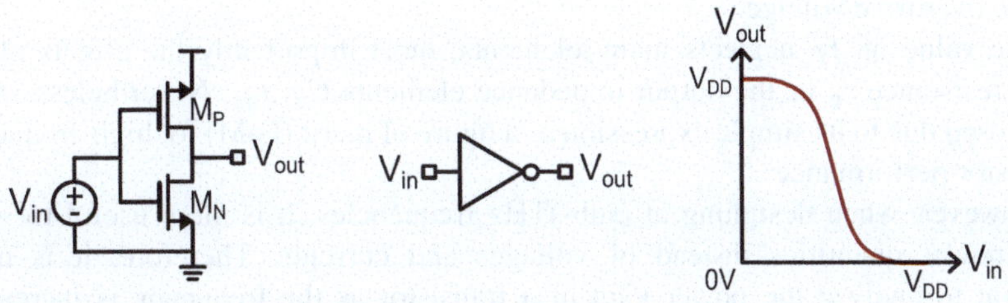

Fig. 3.19 Schematic, symbol, and transfer curve of a CMOS inverter

inverts the digital input signal. As illustrated by the transfer curve in Fig. 3.19, the signal transition happens very steeply, indicating a high robustness against moderate levels of noise in the input. Furthermore, while the inverter draws current from the supply while providing a signal transition, it does not draw any current when the input is static, since either the NMOS or PMOS is turned off.

The inverter can be extended toward more complicated logic functions, for example, to OR and AND gates, to adders and multipliers, and finally to extremely complicated processors and algorithm accelerators, all with similar characteristics as the CMOS inverter. These desirable characteristics of CMOS gates (high output swing, large noise margin, and no static power consumption) have enabled the realization of arbitrary logical functions at miniaturized scale with a low power consumption which has been the main driving force of the rise of CMOS technologies.

3.2.3 *MOSFETs at High Frequencies*

3.2.3.1 Transistor Gain: f_T and f_{max}

Compared to expression 3.96, the maximum transistor gain degrades with increasing frequency. This is mainly the result of the parasitic transistor elements, shown in Fig. 3.17, which filter out useful signal at high frequencies.

A first way to quantify the high-frequency gain performance of a transistor is through the transition frequency f_T, which is the frequency at which the current gain becomes unity. The current gain is calculated by providing a short circuit between the drain and the source of Fig. 3.17 and ignoring the series resistances and feed-forward current contribution through C_{gd}. Equating this current gain to unity yields the following expression for f_T [7]:

$$f_T = \frac{g_m}{2\pi(C_{gs} + C_{gd})} = \frac{3\mu(V_{GS} - V_T)}{4\pi L^2} . \tag{3.99}$$

The last part of the equation is valid in the strong-inversion region, where $C_{GS} = (2/3)WLC_{ox}$ and $C_{GD} \approx 0$ [69]. In contrast to lower frequencies (Eq. 3.96), a large high-frequency gain is obtained by minimizing the gate length L and choosing a large overdrive voltage.

The value of f_T neglects many elements, most importantly the effects of the input resistance r_g or the output impedance elements C_{ds},r_{ds}. Nevertheless, f_T is often used due to its simple expression as a figure of merit (FoM) for high-frequency transistor performance.

However, when designing at (sub-)THz frequencies, it is more useful to work with power quantities instead of voltages and currents. Therefore, it is more relevant to analyze the power gain of a transistor as the frequency is increased. The power gain can be calculated based on the schematic of Fig. 3.17 again, with some simplifying assumptions, e.g., neglecting the feedback through C_{gd} and series elements r_s, r_d. The maximum power gain can then be approximated, assuming a perfect conjugate match at the input and output:

$$G_{\max} \approx \frac{r_{ds}}{4r_g}\left(\frac{g_m}{2\pi f(C_{gs}+C_{gd})}\right)^2 = \frac{r_{ds}}{4r_g}\left(\frac{f_T}{f}\right)^2 . \tag{3.100}$$

From this, the frequency at which the power gain becomes unity, called the maximum oscillation frequency or $f_{\max}$, is given by

$$f_{\max} = \frac{f_T}{2}\sqrt{\frac{r_{ds}}{r_g}} . \tag{3.101}$$

Next to the parasitic capacitances also effecting f_T, the gate resistance plays an important role for the power gain at high frequencies. A more elaborate expression for $f_{\max}$ can be found including C_{gd} and the series resistance of the channel r_{ch} and source r_s [70]:

$$f_{\max} = \frac{f_T}{2\sqrt{r_g \frac{g_m C_{gd}}{C_{gs}+C_{gd}} + \frac{r_g+r_{ch}+r_s}{r_{ds}}}} , \tag{3.102}$$

indicating the importance of all capacitive and resistive parasitic elements, resulting from the intrinsic device and from the physical layout and interconnect, as will be discussed later.

3.2.3.2 Transistor Noise: $F_{\min}$

As discussed extensively in Chap. 2, noise is one of the main limiting factors of performance in communication systems. Apart from the noise present in the environment, additional noise is generated by the transistors from various sources. As described in Sect. 2.3.2, noise is commonly described using a PSD (expressed in V^2/Hz or A^2/HZ) which can be integrated over a certain frequency range to quantify the amount of noise power in that band.

Figure 3.20 shows a small-signal model of a MOS transistor with the most important parasitic elements and noise sources. Firstly, thermal noise is generated by the gate resistance and can be expressed as

$$\overline{V_{ng}^2} = 4kTr_g . \tag{3.103}$$

Secondly, due to the resistive nature of the channel, a current noise is generated described by

$$\overline{I_{nd}^2} = 4kT\frac{\gamma}{\alpha}g_m , \tag{3.104}$$

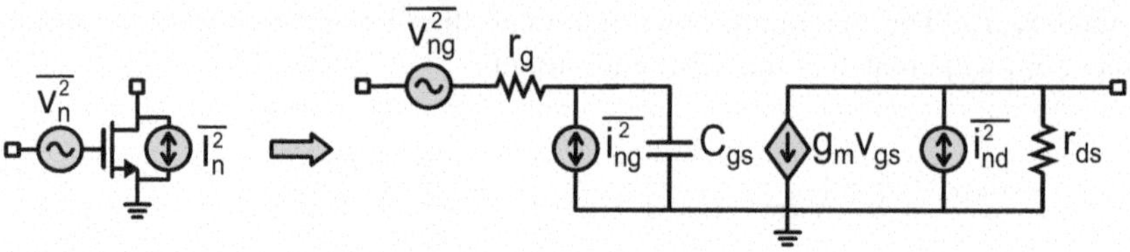

Fig. 3.20 Small-signal MOS transistor model with most important parasitic elements and noise sources

where γ and α are technology parameters [70], not further discussed here. Lastly, the third noise source is the drain-induced gate noise, which is the noise current through the channel which couples to the gate at higher frequencies, described by Lee [7]

$$\overline{I_{ng}^2} = 4kT\delta g_g \ , \tag{3.105}$$

where δ is another technology parameter and g_g is expressed as [7]

$$g_g = \frac{\alpha\omega^2 C_{gs}^2}{5g_m} \ . \tag{3.106}$$

Note that I_{ng}^2 and I_{nd}^2 are correlated (i.e., not independent) noise sources, which should be included in noise calculations.

From these expressions and the small-signal model, the equivalent input voltage and current noise source as shown in Fig. 3.20 can be derived, neglecting the output resistance r_{ds}:

$$\overline{V_n^2} = \overline{V_{ng}^2} + r_g^2\overline{I_{ng}^2} + \left[\frac{1 + s\, r_g C_{gs}}{g_m}\right]^2 \overline{I_{nd}^2} \tag{3.107}$$

$$\overline{I_n^2} = \overline{I_{ng}^2} + \left[\frac{sC_{gs}}{g_m}\right]^2 \overline{I_{nd}^2} \ , \tag{3.108}$$

where $s = j\omega$ is the Laplace variable, often used for phasor calculations.

Finally, assuming a perfect noise match, the minimum noise $F_{\min}$ of a MOS transistor can be expressed as follows:[2]

$$F_{\min} = 1 + 2\frac{\gamma}{\alpha}\frac{f}{f_T}\sqrt{1 + \frac{\alpha}{\gamma}\, g_m r_g} \ . \tag{3.109}$$

[2] How to match for noise will be discussed in the final section of this chapter.

This calculation neglects the drain-induced gate noise (to make the calculation bearable) as it is not a dominant factor in the overall noise. The expression shows an increase in transistor noise as the frequency increases and describes how the noise performance improves as the technology advances (i.e., f_T increases).

3.2.4 *Transistor Optimization*

As indicated by the derived expressions of f_T, $f_{\max}$, and $F_{\min}$, the transistor performance at high frequencies heavily depends on the mitigation of parasitic elements. These parasitics have contributions not only at the device level but also at the layout and interconnect level. Since both aspects are heavily interdependent, transistor optimization must happen at both levels simultaneously.

At the device level, the gate length is usually kept as small as possible to minimize the amount of parasitic capacitance for a given amount of output current or transconductance. Next, the width is typically designed to deliver a specified amount of output current or power. Note that a larger transistor width, for obtaining a larger output power, inevitably increases the parasitic elements and thus degrades transistor gain. Finally, the transistor bias ($V_{GS} - V_T$) is optimized for gain or noise, while taking power efficiency into account.

At the interconnect level, the layout for a given transistor size should be drawn to minimize degrading layout parasitics as much as possible. Minimizing the parasitic capacitance can be obtained by avoiding large overlap between the terminals, while series resistance can be decreased by ensuring short connections. To avoid layouts with a high aspect ratio, a transistor is usually broken up into several fingers and combined as shown in Fig. 3.21 where a transistor of width W is divided into N_f fingers of width $W_f = W/N_f$. This significantly reduces the series resistance of otherwise long connections and provides more layout flexibility. However, more optimizations are required to achieve adequate performance at (sub-)THz frequencies. Especially the gate resistance, which serves as an important bottleneck for high-frequency performance, is a more complex problem that necessitates a more thorough investigation.

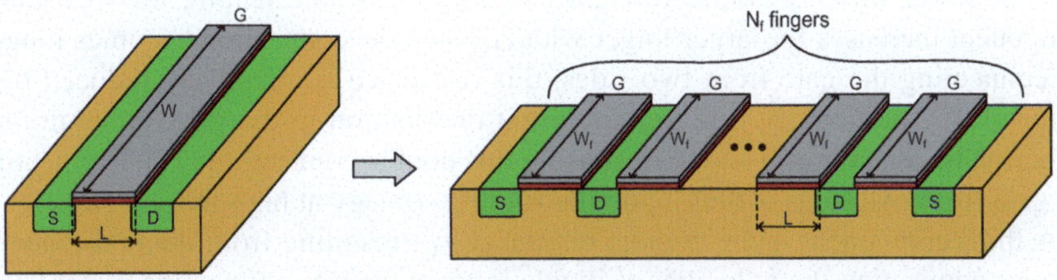

Fig. 3.21 Breaking up a single MOS transistor into several fingers

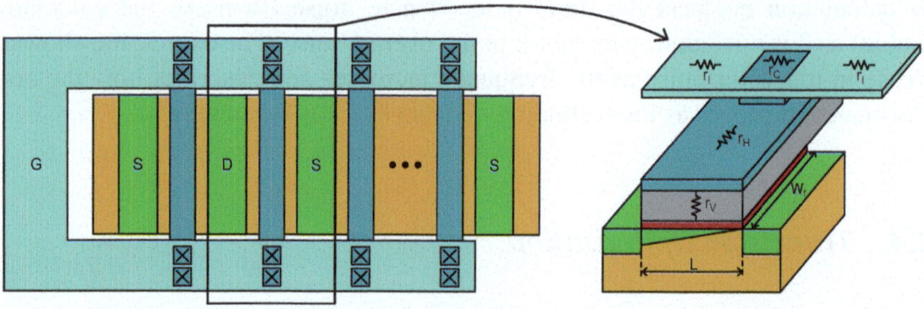

Fig. 3.22 Transistor layout with more in-depth illustration of the gate connection and various contributions to the gate resistance

3.2.4.1 Gate Resistance Optimization

Figure 3.22 depicts a schematic top-view layout of a MOS transistor showing the gate connections, as well as a 3D zoomed-in view of a single gate. The gate connections are first routed outward using a low-resistance layer, often made up of polysilicon. Outside of the channel, this layer is connected to a metal layer that combines the connections from both sides. The total gate resistance thus consists of several contributions as illustrated in Fig. 3.22: the vertical gate resistance r_V, the horizontal gate resistance r_H, the contact resistance r_C, and the interconnect resistance r_I, which can be expressed as

$$r_g = r_V + r_H + r_C + r_I \; . \tag{3.110}$$

First, the vertical gate resistance r_V is determined by the resistivity of the polysilicon material, the layer thickness, and the vertical cross-section of the gate finger (i.e., area from top view). Since the area of a finger equals $W_f \cdot L$, the vertical gate resistance increases as the number of fingers is increased, since the width per finger W_f is decreased.

Second, the horizontal gate resistance r_V is calculated by the sheet resistance of the routing layer and the width and length of the connection. In some processes, a highly conductivity silicide layer is often added on top of the polysilicon layer to lower this horizontal resistance. For a fixed gate length, this resistance component increases for larger finger widths, since the connection becomes longer. By connecting the gate from two sides, this resistance is effectively reduced by a factor of 4. Additionally, due to the current distribution in the gate, an additional reduction by a factor of 3 is obtained using this double connection [71]. Connecting the gate from two sides is thus a popular routing strategy at high frequencies.

A third component is the contact resistance r_C, resulting from the resistance of the vias connecting the polysilicon/silicide layer to the interconnecting metal layer. This resistance can be minimized by providing as many vias as possible between the layers.

Fourth, the metal layer interconnecting all gate connections adds series resistance r_I which depends on the resistivity of the metal and grows with a number of fingers as the distance between the fingers increases for a total width W. Again, by connecting from two sides, this resistance is reduced by a factor of 2.

Lastly, for completeness it should be mentioned that the gate resistance also has a component arising from the non-quasistatic (NQS) effect in the channel. However, this part has a negligible contribution to the overall gate resistance so will not be discussed further.

Knowing how the different gate resistance contributions depend on the physical dimensions of the fingers, the optimal finger width can be determined. Indeed, while a smaller finger width (i.e., a higher number of fingers for a fixed total width) decreases the horizontal resistance, it will increase the vertical and especially interconnect resistance. For a given transistor width W, there exists an optimal finger width W_f determined by a trade-off mostly between the horizontal gate resistance and interconnect resistance. Interestingly, the gate resistance tends to decrease for larger transistor widths. However, due to a more significant increase in parasitic capacitance ($C_{par} \sim W \cdot L$) and decrease in output resistance ($r_{ds} \sim 1/I_{DS}$), the $f_{\max}$ degrades considerably for larger transistors.

3.2.4.2 Layout Optimization and Modeling

With all of the discussed factors in mind, the transistor layout can be optimized to maximize the power gain at high frequencies for a given W, W_f, and L. Figure 3.23 shows the transistor layout used in this work, where mainly the gate resistance and parasitic capacitance are minimized [72]. The gate is brought out to both sides in polysilicon, combined in the lowest metal layer, and brought up to the highest metal layer to be connected to external components. All source connections are connected to a large ground plane in multiple parallel metal layers to minimize the source resistance. The drain connections to higher metal layers are made over the channel to minimize overlap with the gate metal, thus lowering parasitic C_{gd} capacitance. For

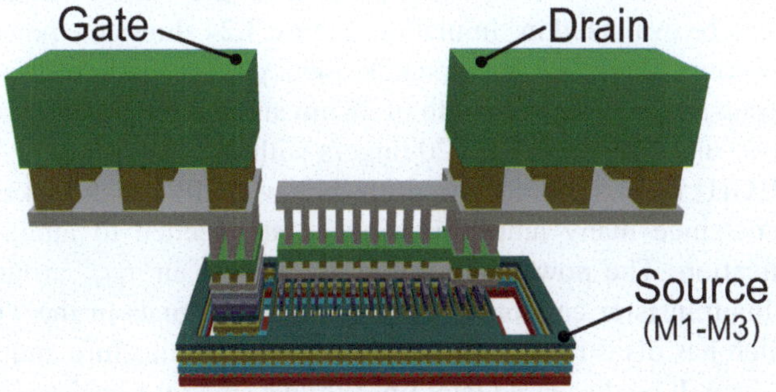

Fig. 3.23 Layout of an NMOS transistor optimized for high-frequency performance

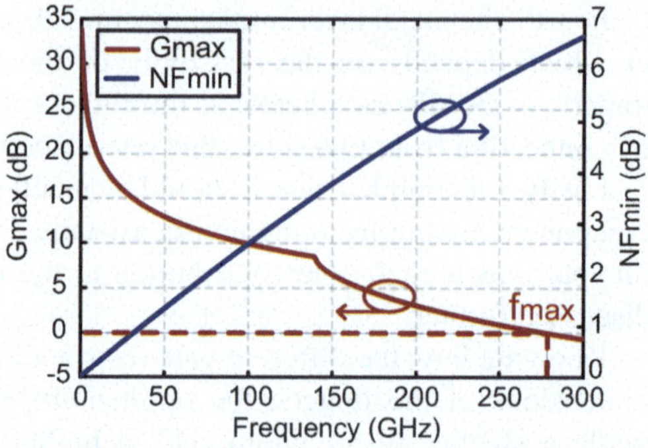

Fig. 3.24 The $G_{\max}$ and $NF_{\min}$ of a single transistor (W/L = 25 μm/28 nm)

larger transistor widths, multiple of these unit transistor cells are placed in parallel, which reduces the overall interconnect loss compared to one large transistor cell with many fingers.

Not only should the size and layout of the transistor be optimized for maximum performance, but the behavior should also be accurately modeled in order to precisely predict its behavior. The intrinsic behavior of the transistor is included in the transistor model provided by the foundry. The parasitics in a transistor cell are usually modeled in two separate steps. At lower metal layers, where the dimensions are rather small, the parasitic resistance and capacitance are modeled using a parasitic extracting tool. This tool models all drawn layers as a series of lumped resistors and capacitors, capturing the loss of and coupling between all metal layers. For the higher metal layers, where the dimensions of interconnects are significantly larger, the influence of parasitic inductance becomes more important. As this is not captured by parasitic extraction tools, these connections must be modeled by an EM solver. Where a 2.5D EM solver (which simplifies the fields along the thickness of the layers) suffices at most frequencies, a 3D solver is necessary at high (THz) frequencies to adequately model interconnects in all dimensions. A full transistor model thus includes the intrinsic transistor model at the device level, a parasitic extraction of the parasitics at lower metal levels, and an EM model of interconnects at the highest metal layers.

Having fully optimized the layout and modeled the whole transistor cell, the performance can be evaluated in simulation. Figure 3.24 shows the power gain $G_{\max}$ and minimum noise figure $NF_{\min}$ versus frequency for a single transistor in 28 nm CMOS. The transistor has a gate length of 28 nm and the total width around 25 μm, divided into two unit cells each with 20 fingers with $W_f = 630$ nm.

Above 100 GHz the maximum gain drops below 10 dB which makes the circuits very inefficient since many amplifying stages are needed to attain the desired signal amplification. The power gain drops to 0 dB at an $f_{\max}$ around 280 GHz. As a result, this transistor cannot be used to amplify signals in the THz range (> 300 GHz) which has disastrous consequences for the architecture and efficiency of THz systems, as will be discussed later on. While a smaller transistor size can offer a slightly higher $f_{\max}$, its low current drawing capability limits its use in practical circuits.

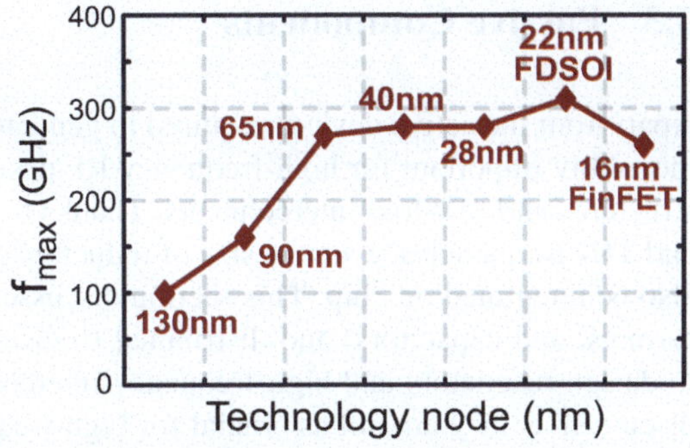

Fig. 3.25 Simulated transistor f_{max} across different CMOS technology nodes

Furthermore, the minimum noise figure clearly increases with frequency, meaning more noise will inevitably be generated by the circuits as the frequency is increased.

3.2.5 *Effects of Scaling*

Technology scaling has been the largest reason for performance improvements for the last decades. Is this something we can continue to rely on? Well, unfortunately not. Since the main focus of technology scaling has been on improving the performance and the density of digital logic, it has not been as beneficial for RF performance.

Figure 3.25 shows the simulated f_{max} of a single transistor across various CMOS technology nodes. Clearly, the f_{max} has saturated around 300 GHz and has even decreased in the most advanced processes. Indeed, as the gate length has continued to scale down, the maximum supply voltage has decreased as well, resulting in limited voltage headroom and hence limited overdrive voltage. A lower supply voltage not only restricts improvements in transistor speed but also directly degrades the achievable linearity and output power of amplifiers.

Furthermore, the most advanced CMOS technologies started implementing a new type of transistor structure which extend the channel upward into a 3D structure called a fin field-effect transistor or FinFET [73]. Compared to a traditional planar transistor, the FinFET provides better electrostatic control over the gate while improving density. Even though this technique has been beneficial for digital circuits, it has not been as advantageous for the RF performance. Indeed, due to the 3D fin structure both the gate resistance and parasitic capacitance is larger than for planar nodes [74]. While the larger output resistance (thanks to the better channel control) partially offsets the aforementioned effects, the f_{max} clearly suffers from this digital-focused technique. Future transistor structures that are currently being implemented in the latest state-of-the-art CMOS nodes, such as gate-all-around (GAA) transistors, will only exacerbate the challenges posed for RF designers.

3.3 Passive Components

Apart from the active devices required to generate signal power, passive devices are incredibly important for high-frequency RF design as it heavily relies on resonator structures and low-loss interconnects. Thanks to the small wavelength at mm-Wave and THz frequencies, even the size of inductors and transformers is small enough to also be integrated on chip. This section discusses passive lumped (inductors, transformers, and capacitors) and distributed (transmission lines) components required to design resonators and high-frequency interconnects. Resistors are not explicitly discussed as they are not as crucial for high-frequency performance. Since most of these elements are implemented in the metallization layers of the technology, the metal stack and its evolution with technology scaling is discussed first.

3.3.1 Metal Stack and Effects of Scaling

Figure 3.26 shows the cross-section of several CMOS technologies, consisting of a front end of line (FEOL) and a back end of line (BEOL). The FEOL includes the silicon substrate where all active components are implemented, as shown previously in Fig. 3.14. The BEOL contains all metal layers, separated by oxide layers, that can be connected using vias. Since the passive components, which are implemented in these metal layers, play a crucial role in high-frequency circuits, the BEOL forms

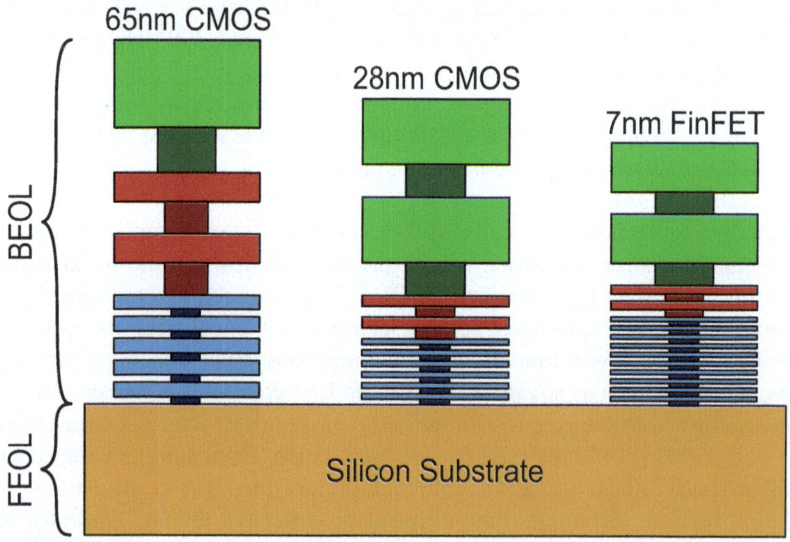

Fig. 3.26 Illustration of how the back-end metallization scales with several CMOS technologies

an important specification when choosing the technology in addition to the active device performance.

The metallization BEOL is illustrated for several CMOS technologies in Fig. 3.26, only showing the polished layers. Usually, a thick redistribution layer is added on top but is not commonly used for high-frequency passives due to its high surface roughness. It should be noted that this figure is not to scale, as in reality the silicon substrate has a substantially greater thickness than the shown metal layers. For instance, in 28 nm CMOS the substrate ($\approx 300\,\mu$m thick) is about 60 times as thick as the metal stack. Furthermore, the drawn dimensions aim to give a rough indication of general trends but do not represent exact numbers, as such technology-specific information is highly sensitive.

Most metal stacks are optimized for digital systems, where a large number of metal layers are needed to route the many interconnects and a few thicker metals are used for power distribution. For RF design, however, thick top metals are desired to implement low loss inductors and transmission lines. Therefore, most foundries provide multiple metallization options in the same technology, offering a trade-off between digital and RF performance, for example, by replacing thinner metal layers by one or more very thick top layers.

Still, since the advancements of CMOS have largely been driven by digital system performance, technology scaling has evolved in that light with several important consequences for RF circuits. As visible in Fig. 3.26, the number of layers has increased substantially in order to support the growing complexity in chip interconnects. Furthermore, the thickness and spacing of the metal layers have reduced significantly to provide denser logical gates with a finer pitch. The thinner metals lead to a higher resistance and therefore larger loss in high-frequency passives and interconnects. Additionally, since the top metals are closer to the substrate, more coupling to the substrate and thus higher losses are expected. Finally, also the active devices suffer since the via resistance is increased (especially at lower levels) and a higher number of vias must be traversed from the substrate to reach the upper layers. Both increase the resistance in the gate connection, degrading the transistor $f_{\max}$.

3.3.2 Inductors

An inductor is a component that stores energy in its magnetic field generated by flowing current. Inductors are often realized as metal spirals, as the mutual coupling between the different turns leads to a higher inductance compared to a straight line with the same length. Inductors are an essential part of resonators and matching networks and are used to create a high impedance in the supply path. To model the inductive effects of inductor, EM simulator tools must be used as such effects are not modeled by parasitic extractor tools.

3.3.2.1 Inductor Model

At sufficiently low frequencies, an inductor can be modeled as an ideal inductor L with a series resistance R_s, yielding the following impedance:

$$Z_L = j\omega L + R_s \,. \tag{3.111}$$

From this complex impedance of an inductor, corresponding to Z_{11} from a 1-port Z-parameter simulation, the inductance is calculated as

$$L = \frac{\mathrm{Im}\{Z_{11}\}}{\omega} \,, \tag{3.112}$$

while the series resistance is extracted as

$$R_s = \mathrm{Re}\{Z_{11}\} \,. \tag{3.113}$$

The loss of an inductor is commonly quantified with the quality factor or Q-factor of the component, which is defined for an inductor as the ratio of the amount of energy stored in the magnetic field (the useful energy) over the amount of energy dissipated each period at frequency ω, resulting in the following expression:

$$Q_L = \frac{\mathrm{Im}\{Z_{11}\}}{\mathrm{Re}\{Z_{11}\}} = \frac{\omega L}{R_s} \,, \tag{3.114}$$

which generally improves with increasing frequency.

However, a physical inductor also has some parasitic capacitance C_p originating from the capacitance between the metals or to the substrate. This capacitance, although usually quite small, will form an LC resonator[3] with the designed inductance. The frequency at which resonance occurs is called the self-resonance frequency (SRF) and can be expressed as follows:

$$SRF = \frac{1}{2\pi\sqrt{LC_p}} \,. \tag{3.115}$$

Above this frequency, the inductor starts behaving more like a capacitor, since the impedance is now mostly determined by the parasitic capacitance instead of the inductance.

Furthermore, at higher frequencies the inductor losses start to increase because of several effects. First of all, at high frequencies, the current prefers to flow at the surface of a conductor, as shown in Fig. 3.27, showing an exponentially decaying current distribution inward from the surface [67]. As a result, the current flows

[3] Resonators will be discussed more in depth in the last section of this chapter.

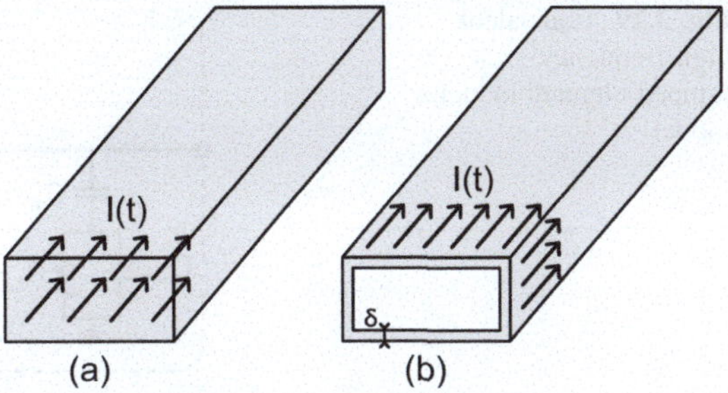

Fig. 3.27 Current distribution in a conductor at low (**a**) and high (**b**) frequencies

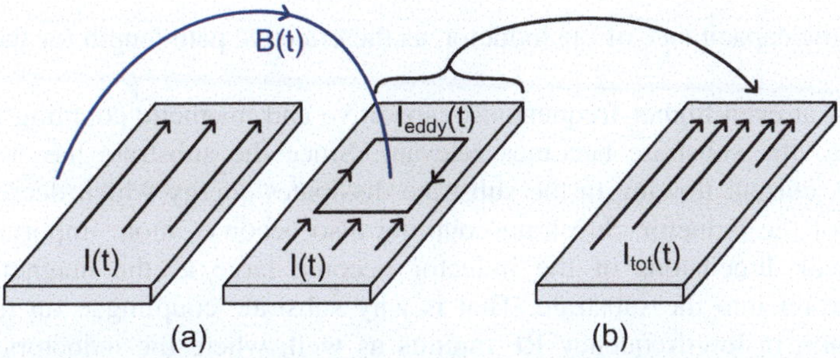

Fig. 3.28 Generation of Eddy currents in adjacent turns (**a**) and resulting current crowding effect (**b**)

through a smaller effective cross-section, increasing the resistance it faces. This effect is often quantified using the "skin depth" δ, which is defined as the depth below the surface where the current density has dropped to $1/e \approx 37\%$ of its value at the surface:

$$\delta = \sqrt{\frac{2}{\omega\mu\sigma}}\,, \tag{3.116}$$

where μ and σ represent the permeability and conductivity, respectively, of the conductor. For example, copper, with a conductivity of $5.8 \cdot 10^7$ S/m, has a skin depth of 2.1, 0.66, and 0.2 μm at 1, 10, and 100 GHz, respectively.

In spiral inductors, the proximity of neighboring turns leads to the current flow concentrating at the edge of the trace [60]. This effect, called "current crowding," arises from loops of current being induced in adjacent turns by the generated magnetic field. These current loops, called "eddy currents," add to the primary current at one edge and subtract from the main current at the other edge of the trace, resulting in a more "crowded" current density at one of the edges, as illustrated in Fig. 3.28. This effect becomes more pronounced at higher frequencies since the induced currents increase with frequency. Note that both of these effects (skin effect and current crowding) not only increase the loss but also influence the inductance

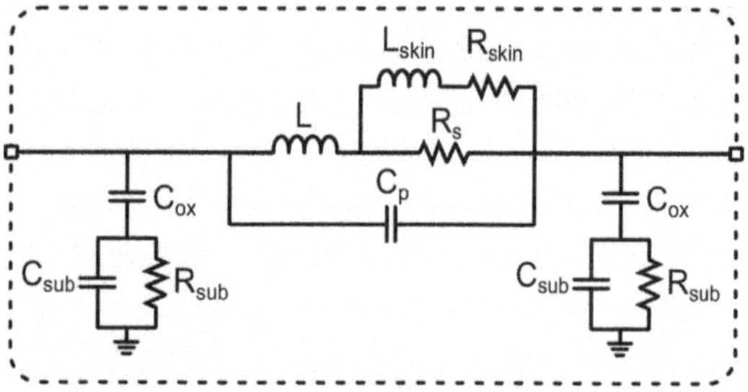

Fig. 3.29 Equivalent high-frequency lumped-element inductor model

and parasitic capacitance of the inductor, as the effective path length for the current changes.

Finally, at even higher frequencies, capacitive and magnetic coupling from the inductor to the substrate becomes relevant. Since the substrate has a nonzero resistance, current flowing in the substrate dissipates energy which decreases the Q-factor of the inductor. Substrate coupling also becomes more important when the physical dimensions of the inductor become large as the magnetic fields extend further into the substrate. That is why substrate coupling is an important phenomenon in low-frequency RF circuits as well, where the inductors take up significant area.

Figure 3.29 shows an equivalent lumped-element schematic of an inductor that aims to model the discussed nonidealities at low and higher frequencies. While such a model provides some insight, it fails to capture the inductor behavior over a broad frequency range, and its component values are challenging to extract. Therefore, inductors are better modeled by extracting their frequency-dependent impedance obtained from rigorous EM simulations.

3.3.2.2 Implementation

To minimize the metal loss, inductors should be implemented in the thickest (so highest) metal layers in the BEOL. In terms of geometric shape, the inductor can be drawn in a loop or spiral shape with one or multiple turns to increase its inductance within a given area. Since the necessary inductance is usually quite small at high frequencies, a single turn often suffices. Due to strict design rules imposed by the foundry, all bends must form angles that are multiples of 45 ° resulting in an octagonal shape as shown in Fig. 3.30 for a single-turn and a double-turn inductor.

A popular strategy to diminish substrate losses is placing a ground shield below the inductor, effectively shielding the substrate from the magnetic fields generated by the inductor. To limit the losses caused by eddy currents in this ground plane, the ground shield is often patterned with cut-outs, breaking up long loops of current in the plane. Since substrate coupling is not the dominant source of loss at (sub)-THz frequencies, such ground shields are rarely used at these frequencies as it lowers the inductance and Q-factor of the inductor (due to eddy currents). It also provides a

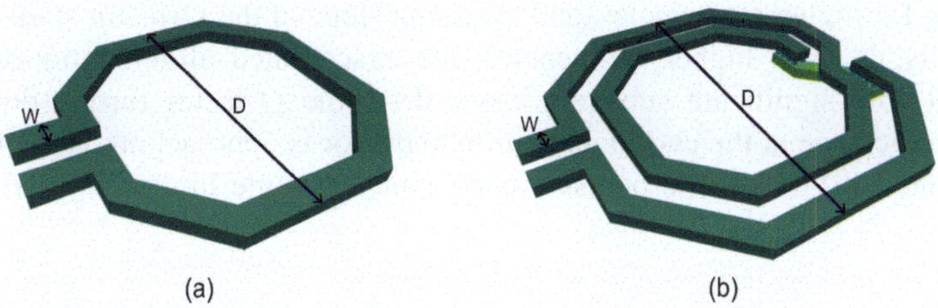

Fig. 3.30 Examples of a single-turn (**a**) and a double-turn (**b**) spiral inductor

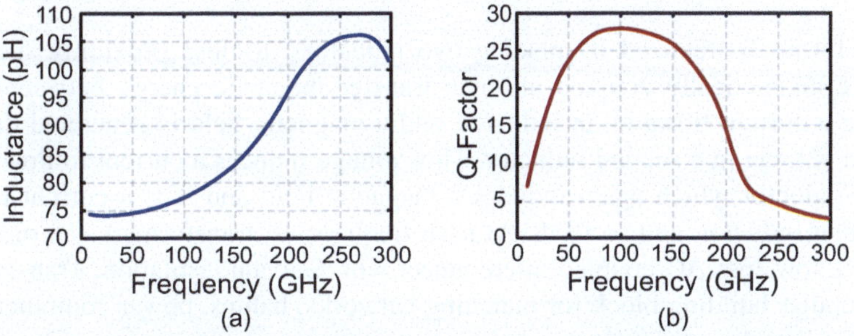

Fig. 3.31 Simulated inductance (**a**) and Q-factor (**b**) of a single-turn spiral inductor with 40 μm diameter and 4 μm trace width

larger parasitic capacitance between the inductor and the shield, further decreasing the Q-factor and SRF.

The preferred inductor topology (i.e., octagonal shape in top metals without a shield) should then be sized, meaning that the optimal diameter, trace width, number of turns, etc. should be chosen for a desired inductance value while maximizing the Q-factor and SRF. For example, increasing the diameter will increase not only the inductance but also the parasitic capacitance thus reducing the SRF. The Q-factor can be increased slightly by making the traces wider and hence reducing the series resistance, but in the process that increases the parasitic capacitance and thus lowers the SRF. Furthermore, wider traces cause the inductance to drop, which requires a larger diameter to compensate, increasing the series resistance and thus partly offsetting the improvements in Q-factor. Besides, it should be noted that increasing the trace width way past the skin depth of the conductor will yield only limited improvements. Lastly, for large inductance values, the number of turns can be increased, leading to a smaller necessary area for the desired inductance. Clearly, optimizing an inductor is an iterative process in which a balance must be found between all of the discussed effects.

Figure 3.31 shows the simulated inductance and Q-factor of a single-turn spiral inductor with a diameter of 40 µm and a trace with of 4 µm. At low frequencies (< 50 GHz), the inductance stays relatively constant, and the Q-factor rises linearly with frequency. Due to the skin effect and current crowding, the loss starts increasing for higher frequencies, and the rising slope of the Q-factor starts leveling off. Finally, at even higher frequencies, the exacerbated aforementioned effects combined with significant substrate losses drive the Q-factor further down. The drop in inductance at the end hints that the inductor is approaching its SRF, which lies beyond 300 GHz as the parasitic capacitance is quite limited in a single-turn inductor.

3.3.3 *Transformers*

A transformer is realized by placing two inductors in close proximity such that they are magnetically coupled and can transfer magnetic energy between them. Assume a current flows in an inductor, and a magnetic field is generated that can create a voltage in a second inductor. This voltage induces a current to flow in the second inductor which again induces a magnetic field and thus a current to flow in the first inductor, and so forth. At high frequencies, transformers can provide a compact, low-loss AC-coupled interconnect with Galvanic isolation. Transformers are a popular building block for matching networks, baluns, power combiners, and splitters, all with convenient way to bias the interconnecting circuits.

3.3.3.1 Transformer Model

A transformer can be modeled as two magnetically coupled inductors, or windings, as shown in Fig. 3.32. The primary winding at the input has an inductance L_p and a quality factor Q_p, while the secondary winding at the output has an inductance L_s and a quality factor Q_s. The coupling is described by the magnetic coupling coefficient k which ranges from 0 to 1 and is defined based on the mutual inductance of the windings M:

$$M = k\sqrt{L_p L_s}\ , \tag{3.117}$$

which leads to the equivalent T-model in Fig. 3.32. As the windings are more tightly coupled, the mutual inductance increases, and the magnetic coupling coefficient approaches closer to a value of 1. Denoting the input as port 1 and the output as port 2, the two-port Z-parameters of this equivalent model can be derived:

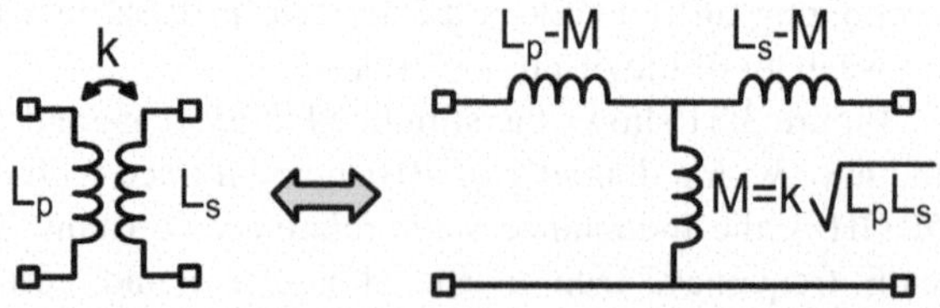

Fig. 3.32 Schematic representation of a transformer and its equivalent model

$$\begin{bmatrix} Z_{11} & Z_{12} \\ Z_{21} & Z_{22} \end{bmatrix} = \begin{bmatrix} R_p + j\omega L_p & -j\omega M \\ -j\omega M & R_s + j\omega L_s \, . \end{bmatrix} \tag{3.118}$$

Conversely, the five parameters describing a transformer can be extracted from its EM simulated two-port Z-parameters as follows:

$$L_p = \frac{\text{Im}\{Z_{11}\}}{\omega} \,, \quad L_s = \frac{\text{Im}\{Z_{22}\}}{\omega} \tag{3.119}$$

$$Q_p = \frac{\text{Im}\{Z_{11}\}}{\text{Re}\{Z_{11}\}} \,, \quad Q_s = \frac{\text{Im}\{Z_{22}\}}{\text{Re}\{Z_{22}\}} \tag{3.120}$$

$$k = \sqrt{\frac{\text{Im}\{Z_{21}\} \cdot \text{Im}\{Z_{12}\}}{\text{Im}\{Z_{11}\} \cdot \text{Im}\{Z_{22}\}}} \,. \tag{3.121}$$

From the Z-parameters and ignoring the losses for now, it can be shown that the input impedance $Z_1 = V_1/I_1$ is transformed into an output impedance $Z_2 = V_2/I_2$ as follows [67]:

$$Z_1 = k^2 \frac{L_p}{L_s} Z_2 \,, \tag{3.122}$$

indicating that an impedance transformation can be obtained which in an ideal transformer ($k = 1$) is determined by the ratio of the primary and secondary inductances.

In terms of loss, all effects discussed for the inductor apply to both the primary and secondary inductors, such as resistive metal loss, skin effect, current crowding, and substrate coupling. An additional loss will be incurred due to the limited magnetic coupling between the windings as part of the magnetic energy is not transferred and thus lost.

Based on the inductor Q-factors and coupling coefficient, the passive efficiency η_{max} of the transformer can be expressed as follows [75]:

$$\eta_{\text{max}} = \frac{1}{1 + 2\sqrt{\left(1 + \frac{1}{Q_p Q_s k^2}\right)\frac{1}{Q_p Q_s k^2}} + \frac{2}{Q_p Q_s k^2}} \,, \tag{3.123}$$

which is the minimum insertion loss of a transformer-based network. Clearly, a high magnetic coupling and low-loss inductors are desired to minimize the loss in the transformer.

Furthermore, the windings are not only magnetically coupled but also capacitively, resulting in significant inter-winding capacitance that, together with the intra-winding capacitance, substantially reduces the SRF compared to a single inductor coil. All of these effects can be combined into a single equivalent lumped-element model, as shown in Fig. 3.33. As this model is very complex and only covers

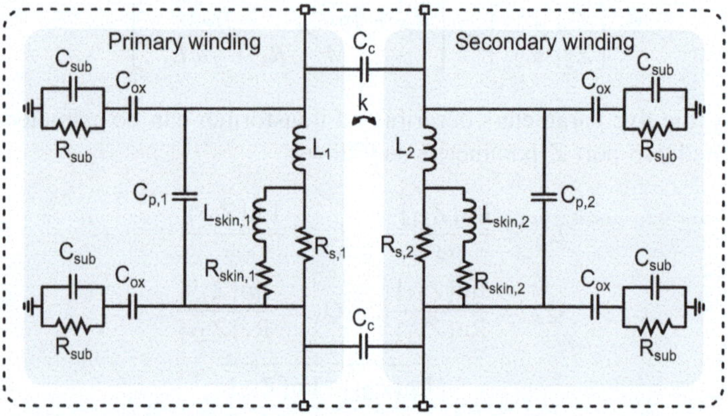

Fig. 3.33 Equivalent high-frequency lumped-element model of a transformer

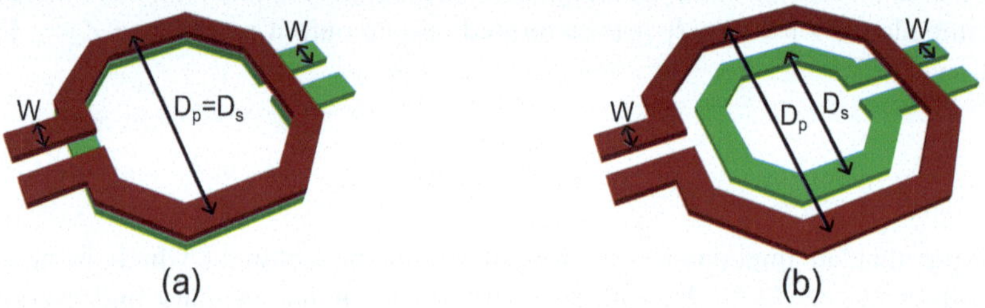

Fig. 3.34 Examples of a single-turn stacked transformer with identical (**a**) and different (**b**) diameters for the primary and secondary windings

the behavior over a limited frequency range, impedance parameters derived from EM simulations are mostly used, similar to an inductor.

3.3.3.2 Implementation

Starting from the basic single-turn octagonal inductor topology, transformers can be implemented in several ways. The most straightforward way is to stack two inductors in different metal layers, as exemplified in Fig. 3.34a. To maximize the coupling, the windings should overlap maximally, and neighboring metals should be used. Do note that maximizing the magnetic coupling also inevitably maximizes the capacitive coupling, hence resulting in a low SRF.

Alternatively, a planar transformer realizes both windings in the same layer. Since one of the inductors must be placed within the other one, the magnetic coupling between the windings is lower compared to a stacked transformer, and as a result the loss is higher. Nevertheless, this topology can be useful in technologies where only one thick metal layer is available [76].

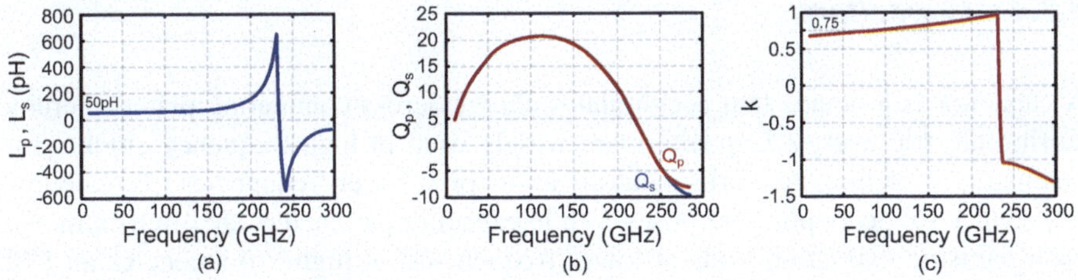

Fig. 3.35 Simulated inductances (**a**), Q-factors (**b**), and coupling factor (**c**) of a single-turn stacked transformer with 30 µm diameter and 3 µm trace width

Sticking to the stacked transformer, several other design choices can be made. Like for a single inductor, the diameter and trace width can be chosen to achieve the desired inductance while maximizing the Q-factor. Furthermore, as transformers are often used as part of a matching network, a certain impedance transformation from input to output is sometimes desired. As demonstrated by Eq. 3.122, an impedance transformation can be attained by sizing the primary and secondary inductors unequally. For instance, the transformer shown in Fig. 3.34b realizes an impedance downconversion, as the secondary winding is smaller than the primary. Note that, since the inductors do not completely overlap in this case, the magnetic coupling will be lower, which increases the transformer insertion loss and reduces the impedance transformation ratio (Eq. 3.122). In general it can be stated that a higher impedance transformation comes at the cost of a higher transformer loss, due to a lower magnetic coupling.

In some matching or filter networks, a lower magnetic coupling might be desirable. In that case, the windings can be offset relative to each other to reduce the overlap and hence the mutual coupling. Since this also reduces the inter-winding capacitance, the SRF typically increases as well. How to choose the optimal transformer parameters based on the matching requirements, as well as strategies to optimally implement the needed transformer, will be described in the next section.

Figure 3.35 shows the extracted parameters from an EM simulated transformer where the inductors with 30 µm diameter and 3 µm trace width are fully aligned. The inductance values, identical for the primary and secondary winding, stay relatively constant for low frequencies and start increasing rapidly as the frequency approaches the SRF. Above the SRF (which is around 230 GHz in this case), the inductance becomes negative indicating that the transformer behaves capacitively rather than inductively. The Q-factors follow a similar behavior as described for the inductor, even becoming negative at frequencies above the SRF as the used definition for magnetic energy does not make sense for capacitive component. The magnetic coupling is also frequency-dependent, but, like the Q-factor, its physical meaning becomes less obvious close to the SRF as the transformer starts to rely mostly on capacitive rather than magnetic coupling. Note that a transformer can still be used at frequencies above the SRF, but as a capacitive rather than an inductive coupling device. In that situation other models, like capacitively coupled lines, might be more accurate and useful to describe its operation.

3.3.4 *Capacitors*

A capacitor is a device that accumulates charge across conducting plates, thereby storing electric energy. Capacitors are widely used in high-frequency circuits for resonators, matching networks, neutralization, or at lower frequencies to create low impedance for decoupling purposes. While capacitor parameters are often extracted by a parasitic extraction tools at lower frequencies, at higher frequencies an EM simulation tool is required to accurately model the inductive effects.

3.3.4.1 Capacitor Model

The basic capacitor model, valid at low frequencies, includes an ideal capacitor C with a series resistance R_s, resulting in the following impedance:

$$Z_C = \frac{1}{j\omega C} + R_s \,. \tag{3.124}$$

The capacitance can thus be derived from the impedance Z_{11} as

$$C = -\frac{1}{\text{Im}\{Z_{11}\}\,\omega} \,, \tag{3.125}$$

which is generally frequency-dependent due to inductive effects. The series resistance can be derived from the real part of this admittance:

$$R_s = \text{Re}\{Z_{11}\} \,. \tag{3.126}$$

Similar to the inductor, the Q-factor quantifies how much energy is stored in the electric field versus dissipated each cycle. The Q-factor of a capacitor can thus be expressed as

$$Q_C = \frac{|\text{Im}\{Z_{11}\}|}{\text{Re}\{Z_{11}\}} = \frac{1}{\omega R_s C} \,, \tag{3.127}$$

showing a decrease in Q-factor as the frequency increases.

To model the behavior at higher frequencies, some parasitic series inductance must be added originating from the feed lines or plates themselves. Assuming a series inductance L_s, the SRF of the capacitor can be defined similar to the inductor:

$$SRF = \frac{1}{2\pi\sqrt{L_s C}} \,, \tag{3.128}$$

above which the capacitor starts behaving like an inductor. Figure 3.36 shows a more complex equivalent lumped-element model of the capacitor valid at higher

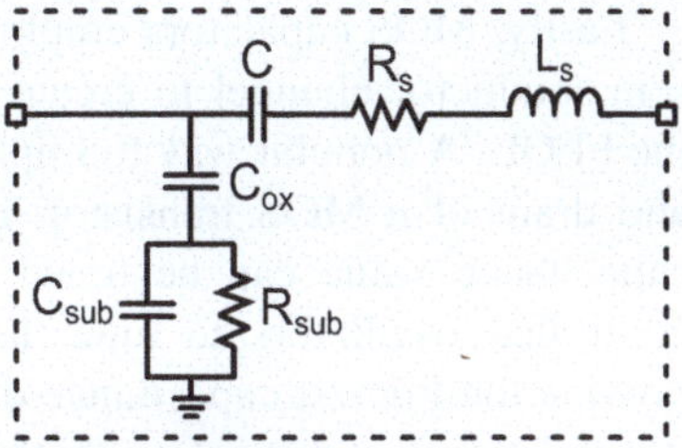

Fig. 3.36 Equivalent high-frequency lumped-element capacitor model

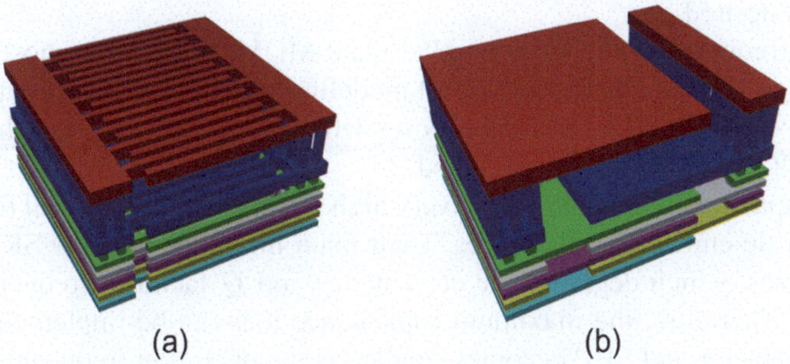

Fig. 3.37 Examples of a multi-finger (**a**) and parallel-plate (**b**) MOM capacitor

frequencies, where also substrate coupling and losses are included. However, compared to inductors and transformers, this model is still relatively simple.

3.3.4.2 Implementation

Capacitors can be divided between variable and fixed capacitors. The capacitance of variable capacitors can be tuned, while fixed capacitors have a fixed value. A further distinction can be made based on their physical composition, resulting in three main types: metal-oxide-metal (MOM) capacitors, metal-insulator-metal (MIM) capacitors, and metal-oxide-metal (MOS) capacitors.

MOM capacitors are implemented in the BEOL by creating overlap between various metal layers. The overlap can be created in multiple ways, most commonly as a parallel-plate capacitor or a multi-finger capacitor, both shown in Fig. 3.37. The multi-finger approach achieves a higher capacitance density since it features side-to-side capacitance between the fingers in addition to the vertical plate-to-plate capacitance also present in the parallel-plate implementation. However, the thin finger structure is more difficult to model and takes significantly longer to EM simulation.

MIM capacitors, on the other hand, contain a special dielectric between the metal plates with a higher permittivity to increase the capacitance density and Q-factor compared to MOM capacitors. They do, however, require an additional process step that increases the chip cost and is not available in every technology.

Lastly, MOS capacitors employ the oxide between the gate connection and the semiconductor channel to create a capacitance and are therefore implemented in the FEOL. A popular way to implement a MOS capacitor is by shorting the source and drain of a MOS transistor. By varying the voltage across the terminals, the capacitance value can be tuned which is very useful in, for example, voltage-controlled oscillators to tune the oscillation frequency. While MOS capacitors provide the highest capacitance density, they suffer from very poor quality factors which limit their use at very high frequencies to only situations where capacitance tunability is desired.

At high frequencies, the fixed parallel-plate MOM capacitor is mostly adopted due to its high Q-factor, reliable and fast modeling, and compatibility with standard CMOS processes. For decoupling purposes where capacitance density has priority over loss, MOS capacitors are employed.

MOM capacitors inherently have a very high Q-factor, which is ideal for low-loss resonator or de-embedding structures. Their main limitation is a low SRF for large capacitor sizes, which degrades the capacitance and Q-factor at frequencies close to the SRF. Therefore, the maximum capacitance that can be implemented with a sufficient Q-factor and SRF becomes smaller as the operation frequency increases. However, the amount of required capacitance typically decreases with frequency; hence capacitors rarely form a performance bottleneck.

Still, the physical size can be reduced by following a couple of recommendations but will inevitably increase when a larger capacitance is required. First, as many metal layers should be used as possible, especially the lower metals, to maximize the capacitance per unit area. Indeed, as is clear from Fig. 3.37, the highest metals have a much larger spacing and thus contribute much less to the overall capacitance. Second, for very large capacitance values, the shape should be drawn such that a long, thin structure is avoided which introduces significant series inductance that limits the SRF. It is also important to include the feed lines and interconnect of the capacitor in the model at high frequencies, as they can also add unwanted inductance.

Figure 3.38 shows the simulated capacitance and Q-factor of a 5 μm wide and 4 μm long parallel-plate MOM capacitor. For this moderate capacitance (about 28 fF at lower frequencies), the SRF is large (>300 GHz) and the maximum Q-factor very high (≈160). The effects of approaching the SRF are, however, visible at the highest frequencies, being an increase in apparent capacitance and decrease of the Q-factor.

3.3.5 *Transmission Lines*

Long interconnect lines at high frequencies, i.e., where the length becomes a meaningful fraction of the wavelength, can be modeled as transmission lines. Furthermore, transmission lines form the basis of many microwave or RF structures, such as power splitters/combiners, hybrids, filters, and many others.

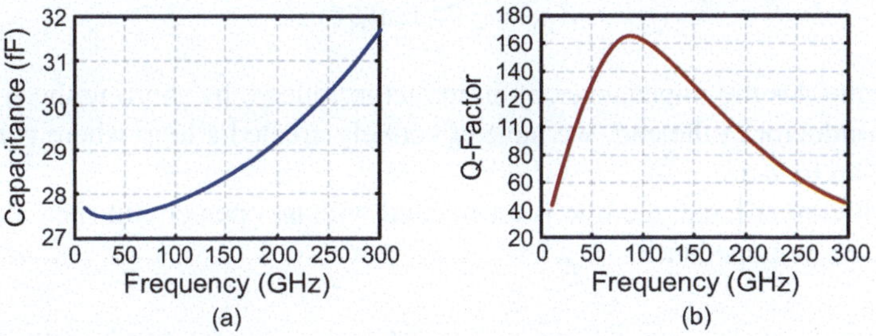

Fig. 3.38 Simulated capacitance (**a**) and Q-factor (**b**) of a parallel-plate MOM capacitor with 5 μm width and 4 μm length

3.3.5.1 Transmission Line Model

The transmission line model has already been extensively studied in Sect. 3.1.4 as it forms the foundation of high-frequency circuit design. A transmission line is modeled as distributed component consisting of a cascade of infinitesimally short line sections. By characterizing each section with unit lumped-element component values (R, L, G, C), the behavior of a propagating traveling wave emerges. The most important characteristics of a transmission line, based on these unit elements, are now summarized:

$$Z_c = \sqrt{\frac{R + j\omega L}{G + j\omega C}} \approx \sqrt{\frac{L}{C}} \tag{3.129}$$

$$\gamma = \alpha + j\beta \approx j\omega\sqrt{LC} \tag{3.130}$$

$$v_p = \lambda \cdot f \approx \frac{1}{\sqrt{LC}}, \tag{3.131}$$

which correspond to the characteristic impedance, propagation constant, and phase velocity of the transmission line. The approximated expressions are more useful and insightful while designing and are valid for most lines that exhibit a moderate amount of attenuation. No complicated equivalent circuit is required to model the behavior at high frequencies, like for the previous lumped components, as this is inherently included by the distributed way of modeling.

As analyzed in Sect. 3.1.4, when a transmission line is terminated with a load impedance $Z_L \neq Z_c$, that impedance is transformed by the line of length ℓ to an input impedance equal to

$$Z_{\text{in}} = Z_c \frac{Z_L + jZ_c \tan(\beta\ell)}{Z_c + jZ_L \tan(\beta\ell)}, \tag{3.132}$$

which equals Z_c when the line is matched to its characteristic impedance. If the line is shorted (i.e., $Z_L = 0$), the input impedance reduces to

$$Z_{\text{in}} = jZ_c \tan(\beta\ell) , \tag{3.133}$$

which resembles the impedances of an inductor. Indeed, by shorting the signal and ground conductor at the end, we have effectively created a loop where current can flow, like an inductor.

If, on the other hand, the line is terminated with an open circuit (i.e., $Z_L = \infty$), the lines see an impedance:

$$Z_{\text{in}} = \frac{Z_c}{j \tan(\beta\ell)} , \tag{3.134}$$

indicating that the line behaves like a capacitor. Again, leaving two conductors open effectively creates two parallel plates that act as a capacitor.

Using these last two equations, a transmission line can be used to implement a certain inductance or a capacitance, as long as $\ell < \lambda/4$. For longer lengths, the transmission line changes periodically between inductive and capacitive behaviors with an increasing length. The range of possible values is determined by the magnitude of Z_c and is pretty limited compared to dedicated inductor/capacitor components. Furthermore, for the same component value, a transmission line will take up considerably more area compared to its lumped-component variant. However, in some situations at high frequencies, a transmission line might have lower losses and be easier to model than a lumped inductor or capacitor and is therefore sometimes adopted instead.

3.3.5.2 Implementation

Many different types of transmission lines exist, each with different characteristics (i.e., R, L, G, C per unit length). To support a single TEM mode, for single-ended signals, two conductors are required. Figure 3.39 shows the three most commonly used types of transmission lines in integrated circuits.

A microstrip line is depicted in Fig. 3.39a, consisting of a signal conductor (usually in the top metal layer for minimal loss) and a ground plane underneath.

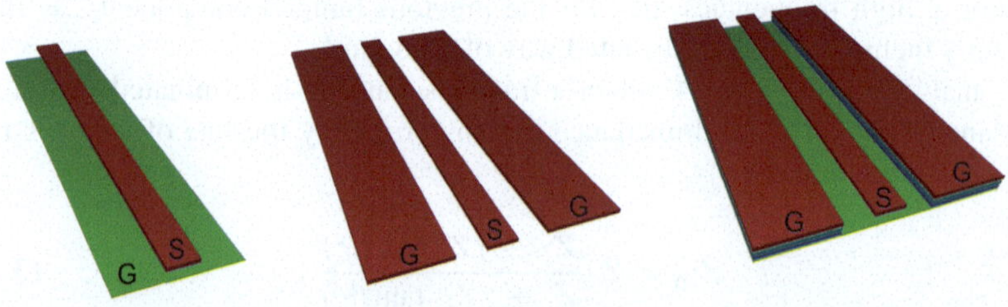

Fig. 3.39 Various transmission line types: microstrip line (**a**), coplanar waveguide (**b**), and grounded coplanar waveguide (**c**)

The characteristic impedance is determined by the width of the line and the spacing between the two conductors. For instance, a wider signal trace diminishes the line inductance but increases the capacitance between the traces, both reducing the characteristic impedance of the line. A coplanar waveguide (CPW) is shown in Fig. 3.39b, which contains a signal trace in the center with two ground traces on both sides. The characteristic impedance is now mainly determined by the width of the center conductor and the spacing to its neighboring traces. Decreasing the spacing leads to an increase in line capacitance and hence a decrease in characteristic impedance. A combination of the two types is depicted in Fig. 3.39c, called a grounded coplanar waveguide (GCPW), where the ground traces are connected to a ground plane in a lower layer.

For differential signaling, an extra conductor is added to support an extra mode. While the same line types can be used from Fig. 3.39, an extra degree of freedom is now available to tune the characteristics of the two propagating modes, which are typically divided between a differential-mode and a common-mode, more information on that in the next section. While two conductors are sufficient to fully characterize the differential-mode of the propagating signal, it is still recommended to add a ground plane to fully model the common-mode as well. After all, the return current in common-mode will inevitably find a path to flow whether you explicitly model it or not. To avoid unexpected results, it is good practice to always model the path of the return current when simulating transmission lines. However, for short interconnects (relative to the wavelength), this is less important as the transmission line behavior is not as pronounced and hence neither are deviations on that behavior.

As a final side note, it should be emphasized that the wavelength of the propagating signal does not correspond to the wavelength of an EM wave in free space but is reduced since most of the wave is propagating in a material with a higher permittivity. Suppose a medium with relative permittivity ϵ_r, the wavelength reduces by a factor of $\sqrt{\epsilon_r}$. Usually, however, the wave propagates in multiple materials with different permittivities. For instance, most of the signal propagates in vacuum ($\epsilon_r = 1$) and the dielectric material between the metal layers, like SiO_2 ($\epsilon_r = 3.9$), while part of the signal might propagate through the silicon substrate ($\epsilon_r = 11.9$). The effective (or average) permittivity that determines the physical wavelength is then a weighted sum of the different material permittivities based on what proportion of the overall electric field propagates in it. Or the wavelength can simply be deduced from the simulated propagation constant.

3.4 Conclusion

This chapter gave an overview of the fundamental concepts in electromagnetism, describing the notion of electromagnetic waves, both in free space and within dielectric waveguides. From electromagnetism, basic circuit theory and transmission line theory were derived and described as good foundation for high-frequency circuit design.

The behavior of the MOS transistor was then discussed, with a large emphasis on its high-frequency characteristics. Several guidelines were introduced to optimize its performance. Finally, the primary passive components were introduced: the inductor, the transformer, the capacitor, and the transmission line. Their behavior was analyzed at low and high frequencies, as well as the implementation in CMOS technologies.

The analyses have shown that, despite the many benefits of CMOS, the performance of the active and passive devices dramatically deteriorates as the operation frequency increases. Furthermore, technological advancements in CMOS will not benefit the performance above 100 GHz.

Chapter 4
High-Frequency Circuit Design

The previous chapter has illustrated how both the active and passive components have sub-optimal performance at higher frequencies that only degrades as CMOS nodes are scaling down. Meticulous design techniques are hence necessary to realize high-frequency circuits with maximal performance and efficiency. This chapter therefore discusses the basics of impedance matching, resonators and matching networks which are crucial in RF circuit design. Then, the two primary building blocks for high-frequency amplifiers above 100 GHz in this book are detailed: the neutralized pseudo-differential pair and transformer-based matching network. Finally, the design of various amplifier-based circuits are discussed.

4.1 Matching Fundamentals

4.1.1 S-Parameters

RF circuits are mostly analyzed and measured by means of incident and reflected waves with units expressed in power quantities rather than voltages and currents. For this reason, multi-port networks at high frequencies are often described using scattering parameters or S-parameters, which describe a network by means of the incident and reflected voltage waves at each port. The S-parameters are more formally defined as follows [65]:

$$S_{ij} = \left.\frac{V_i^-}{V_j^+}\right|_{V_k^+=0 \text{ for } k\neq j}, \tag{4.1}$$

C. D'heer, P. Reynaert, *THz and Sub-THz CMOS Electronics for High-Speed Telecommunication*, Analog Circuits and Signal Processing,
https://doi.org/10.1007/978-3-031-64439-9_4

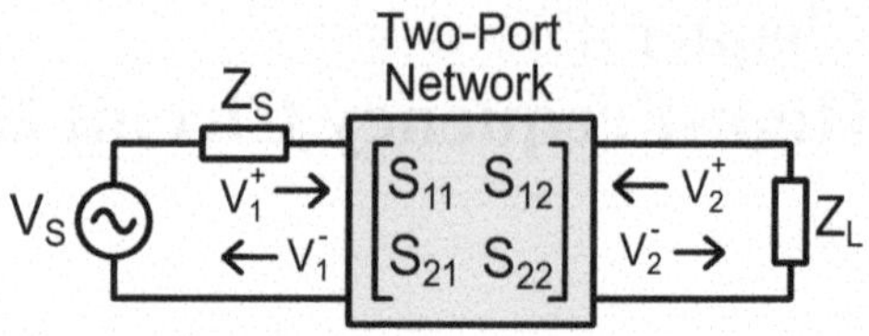

Fig. 4.1 A generic two-port network with incident and reflected waves at the input and output

where V_i^- is the reflected wave at port i while driving port j with an incident wave V_j^+. Note that every other port $k \neq j$ should be terminated in matched loads to avoid reflections and thus incident waves into the network. One of the reasons why measuring S-parameters is much more practical at high frequencies is that matching all the ports not under test is much easier than shorting them (to measure Y-parameters) or providing an open (to measure Z-parameters).

Figure 4.1 illustrates a two-port network characterized by S-parameters with a source and a load. The incident and reflected voltage waves can then be expressed as

$$\begin{bmatrix} V_1^- \\ V_2^- \end{bmatrix} = \begin{bmatrix} S_{11} & S_{12} \\ S_{21} & S_{22} \end{bmatrix} \begin{bmatrix} V_1^+ \\ V_2^+ \end{bmatrix} . \tag{4.2}$$

S_{11} and S_{22} correspond to the reflection coefficients Γ_S and Γ_L at the input and output, respectively, as defined in Eq. 3.82. Assuming the network represents an amplifying device, S_{21} can be interpreted as the forward gain from input to output, while S_{12} denotes the reverse isolation between the ports. Also, the S-parameters are often expressed in decibel units as follows:

$$S_{ij|\mathrm{dB}} = 20 \log(|S_{ij}|) . \tag{4.3}$$

4.1.2 Types of Matching

Matching, or impedance matching, is a technique where an impedance is designed to optimize a specific metric or behavior. In the context of transmission lines, matching the load impedance to the characteristic impedance of the line was necessary to avoid reflections at the load interface. This is called "reflectionless matching." However, reflections are not always problematic, for instance, when the physical distance between the source and load is significantly smaller than the signal wavelength (see Sect. A.4). Moreover, there are several other metrics that should be optimized in various situations, requiring a different matching strategy.

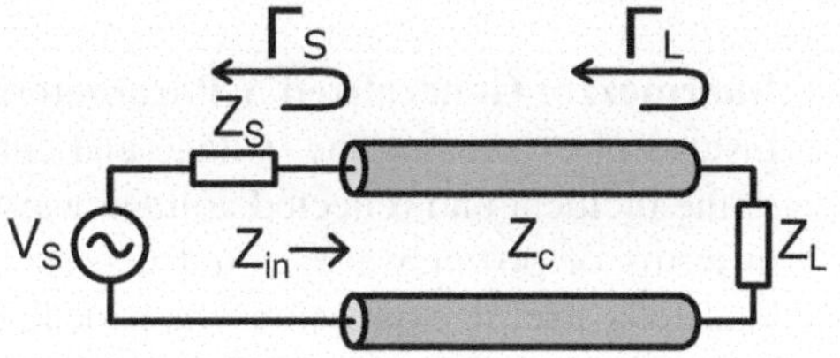

Fig. 4.2 Generic transmission line circuit with a mismatched source and load

4.1.2.1 Conjugate Matching

As stated before, RF systems are mostly concerned about power, rather than voltage or current. Consider the generic transmission line circuit in Fig. 4.2 driven by a source V_S with source impedance Z_S, and loaded by Z_L. A relevant question is: What impedance Z_{in} will result in maximal power transfer from the source into the line? It can be easily proven ([65]) that Z_{in} must be equal to the complex conjugate of the source impedance:

$$Z_{\text{in}} = Z_S^* \,. \tag{4.4}$$

This condition is called conjugate matching or gain matching and results in maximum power transfer from the source into the line equal to

$$P_{\text{in}} = \frac{V_{S,\text{rms}}^2}{4R_S} = P_{\text{avs}} \,, \tag{4.5}$$

which defined as the power available from the source P_{avs}.

The same condition holds at the load interface, where Z_L must be conjugately matched to Z_c for maximum power transfer from the line into the load. Note that the reflection coefficients Γ_S, Γ_L may be nonzero in this situation and leading to a standing wave on the line. Indeed, in some cases the power in multiple reflections may add constructively leading to a greater power delivery to the load than would be delivered if there were no reflections. These conditions apply for any combination of source and load, for example, between two amplifier stages where the one desires to transfer as much power from one stage into the next one. Evidently, in case all impedances are real, conjugate matching reduces to the same condition as reflectionless matching.

Intermezzo: Generalized S-Parameters

Instead of expressing the voltage and current on a transmission line in terms of the incident and reflected voltage wave, this can equivalently be expressed in terms of power waves. First described by Kurokawa [77], power waves lead to a useful expression when dealing with power transfer and a more generalized description of S-parameters. The incident and reflected power wave amplitudes, a and b, are defined as a function of the total voltage and current as follows:

$$a = \frac{V + Z_R I}{2\sqrt{R_R}} \tag{4.6}$$

$$b = \frac{V - Z_R^* I}{2\sqrt{R_R}}, \tag{4.7}$$

where $Z_R = R_R + jX_R$ is the reference impedance and can be a complex number. The power delivered to a load can then be expressed as the difference between the incident and reflected power wave [65]:

$$P_L = \frac{1}{2}|a|^2 - \frac{1}{2}|b|^2 \tag{4.8}$$

and is valid for any complex reference impedance Z_R, as opposed to only real impedances Z_c like in expression 3.88. Furthermore, the reflection coefficient Γ_p for the reflected power wave has the following expression [65]:

$$\Gamma_p = \frac{b}{a} = \frac{Z_L - Z_R^*}{Z_L + Z_R}, \tag{4.9}$$

which corresponds to the conjugate matching condition instead of the reflectionless matching condition, but both are identical in the case of a purely real reference impedance. Confusingly, many circuit simulators employ this definition of generalized S-parameters under the label of "regular" S-parameters. Optimizing for minimal S_{11} thus results in a conjugate match rather than a reflectionless match at the input of the block. As conjugate matching is usually the desired outcome, this is not a problem, yet important to be conscious of.

To make matters more complicated, neither reflectionless matching nor conjugate matching necessarily leads to the highest efficiency. In the case of a conjugate match, 50% of the power is dissipated in the source impedance Z_S, while only 50% is transferred to the load. Increasing the load resistance will result in a higher efficiency since a lower fraction of the total power is dissipated in the source resistance. However, the total amount of power generated and dissipated in the load will be lower compared to the conjugate match case. If a certain power specification is to be met, more power gain stages will be needed to compensate the reduction in power transfer, which will mostly offset the improvement in efficiency. Therefore, in high-frequency circuits, matching will be performed for maximal power instead of efficiency. Furthermore, the voltage source V_S and its impedance Z_S together represent a power source and should be viewed as such. The power dissipated in Z_S is not real power that is lost; rather, it is a necessary part of the equivalent representation of a power source. It is therefore better to interpret the source as a power source with maximum available power of P_{avs} with internal impedance Z_S.

4.1.2.2 Large-Signal Power Matching

When designing a power amplifier (PA) at the output of the TX, the goal is to generate and transfer as much power as possible to into the load. Due to the nonlinear behavior of the PA, conjugately matching the output of the PA, which leads to maximum power gain, does not automatically lead to maximum power generation by the PA. Indeed, to maximize the output power of the PA, a certain optimum load impedance must be provided to the PA, which is called power matching or large-signal matching [78]. This impedance is generally different from the complex conjugate of its small-signal output impedance that is provided in the case of gain matching. This large-signal impedance is typically found by performing a loadpull simulation, where the PA is terminated by various load impedances for which the output power is evaluated. When the load is optimized for output power, an inevitable penalty in power gain is suffered compared to a conjugate match, as visualized in Fig. 4.3, where a conjugately matched amplifier would deliver lower output power due to early compression. Note that for small-signal operation, i.e., the amplifier behaves linearly, a gain and power match yields the same load impedance.

4.1.2.3 Noise Matching

Finally, when designing the low-noise amplifier (LNA) at the input of the RX, the main concern is to optimize the SNR and thus minimize the noise figure of the

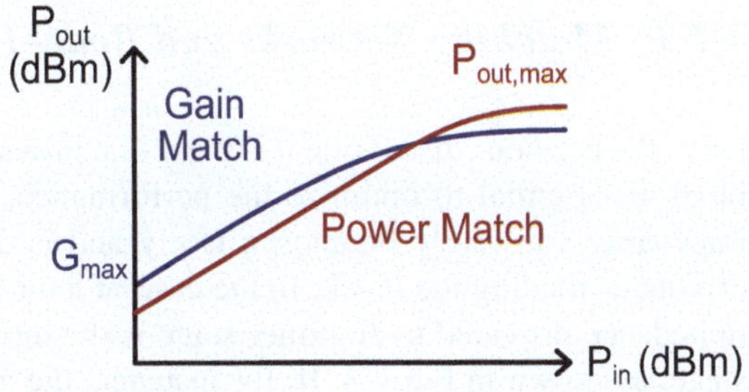

Fig. 4.3 Difference between a gain match and a power match in a power amplifier

LNA. From two-port noise theory [7], it can be shown that the NF of an amplifier is minimized for a certain source impedance at its input. The amplifier is then said to be noise matched at the input. If the source resistance ($R_s = 1/G_s$) deviates from the optimal resistance ($R_{\text{opt}} = 1/G_{\text{opt}}$), the noise figure increases according to [67]:

$$F = F_{\text{min}} + R_n R_s |G_{\text{opt}} - G_s|^2 \,, \tag{4.10}$$

where R_n is the noise sensitivity parameter, quantifying the degradation in noise figure for a certain deviation from the optimal source impedance. The optimal noise impedance is determined by the relative magnitude of the equivalent voltage and current noise sources of the amplifier and is influenced by the size and bias of the amplifier. Like for the PA, designing the source impedance for the LNA comes with an inherent trade-off between noise (noise matching) and power gain (conjugate matching). However, various design techniques can be employed to bring the necessary source impedance for both goals closer together and hence minimize the degradation in both NF and power gain.

Intermezzo: What Is Up with 50 Ω?
In the RF world, many instruments, cables, or modules are standardized to provide an impedance of 50 Ω. This exact value is not chosen randomly but is the result of a deliberate trade-off between various factors. First, it can be shown that a characteristic impedance of 30 Ω results in maximum power-handling capability in coaxial cables [7]. However, minimizing the cable attenuation yields an optimum impedance of 77 Ω [7]. As a compromise between power-handling and attenuation, 50 Ω was chosen as the golden standard. Some applications where power levels are low, like audio and video cables, use 75 Ω instead to solely minimize cable loss. Of course, it should be noted that a 50 Ω interface is only required when interfacing with standardized external components. Within an integrated circuit, the impedances can be chosen freely to optimize various other metrics, such as gain, power, or noise, as discussed in this section.

4.1.3 Matching Networks and Bode–Fano Limit

From the previous discussion it is clear that providing the desired impedance to each block is essential to optimize the performance. Unfortunately, the source or load impedance can rarely be chosen freely and is determined by the specific circuits driving or loading the block. In the case of a multistage amplifier network, the load impedance provided to amplifier stage is the input impedance of the next amplifier stage, as shown in Fig. 4.4. If, for instance, the amplifier circuit is implemented by

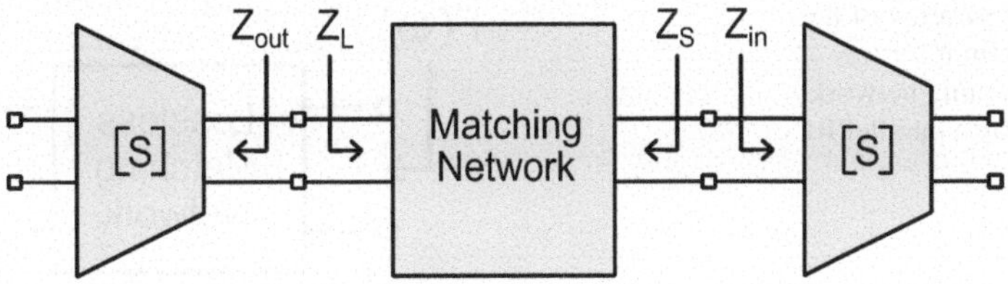

Fig. 4.4 General framework of a matching network interfacing two amplifier stages

means of an NMOS transistor, the input and output impedances are determined by different elements, see the small-signal model in Fig. 3.17, approximately like:

$$Z_{\text{in}} \approx r_g + \frac{1}{j\omega C_{gs}} \tag{4.11}$$

$$Z_{\text{out}} \approx r_{ds} \parallel \frac{1}{j\omega C_{ds}} . \tag{4.12}$$

Since both impedances are capacitive in nature, they obviously do not automatically satisfy the condition for conjugate matching. For this reason, a matching network should be inserted that transforms the input impedance of the second stage into the desired load impedance for the first stage, e.g., the complex conjugate of its output impedance. This network must not only compensate the reactive part, but also provide an impedance up- or downconversion to ensure conjugate matching. The matching network can be any passive network with a frequency-dependent characteristic that thus functions as a filter. If the network is lossless, conjugate matching the output of the first stage ($Z_L = Z^*_{\text{out}}$) will automatically lead to a conjugate match at the input of the second stage ($Z_S = Z^*_{\text{in}}$). However, when losses are included in the network, this is not necessarily the case, and more complex equations are required to simultaneously conjugately match both sides [65]. Later in this section different possible implementations will be discussed.

So far, matching has been discussed without specifying a particular frequency or bandwidth. However, circuits must exhibit the desired performance over a certain bandwidth in order to process modulated signals. Since both the input and output impedances of circuits vary over frequency (see Eq. 4.11), the matching network must have a particular frequency response in order to satisfy the desired matching conditions over the specified bandwidth. Even though the challenges and design guidelines related to this problem are highly circuit-specific, Bode and Fano were able to formulate an upper bound for how well a certain load can be matched. For example, supposing the load can be represented by an equivalent parallel RC load, as depicted in Fig. 4.5, the Bode–Fano criterion states that [79, 80]

$$\int_0^{\infty} \ln\left(\frac{1}{|\Gamma(\omega)|}\right) d\omega \leq \frac{\pi}{\text{RC}} . \tag{4.13}$$

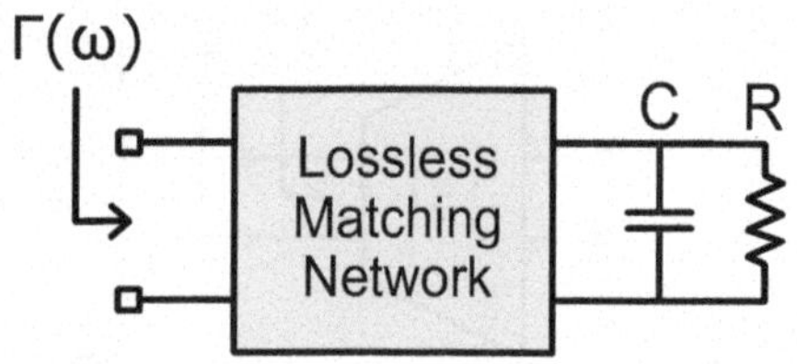

Fig. 4.5 Illustration of the Bode–Fano limit for a lossless matching network terminated by a parallel RC load

The frequency range over which this load can be adequately transformed into a certain impedance is thus limited by the characteristics of the load. More specifically, a load with a higher Q-factor (remember the expression for the Q-factor a lossy capacitor in Eq. 3.127) will be more difficult to match. Furthermore, this expression unveils a fundamental trade-off between the bandwidth and quality of the match, i.e., $|\Gamma(\omega)|$. A more strict specification on $|\Gamma(\omega)|$ will limit the bandwidth over which it can be achieved.

As an example, let us specify $|\Gamma(\omega)| = -10\,\text{dB}$, which reduces the left-side integral to approximately $1.1 \cdot \Delta\omega$. Using the expression for the Q-factor of a parallel RC load ($Q_p = \omega_0 R_p C_p$), the Bode–Fano criterion reduces to the following expression:

$$\frac{\Delta\omega}{\omega_0} \approx \frac{3}{Q}, \tag{4.14}$$

where $\Delta\omega/\omega_0$ represents the fractional bandwidth around a center frequency ω_0. Using this specification, a load with Q-factor of 10 can be matched over a maximum fractional bandwidth of about 30%. Doubling the Q-factor of the load effectively halves the achievable bandwidth.

While the Bode–Fano limit provides a theoretical limit on the obtainable matching bandwidth, it does not say anything on how to implement the actual matching network, which is the focus of the rest of this subsection.

4.1.4 RLC Resonators

4.1.4.1 RC Low-Pass Circuit

Resonators are the key to high-frequency RF circuit design. To understand why, take the circuit in Fig. 4.6a, a common-source amplifier with a resistive load and thus a voltage gain of $g_m \cdot Z$, where Z is the frequency-dependent load of the amplifier. Assuming a constant transistor transconductance over frequency, the frequency behavior of the circuit can be analyzed by solely looking at the load, or transimpedance, of the common source. The transimpedance is comprised of the load resistor R parallel with the total parasitic capacitance at the output (from the current and next stage), expressed as

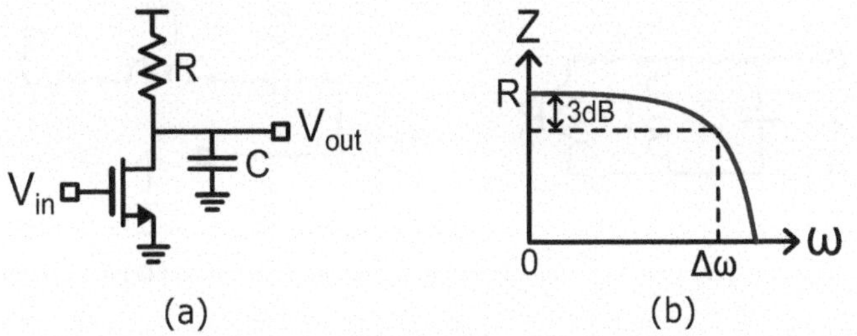

Fig. 4.6 Schematic of a resistively loaded common-source amplifier (**a**) and corresponding transimpedance response over frequency (**b**)

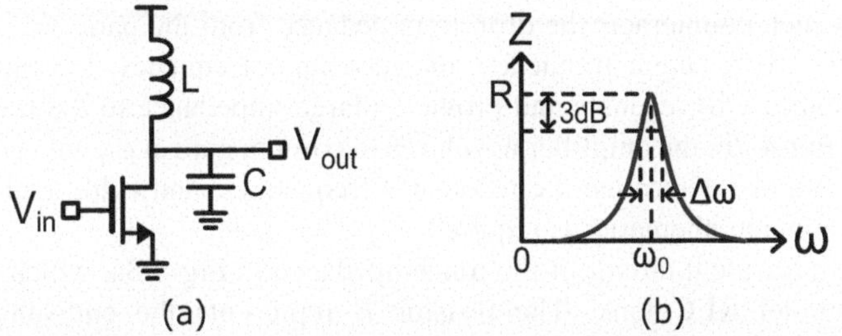

Fig. 4.7 Schematic of an inductively loaded common-source amplifier (**a**) and corresponding transimpedance response over frequency (**b**)

$$Z = R \parallel \frac{1}{j\omega C} = \frac{R}{1 + j\omega RC}, \tag{4.15}$$

and plotted on Fig. 4.6b demonstrating a low-pass behavior. The 3-dB bandwidth is the frequency where this transimpedance drops 3 dB, or a factor of $\sqrt{2}$, relative to its maximum (DC) value of R, which can be calculated to equal:

$$\omega_{3\text{dB}} = \Delta\omega = \frac{1}{\text{RC}}. \tag{4.16}$$

This shows that there is a trade-off between gain and bandwidth, and even for low values of R, the unavoidable parasitic capacitance at the output results in achievable bandwidths in the low-GHz range. Even though this bandwidth can be extended into the 10–20 GHz in advanced technologies with inductive peaking techniques, this topology does not support operation above 100 GHz.

4.1.4.2 Parallel RLC Circuit

This is where a resonator-based approach comes to the rescue. Consider the circuit in Fig. 4.7a, where the common-source stage is loaded with an inductor instead of

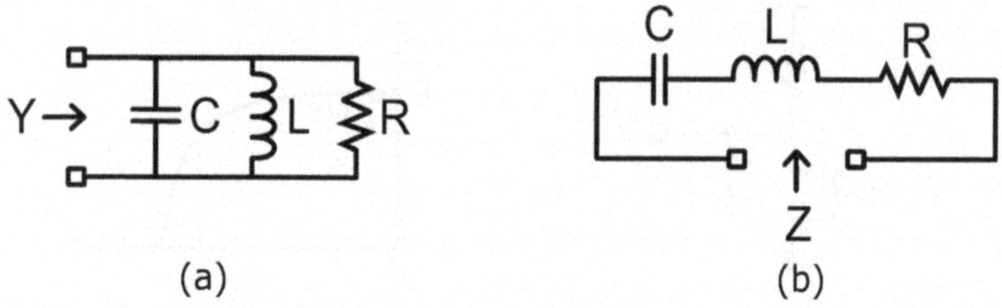

Fig. 4.8 Schematic of a parallel (**a**) and series (**b**) RLC circuit

a resistor. While the inductor provides a short at DC, its impedance increases with frequency which counteracts the drop in impedance from the parasitic capacitance at the output. At a single frequency, the resonance frequency, the inductor, and total capacitance will resonate and provide a large impedance to the transistor. As depicted in Fig. 4.7b, the amplifier now has a maximum gain at ω_0 with certain 3-dB bandwidth $\Delta\omega$. To determine the resonance frequency, bandwidth, and maximum gain value, some mathematics is required.

Take the equivalent circuit of the transimpedance in Fig. 4.8a, which effectively forms a parallel RLC tank. The resistor R represents the equivalent parallel resistance arising from the resistive losses of the inductor L and capacitance C, and hence L and C can be considered lossless. The admittance of this circuit can be expressed as

$$Y = j\omega C + \frac{1}{j\omega L} + \frac{1}{R} = j\omega C\left(1 - \frac{1}{\omega^2 LC}\right) + \frac{1}{R}\,. \tag{4.17}$$

By inspecting the network, it can be observed that the admittance approaches infinity near DC (dominated by the short of the inductor) and at very high frequencies (shorted by the capacitor). However, both terms cancel out at $\omega = \omega_0$:

$$\omega_0 = \frac{1}{\sqrt{LC}}\,, \tag{4.18}$$

which is called the resonance or resonant frequency. At that frequency, the capacitance and inductance resonate each other out, leading to a minimal value in the admittance, or maximal value in the impedance, which is purely real and determined by the value of R, as illustrated by Fig. 4.7b. The Q-factor of this RLC tank at resonance is defined as

$$Q = \omega_0 CR = \frac{R}{\omega_0 L} = \frac{R}{\sqrt{L/C}}\,, \tag{4.19}$$

which has many equivalent expressions by using Eq. 4.18. It can be shown that the 3-dB bandwidth $\Delta\omega$ of the admittance response is given by

$$\Delta\omega = \frac{\omega_0}{Q}\,, \tag{4.20}$$

which shows that the bandwidth is inversely proportional to the Q-factor of the tank and increases proportional to the resonance frequency of the tank.

4.1.4.3 Series RLC Circuit

Figure 4.8b shows the reciprocal of discussed circuit: the series RLC circuit. Similar to the parallel RLC circuit, a resonance condition occurs at $\omega_0 = 1/\sqrt{LC}$ where the inductance and capacitance cancel each other out. Instead of the admittance, now the impedance reaches a minimum value at resonance equal to a value of R. The Q-factor also has a similar expression, but in a reciprocal form compared the parallel case:

$$Q = \frac{1}{\omega_0 CR} = \frac{\omega_0 L}{R} = \frac{\sqrt{L/C}}{R}\,. \tag{4.21}$$

4.1.4.4 Parallel–Series Transformation

Since both the parallel and series representations have their use cases, it is convenient to know how to convert one into the other. First consider the arbitrary impedance in Fig. 4.9a, consisting of a series resistance R_s and reactance X_s which can be transformed into a parallel resistance R_p and reactance X_p as follows [67]:

$$R_p = R_s(1 + Q^2) \approx Q^2 R_s \tag{4.22}$$

$$X_p = X_s(1 + Q^{-2}) \approx X_s\,, \tag{4.23}$$

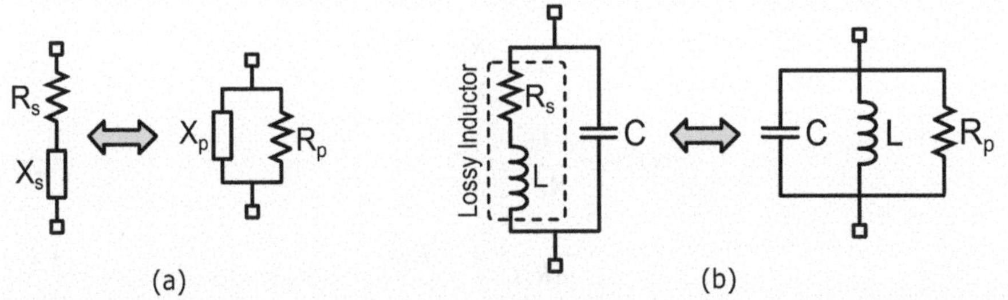

Fig. 4.9 A series to parallel transformation of an arbitrary impedance (**a**) and parallel RLC circuit with inductor loss (**b**)

where the approximated expressions are valid in case of a high-component Q-factor. The reactance can represent a capacitor or inductor component.

Figure 4.9b demonstrates a case where this transformation proves useful. A parallel LC tank is formed with an ideal capacitor and a lossy inductor. As discussed, MOM capacitors can be implemented with very high Q-factors and can thus be considered ideal. Inductors, on the other hand, only have a Q-factor in the range of 10–20. Using the series–parallel transformation, the network can be represented by an ideal LC tank with equivalent parallel resistance R_p calculated as

$$R_p = R_s(1 + Q_L^2) , \tag{4.24}$$

where Q_L is the Q-factor of the inductor as well as of the complete RLC tank since the capacitor is ideal. If the capacitor also introduces extra losses, this will further reduce the overall Q-factor of the tank and thus reduce the equivalent R_p. In general, the total Q-factor of the network is approximately equal to the parallel combination of the individual Qs.

4.1.5 *The Many Faces of Q*

The Q-factor seems to be a very important concept which has been defined in several ways up till now. To avoid confusion, let us summarize the various definitions of Q.

4.1.5.1 Untuned Component

For single components, i.e., a capacitor C or inductor L, the Q-factor signifies the ratio of energy storage to energy loss over one cycle of oscillation. The energy storage happens in the magnetic or electric field, for an inductor or capacitor, respectively. This definition leads to the following expressions for a series or parallel representation:

$$Q = \frac{X_s}{R_s} = \frac{R_p}{X_p} \tag{4.25}$$

$$Q_L = \frac{\omega L}{R_s} = \frac{R_p}{\omega L} \tag{4.26}$$

$$Q_C = \frac{1}{\omega R_s C} = \omega R_p C , \tag{4.27}$$

where it is assumed that $X_s \approx X_p$ and $R_p \approx Q^2 R_s$, as discussed before.

4.1.5.2 Tuned Network

For a tuned RLC network, the Q-factor is defined as the total reactive energy stored in the network relative to the energy dissipated per cycle. In contrast to an untuned component, the total reactive energy now includes the magnetic and electric energy stored in both fields. At resonance, the total reactive energy in a lossless network stays constant and moves between capacitive and inductive energy. With component losses, some of that energy is dissipated in the resistive element each cycle. This leads to the following Q-factor expression for a series or parallel RLC tank:

$$Q = \omega_0 \frac{E_{\text{tot}}}{P_{\text{dis}}} = \frac{R_p}{\sqrt{L/C}} = \frac{\sqrt{L/C}}{R_s} . \tag{4.28}$$

Furthermore, the Q-factor determines the fractional bandwidth of the response:

$$Q = \frac{\omega_0}{\Delta\omega} , \tag{4.29}$$

which is a similar finding as what was derived from the Bode–Fano criterion (Eq. 4.14) which quantifies the bandwidth of a matching network in terms of the Q-factor of the load. From both expressions, it is clear that a broadband response inevitably comes with larger losses and thus lower efficiency in the circuit, a trade-off that will come up many more times in RF circuit design.

Finally, the ratio of the reactive voltage/current to the real voltage/current is also determined by the Q-factor. For a parallel and series tank, respectively:

$$\text{parallel:} \quad Q = \frac{V_L}{V_R} = \frac{V_C}{V_R} \tag{4.30}$$

$$\text{series:} \quad Q = \frac{I_L}{I_R} = \frac{I_C}{I_R} . \tag{4.31}$$

This means RLC resonators can be used to achieve voltage or current gain! Even though these circuits cannot provide power gain, as they are passive networks, the reactive gain can be used to achieve impedance up- or downconversion, as will be discussed in the next subsection.

4.1.6 Matching Network Implementations

The final step in our discussion of impedance matching is to look at various ways to obtain the desired impedance transformation. This transformation typically includes the conversion of a capacitive into an inductive reactance, as well as an up- or downconversion of the resistance. Matching networks can be broadly subdivided into two main types: lumped-element and distributed matching networks.

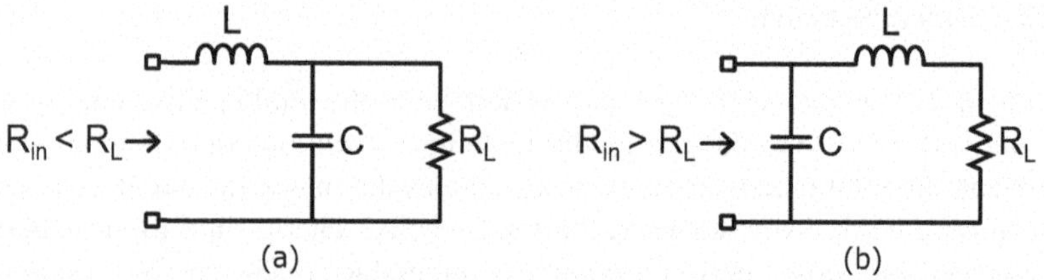

Fig. 4.10 Examples of an L-matching network used to boost (**a**) or lower (**b**) the input impedance

4.1.6.1 Lumped-Element Matching

One way to implement matching networks is using lumped-elements, such as inductors, capacitors, and transformers. Using the property that the voltages or currents are multiplied by Q in resonant RLC networks, an impedance transformation can be obtained. Consider the L-match network in Fig. 4.10a consisting of a series inductor and shunt capacitor. By transforming the parallel capacitor and load resistor into its equivalent series network, the equivalent series resistance is calculated as

$$R_{\text{in}} = \frac{R_L}{1 + Q^2} , \quad (4.32)$$

which equals the input impedance of the whole network at the frequency where the inductance and the transformed capacitance resonate. The L-match network has effectively reduced the load resistance to a lower input resistance.

Alternatively, as depicted in Fig. 4.10b, the series inductance and load resistance can be transformed into its equivalent parallel network, leading to an equivalent parallel resistance of

$$R_{\text{in}} = R_L(1 + Q^2) , \quad (4.33)$$

obtaining an increase in input resistance at resonance.

Looking at these simple examples, a couple of observations can be made. First, to achieve the desired transformation at the frequency of interest, the reactive elements should be chosen such that they resonate (after the necessary transformations) at the desired frequency. Second, in case the load (or desired input impedance) is not purely real but also contains a nonzero reactance, this element can simply be absorbed in the reactance elements of the matching network when doing the necessary calculations. Third, in both networks the inductive and capacitive elements can be interchanged without changing the matching performance at resonance. What combination to choose mostly depends on the DC properties of the network (DC coupled vs. DC blocking) and possible implementation aspects. Lastly, and most importantly, one can observe that the obtained impedance transformation ratio is

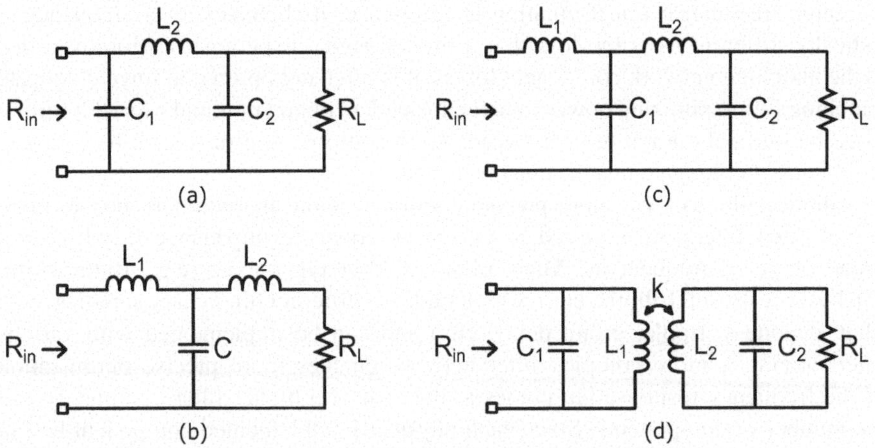

Fig. 4.11 Examples of higher order lumped-element matching networks: Pi-match (**a**), T-match (**b**), multi-section L-match (**c**), magnetically coupled resonator match (**d**)

fully determined by the Q-factor of the network. This means there is no additional freedom to choose the Q-factor and hence bandwidth of the matching network. For additional degrees of freedom, higher order networks are required.

Figure 4.11 shows various examples of high-order lumped-element matching networks. An additional degree of freedom can be obtained by cascading two L-match networks. Combining an upconverting L-section with a downconverting L-section results in a Pi-match network as shown in Fig. 4.11a. Reversing the order gives rise to the T-match network in Fig. 4.11b. In both cases, the two middle elements of the network can be merged resulting in a three-element network. Making the two L-match sections perform downconversion results in a four-element multi-section L-match network as depicted in Fig. 4.11c. In all three cases the desired impedance transformation is performed in two steps, via an intermediate resistance, which brings an extra degree of freedom that can be used to alter the bandwidth of the response. In case the wanted transformation ratio is very high, possible too high for a single L-match section (for instance due to the limited inductor Q-factor), the multi-section L-match network from Fig. 4.11c enables to realize the transformation in two smaller steps. Another reason for wanting two low-Q L-match sections instead of one high-Q one is that the same transformation can be achieved with a lower overall Q-factor and thus more broadband response. More details and design equations can be found in [7, 67].

In addition to controlling the impedance transformation and Q-factor (bandwidth) of the network, the frequency response of the network, i.e., the insertion loss over frequency, is highly important as well. While a decent impedance match is required to obtain a good frequency response, optimizing the loss response demands additional design considerations. For instance, while the inductive and capacitive elements could be mostly interchanged in the discussed networks with the same impedance transformation at resonance, its behavior over frequency is radically different as is the loss that a filtered signal experiences. Moreover, due to the matching network component losses, it is often not optimal to force conjugate matching for maximizing power transfer. Instead, to obtain minimal overall insertion loss, the optimal compromise between component loss and mismatch loss (due to an imperfect conjugate match) must be found.

Additionally, to have more precise control over the frequency response, more complicated filters are required to ensure an adequate impedance match over a broad range of frequencies. Many different filter types exist (e.g., Butterworth, Chebyshev, Bessel, Elliptic, etc.), which trade off different filter characteristics (e.g., filter steepness, ripple, group delay, etc.) and can be implemented with various filter orders. A more complex filter network enables more precise optimization of the frequency response but comes at the cost of a higher filter complexity and the number of components. Since each physically implemented component has an associated loss, increasing the number of components in the network inevitably increases the minimum insertion loss of the filter. This is similar to how more L-match stages enable a lower network Q-factor with more broadband response, but higher loss. At very high frequencies, where the loss of passive components is quite significant, the filter order of practical networks must be limited.

In the end, choosing the topology of the matching network comes down to a trade-off between network bandwidth and insertion loss. More components might enable a better impedance match over a large frequency range but introduce more resistive loss. A highly popular matching network topology at mm-Wave frequencies is the magnetically coupled resonator, as depicted in Fig. 4.11d. Combining a transformer with additional shunt capacitors results in a 4th-order filter response that can achieve highly broadband matching performance with minimal insertion loss [81]. This transformer-based matching network turns out to be highly effective above 100 GHz and will be extensively discussed later in this chapter. For a more in-depth discussion on this interesting topic, the reader is referred to [82].

4.1.6.2 Distributed Matching

Another strategy to implement matching networks is to use distributed components, such as transmission lines. Recall from Sect. 3.3.5 how the input impedance looking into a transmission line changes periodically with its length ℓ:

$$Z_{\text{in}} = Z_c \frac{Z_L + jZ_c \tan(\beta\ell)}{Z_c + jZ_L \tan(\beta\ell)} . \tag{4.34}$$

Furthermore, terminating a transmission line with a short or open leads to an inductive or capacitive input impedance, respectively. Using these facts, impedance transformations can be achieved using transmission lines.

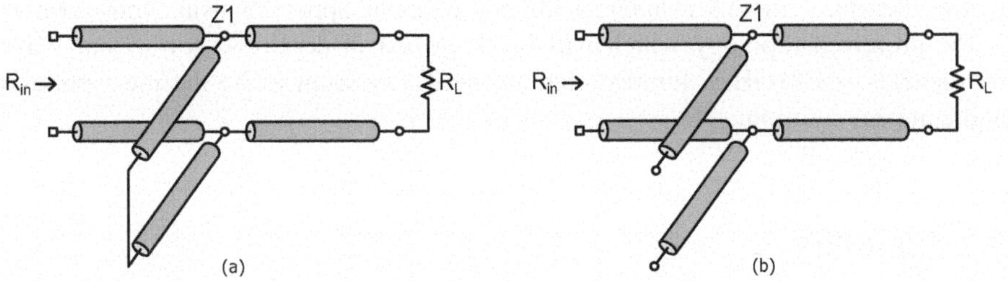

Fig. 4.12 Example of a transmission line-based matching network using a shorted parallel stub (**a**) and open parallel stub (**b**)

For example, consider the matching circuit in Fig. 4.12a. The first section of transmission line transforms the load into another complex impedance according to the above equation, if R_L does not equal the characteristic impedance of the line. By adding a short shunt stub to the line, a positive reactance is added in parallel to that impedance. Depending on whether the impedance at that point is matched to the line, the final piece of transmission line will again transform the impedance into another value. By adequately choosing the length of the lines, any complex load impedance can be transformed into a large range of input impedances.

Depending on the impedances, an open stub may be more appropriate than a shorted stub, as shown in Fig. 4.12b. It is best practice to choose the stub termination that leads to the stub with shortest length. However, at very high frequencies a shorted stub is preferred since providing a short circuit (e.g., by creating a series resonance) is much easier than creating a high-quality open termination. Furthermore, while series stubs can also be used in theory, these are more difficult to implement and thus rarely used in practice. It must be noted that this transmission line-based design can easily be combined with lumped components within the same network. For example, transmission lines are often used in combination with series capacitors, as an effective way to realize AC coupling between adjacent stages. As this type of matching will not be used in this work, no more unnecessary details will be shared. More details on transmission line matching can be found in [65, 67].

Now the question is whether lumped-element or transmission line-based matching networks are superior. At lower frequencies, transmission lines will occupy significantly larger area than lumped-elements and exhibit greater loss. However, as the frequency increases, lumped components become more lossy, and parasitic components start to degrade the expected behavior. In contrast, transmission lines are well modeled at higher frequencies, have a smaller footprint, and show lower losses than their lumped equivalents. Obviously there is a crossover frequency from which point a transmission line approach is more beneficial than a lumped-element approach. The exact frequency is difficult to pinpoint and depends on the application and technology. In the THz range (>300 GHz), transmission lines seem to be the best way to go due to their small size and the difficulty to implement lumped inductors. In the sub-THz range (100–300 GHz), both approaches are feasible and adopted in the literature. In this chapter, a lumped-element approach using transformers is the preferred topology which will be described in depth. At lower mm-Wave frequencies (<100 GHz), lumped-element networks seem to be the most popular and compact solution.

4.2 Stability

Stability is an essential condition for an amplifier, or any other block, to work properly since an instability will lead to the excitation of a large oscillation that disrupts the operation of the circuit. An instability can be best understood in the context of system theory. As introduced in Chap. 2, the behavior of a system or block can be described by a transfer function in the frequency domain, or an impulse response in the time domain. By decomposing the numerator and denominator of the transfer function, the zeros and poles of the system can easily be identified:

$$H(s) = \frac{a_0}{b_0} \cdot \frac{(s+z_1)(s+z_2)\ldots(s+z_m)}{(s+p_1)(s+p_2)\ldots(s+p_n)}, \tag{4.35}$$

where each complex pole ($s = \sigma + j\omega$) leads to term in the impulse response of the form $k\, e^{\sigma t}$. An instability corresponds to a complex pole that has a positive real part ($\sigma > 0$) and thus leads to a transient response that rises exponentially in time. When visualized on a pole–zero plot, see Fig. 4.13, this means the right-half plane (RHP) cannot contain any poles. While the imaginary axis denotes the oscillation frequency, the real part indicates the change of amplitude in time. A negative or zero real part leads to an exponentially decaying or sustained oscillation, respectively, at that frequency ω in the time domain. A RHP pole, on the other hand, manifests itself as an oscillating signal that grows exponentially. The oscillation keeps on growing until the system (i.e., the actual circuit) starts to saturate causing the gain to decrease and hence slow the growth of the oscillation. After a certain settling time, the amplitude of the oscillation will reach a constant value, corresponding to a $\sigma = 1$ of the unstable pole in the (saturated) transfer function. Note that an electronic oscillator works in a similar way, although in that case the instability is the result of a deliberate design choice.

In theory, an oscillation must be triggered by an input signal at that frequency. One might therefore assume that an instability occurring outside of the band of interest is not a problem as there is no input at that frequency. Alas! Since the system is highly sensitive at the unstable frequency (remember that a pole leads to an infinitely high response), even the tiniest of signals, noise for instance, will trigger the oscillation. As every physical circuit is subject to noise, every instability will be in reality be triggered and cause unwanted mayhem. If σ is very small and the amplitude of the oscillation quickly settles to a sufficiently small magnitude such that it has negligible influence on the performance, then the instability theoretically is not a huge problem. However, since the exact value of the pole is often very

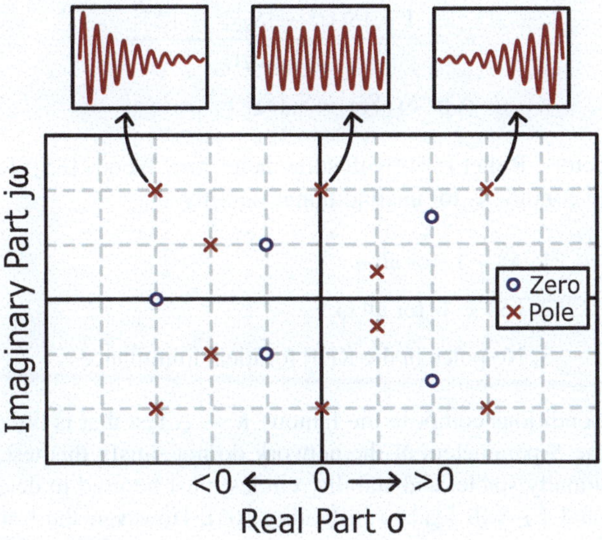

Fig. 4.13 Example of an arbitrary pole–zero plot with visualization how the real part of the pole affects the transient response

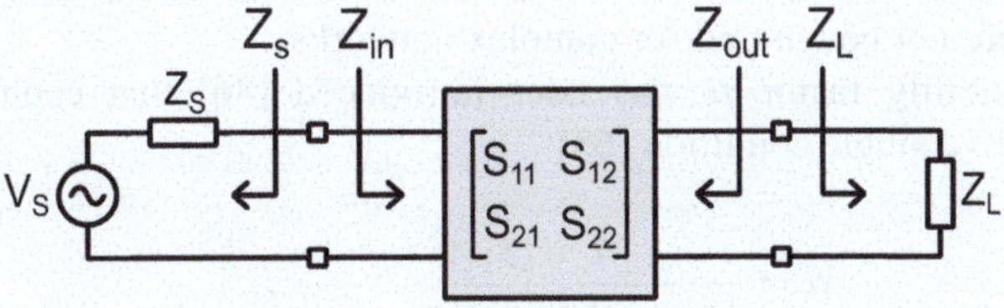

Fig. 4.14 Two-port network with impedances and reflection coefficients

difficult to model accurately, it is better to avoid instability altogether, ideally with a healthy margin. But how can stability be verified in complex networks? A thorough overview of the most popular methods is given in [83], a concise overview will be given here.

4.2.1 Two-Port Verification Criteria

There are many ways to analyze stability in a network. For two-port networks, the simplest method involves the calculation of stability verification criteria. Consider the general two-port network in Fig. 4.14 with specified S-parameters. The Rollet stability factor K of this network can be calculated as follows [65]:

$$K = \frac{1 - |S_{11}|^2 - |S_{22}|^2 + |\Delta|^2}{2|S_{12}S_{21}|} \tag{4.36}$$

$$\Delta = S_{11}S_{22} - S_{12}S_{21} \ . \tag{4.37}$$

Using these factors, Rollet [84] (with some help from Woods [85]) formulated the following three conditions for unconditional stability:

$$1.\ K > 1 \ \text{ for all } \omega \tag{4.38}$$

$$2.\ |\Delta| < 1 \ \text{ for all } \omega \tag{4.39}$$

$$3.\ \text{No poles in the RHP for ideal immittances}\ . \tag{4.40}$$

The first two conditions compose the famous $K - \Delta$ test that is often used by RF designers. If the S-parameters of the network do not satisfy this test, the network is only conditionally stable, and stability circles must be used to determine which values of Γ_S and Γ_L will lead to instability [65]. However, the last condition is often ignored and needs some explanation. It says that no RHP poles (i.e., poles with $\sigma > 0$) are allowed in the system when exposed to an ideal immittance (either a perfect open or short). Even though ideal immittances do not physically exist, this condition can uncover hidden instabilities that otherwise would go unnoticed and should therefore not be ignored for complex networks.

Another stability factor μ was later introduced [86] that captures stability verification into a single condition [65]:

$$\mu = \frac{1 - |S_{11}|^2}{|S_{22} - \Delta S_{11}^*| + |S_{12}S_{21}|} > 1 \ . \tag{4.41}$$

Next to being more convenient to verify than two parameters, the geometric stability factor μ also quantifies how stable a network is, where a large value signifies greater stability.

4.2.2 *Open-Loop Stability Verification*

Alternatively, if the feedback path that is possibly causing an oscillation is known, the loop gain can be analyzed to assess the stability. As illustrated by Fig. 4.15, by breaking up the loop formed by forward amplifier $A(\omega)$ and feedback path $\beta(\omega)$, the loop gain $A\beta(\omega)$ can be calculated. Based on the magnitude and phase response of this loop gain, the stability of the system can be assessed. Barkhausen [87] stated that in order to trigger an oscillation, the loop gain should have a magnitude of one while providing a total phase shift of 360°, effectively creating a system with

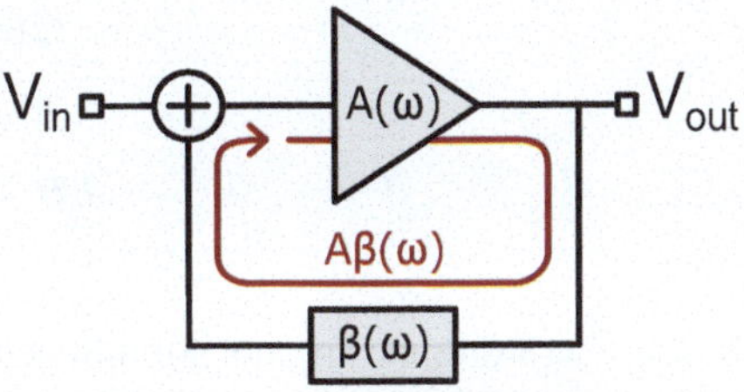

Fig. 4.15 General diagram of a forward amplifier A with feedback factor β to determine to loop gain

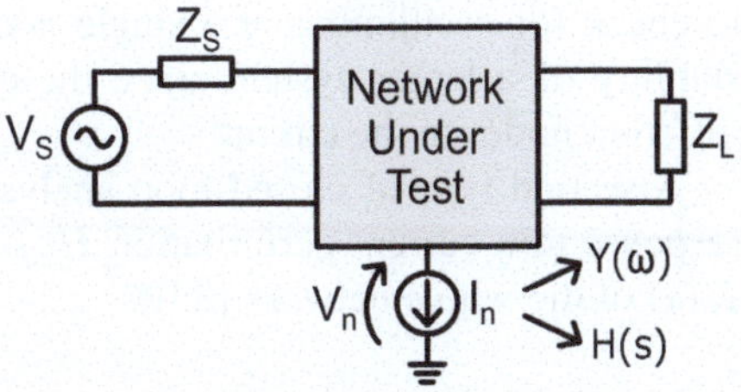

Fig. 4.16 Principle of closed-loop stability verification techniques by inserting an ideal current source to perform an admittance analysis or pole–zero estimation of the network under test

positive feedback. While Nyquist extended this theory with a more graphical method [88], introducing the Nyquist plot, Bode presented a more intuitive representation [89], the well-known Bode plot, with the measurable quantities of gain margin and phase margin. These margins, easily calculated from the magnitude and phase response of the loop gain, quantify how stable the system is. While this method is very intuitive and commonly used in analog design, it is seldom used in RF design. The reason is that a high-frequency RF circuit of decent size has many feedback loops, created by the numerous parasitic elements and couplings originating from the transistors and layout, resulting in a highly complicated interconnected network. Since it is often difficult (or impossible) to identify the exact feedback path that might be causing problems, this type of analysis is not practical.

4.2.3 Closed-Loop Stability Verification

Finally, closed-loop analyses are able to more accurately assess stability in complex systems without needing to identify and break up specific feedback loops. By inserting a small perturbation inside the circuit, typically a small current, its response reveals possible unstable behavior, even when caused by unwanted or hidden parasitic loops. As illustrated in Fig. 4.16, stability can be verified by using an ideal current source and by analyzing either the simulated admittance or the poles and zeros arising from the estimated transfer function $H(s)$.

First, the admittance analysis is a way to quickly verify whether an oscillation can be triggered at a specific node. Concretely, by inserting a small-signal current into a node, the voltage response and corresponding admittance $Y(\omega) = G(\omega) + jB(\omega) = I_n(\omega)/V_n(\omega)$ can be calculated. Kurokawa formulated three conditions that are required for an oscillation to start up at frequency ω_0 [90]:

$$G(\omega_0) < 0 \tag{4.42}$$

$$B(\omega_0) = 0 \tag{4.43}$$

$$\frac{\partial B}{\partial \omega}(\omega_0) > 0\,. \tag{4.44}$$

If these conditions are not simultaneously satisfied at a specific node over all frequencies, no oscillation can be triggered at that node. This is a very fast way to check for oscillations at a single node, but less effective when trying to verify stability of a larger system since these analyses must then be performed on the different nodes in the circuit.

A second type of closed-loop analysis is the pole–zero identification. Here the response to a current perturbation $H(j\omega)$ is used to estimate the transfer function $H(s)$ of the whole network [83]:

$$H(j\omega) \approx \hat{H}(s) = K\frac{\Pi_{i=1}^{m} s + z_i}{\Pi_{i=1}^{n} s + p_i}\,. \tag{4.45}$$

By deriving the different poles p_i from this estimated function, for instance by sing fitting and estimation algorithms, RHP poles and thus unstable behavior can be identified. As opposed to an admittance analysis, a pole–zero identification only has to be performed at a single node in the network, hence providing a more global stability analysis tool. This is based on the fact that any closed-loop transfer function found in a linear system can be expressed with the same denominator, and hence the same poles [91, 92]. One exception is the occurrence of an exact pole–zero cancelation at a certain node in the circuit. To avoid missing a crucial pole, the analysis is best performed on multiple nodes. It must be noted that both closed-loop analyses can be performed with an excitation source V_s to the network, enabling large signal in addition to just small-signal stability analysis.

For a more detailed explanation of the different stability analyses as well as implementation guidelines and examples, the reader is referred to [83].

4.3 Neutralized Pseudo-Differential Pair

This section examines one of the most effective building blocks for amplifiers above 100 GHz: the neutralized pseudo-differential pair. After investigating the most important performance metrics and comparing various amplifier topologies, the design and layout of a neutralized pseudo-differential pair are discussed.

4.3.1 Performance Metrics for Amplifiers

Before being able to optimize an amplifier, we must know what to optimize for. Each specific block usually has multiple specifications that must be designed for simultaneously and that determine the optimal topology and design strategy.

As an amplifier is used to amplify signals, the power gain is an important specification. Different power gain definitions exist for a two-port amplifier depending on whether the mismatch loss as the input or output of the amplifier is included [65]. Where the maximum power gain G_{max} is defined as the power gain with a perfect conjugate match at the input and output, the power gain G includes both the mismatch loss between the source and input, and the output and load of the two ports. While G_{max} provides an indication of how much gain an amplifier block can possibly provide, the actual power gain is what determines the actual amplifier performance. While designing, it is therefore important to also consider the influence of the input/output impedances of the amplifier as opposed to simply maximizing G_{max}.

Within a transceiver system, the general goal is to maximize the signal-to-noise ratio of the signal, as extensively discussed in Chap. 2. This translates into maximizing the output power of the transmitter and minimizing the noise figure of the receiver. The final block of the transmitter, the power amplifier (PA), must therefore be optimized for maximal output power while satisfying the required linearity demands. The first block of the receiver, the low-noise amplifier (LNA), must be designed for a minimal noise figure. Again, for both the PA and LNA, the performance of the output power match and input noise match, respectively, is crucial to obtain the best possible output power and noise performance.

Next, minimizing the power consumption and energy efficiency is a priority for each integrated system, particularly when operating at very high frequencies. While efficiency is an important specification for all blocks, it is especially critical for the most consuming blocks in the system, i.e., the PA, since they dominate the overall power consumption. Two different efficiency metrics are employed in the literature: the drain efficiency (DE) and power added efficiency (PAE). The DE simply links the amount of output RF power that is generated with the amount of DC power that is dissipated, thus assuming the power gain is sufficiently high:

$$\mathrm{DE} = \frac{P_{\mathrm{out,RF}}}{P_{\mathrm{DC}}} . \tag{4.46}$$

However, due to the limited power gin, the DE is not a good indicator of efficiency. Instead, the PAE takes into account how much input power is required to reach the specified output power:

$$\mathrm{PAE} = \frac{P_{\mathrm{out,RF}} - P_{\mathrm{in,RF}}}{P_{\mathrm{DC}}} , \tag{4.47}$$

where P_{DC} should include the power consumption of the driver stages as well. When an amplifier attains a large $P_{out,RF}$ but has very little power gain, this will be reflected in a poor PAE number, as is desired. PAE is thus the preferred efficiency metric at high frequencies.

An important consideration that is often overlooked is stability. If the circuit becomes unstable, an oscillation will build up and corrupt the intended performance, or worst case render the system useless. Any designed circuit must therefore be made unconditionally stable, meaning stability is ensured for every possible combination of source and load impedances to the system. Even though this specification is rather pessimistic, as a practical system will not experience every possible source or load, it is good practice in order to ensure a reliable electronic system. Due to its importance, stability will be more elaborately later on.

Finally, to further improve the reliability, the circuit must be robust against outside disturbances, such as noise or interference signals, and against changes in the environment or fabrication process. Making the circuit robust to process, voltage and temperature (PVT) variations require more intricate design techniques that fall outside the scope of this book. Ensuring the circuit is minimally disturbed by interference signals can be achieved by implementing differential circuits, as discussed next.

4.3.2 *Differential vs. Single-Ended Design*

While the performance of a single transistor was discussed in Sect. 3.2, practical circuits at high frequencies are often implemented in a differential topology. Consider the single-ended and differential common-source circuits in Fig. 4.17,

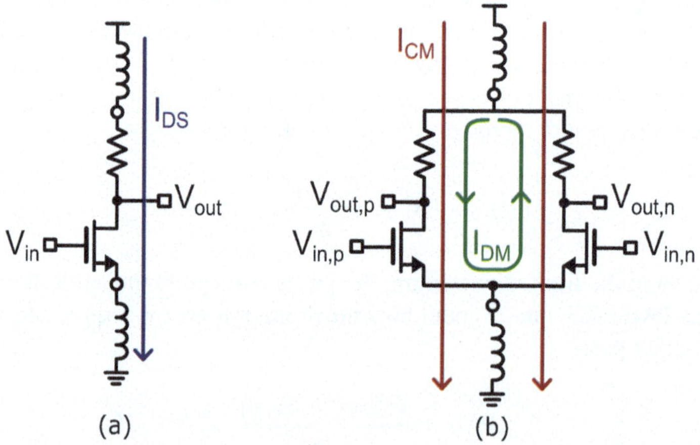

Fig. 4.17 Single-ended (**a**) and differential (**b**) common-source circuit

where the supply and ground interconnects are explicitly drawn as inductors. Since both the input and output of the differential circuit have three pins (including the ground reference), the signals can be decomposed into two different modes: the differential-mode (DM) and the common-mode (CM):

$$V_{\mathrm{DM}} = V_p - V_n \tag{4.48}$$

$$V_{\mathrm{CM}} = (V_p + V_n)/2 \,. \tag{4.49}$$

As visualized in Fig. 4.17b, the differential-mode current flows in a circular path within the circuit, as opposed to the common-mode current or single-ended current in Fig. 4.17a that have to flow through the supply and ground connections. While these connections must always be modeled to some extent, they are much less critical for the performance in differential-mode. This results in a more predictable behavior (since the supply/ground connections are often not trivial to model accurately at high frequencies, especially when they run off-chip) and better performance (as the resistive and inductive nature of these interconnects typically degrades the performance).

Furthermore, any interference signals (e.g., noise, crosstalk, or EMI) usually appear as a CM signal in the circuit, either as a CM input or on the supply rails, ideally leaving the DM performance unaffected. However, even though these CM signals do not appear directly at the DM output, they can affect the operating points of the transistors and thus indirectly affect the differential output. To be fully immune against CM interference at the input or supply rails, a high common-mode rejection (CMR) and power-supply rejection (PSR) are required in the circuit, meaning the gain from these CM nodes toward the differential output is suppressed. For instance, by adding a good current source with a large output impedance between the source connections and the ground node, the operating points of the transistors are fully controlled by these sources and unaffected by CM inputs. Such a truly differential common-source pair with current source often simply referred to as a differential pair. However, since the voltage headroom is very limited in advanced CMOS technologies, such a current source is often removed to maximize the dynamic range of the signal. And besides, at high frequencies the output impedance of the current source inevitably decreases due to its parasitic capacitance to ground, limiting its improvement on the CMR. Two common-source transistors operating differentially without the current source are called a pseudo-differential pair. Despite being interesting and relevant topics, CMR and PSR are not further covered in this book.

Finally, in a differential circuit a higher dynamic range (twice the voltage swing) and thus higher SNR can be attained, however, at the cost of a higher power consumption. To conclude, high-frequency circuits are often implemented in a differential fashion since they are easier to model, less affected by CM interference, and lead to a higher dynamic range than their single-ended counterparts.

4.3.3 Capacitive Neutralization

In Sect. 3.2, the G_{max} of a single transistor was derived assuming a perfect conjugate match at the input and output of the transistor. The maximum power gain of the transistor, or any two-port network in general, can also be expressed as a function of its two-port Y-parameters:

$$G_{max} = \left|\frac{Y_{21}}{Y_{12}}\right|(K - \sqrt{K^2 - 1}) , \tag{4.50}$$

where K represents Rollet's stability factor as discussed before. However, a network can only be simultaneously conjugately matched at the input and output if the network is unconditionally stable [65]. Indeed, a conditionally stable network with $K < 1$ would yield a negative value under the square root in expression 4.50. If the network is not unconditionally stable, the maximum power gain is defined as the max stable gain (MSB):

$$\text{MSB} = \left|\frac{Y_{21}}{Y_{12}}\right| , \tag{4.51}$$

which is equal to G_{max} from expression 4.50 with $K = 1$ or, in other words, the maximum power gain where the network is placed on the edge of unconditional stability. After inspecting the plot of G_{max} of a single transistor in Fig. 4.18, a knee point is observed separating two distinct regions. For frequencies below a f_{stab} = 140 GHz, the transistor is only conditionally stable, thus G_{max} = MSB, while the transistor is unconditionally stable above that frequency. From the small-signal model derived in Sect. 3.2, it is clear that the inherent parasitic capacitance C_{gd} provides a feedback path within the transistor and hence limits the stability as well as the achievable gain at higher frequencies. Luckily, there is an elegant solution to neutralize the feedback effect caused by C_{gd} in a differential topology.

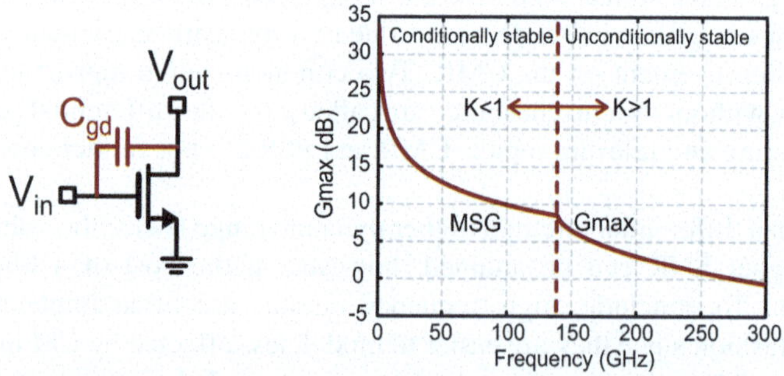

Fig. 4.18 A single-ended common-source amplifier with associated G_{max} curve, showing a conditional and unconditional stability region

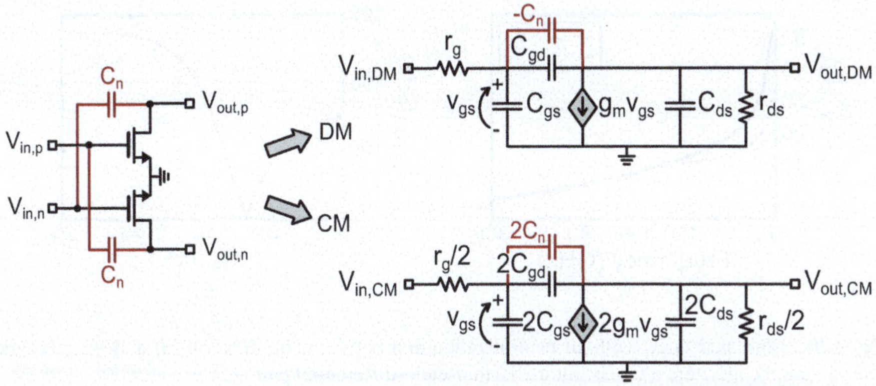

Fig. 4.19 Schematic of a capacitively neutralized pseudo-differential pair with equivalent small-signal model in differential-mode and common-mode

Consider the pseudo-differential pair in Fig. 4.19 with two added capacitors C_n which are cross-connected between the gates and drains of both transistors. This is called capacitive neutralization, and its effect is illustrated by the equivalent small-signal model in differential-mode and common-mode in Fig. 4.19. In differential-mode, the neutralization capacitance appears as a negative capacitance that compensates the effects of C_{gd}. By choosing the value of C_n equal to C_{gd}, the feedback of C_{gd} is canceled, and the amplifier is made unconditionally stable for all frequencies. Mathematically, neutralizing the feedback path in the network, in other words making the network unilateral, leads to an $S_{12} = 0$, which guarantees $K > 1$ (according to Eq. 4.36) for all frequencies. Indeed, as plotted in Fig. 4.20a, $C_n = C_{gd}$ leads to a larger $G_{\max}$ at lower frequencies without showing a kink in the curve. Furthermore, Fig. 4.20b shows the minimal stable frequency f_{stab} versus neutralization capacitance, demonstrating unconditional stability down to DC when C_{gd} is perfectly canceled, which corresponds to $C_n = 6.75$ fF for a 25 μm transistor size.

However, as can be observed in the common-mode equivalent circuit in Fig. 4.19, the neutralization capacitance C_n adds to the degrading effect of C_{gd} and thus worsens the stability in common-mode. Even though the circuit will be used in differential-mode, it can still potentially oscillate in common-mode and negatively affect the differential-mode performance. In order to ensure unconditionally stability in common-mode, large resistors can be placed on the bias lines (CM gate connections) to block CM feedback loops [76]. These resistors kill CM oscillations without interfering with the performance of the amplifier, as no high-frequency signal passes through the resistors. In more complex networks, such as a multistage amplifier chain with shared supply lines, additional CM feedback loops can arise and cause oscillations. These can be mostly suppressed by providing a large impedance, e.g., by means of choke inductors, killing CM gain with minimal current

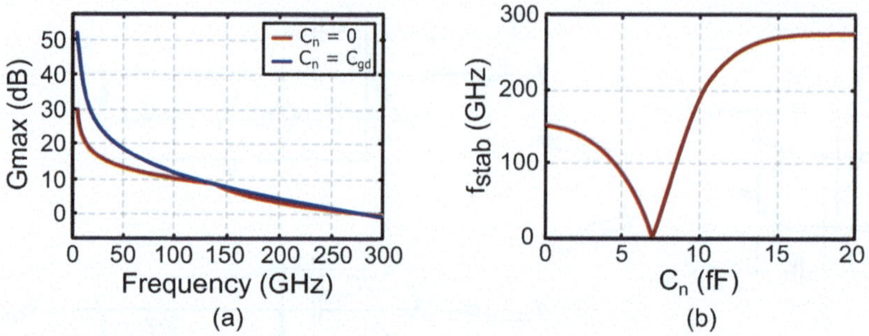

Fig. 4.20 Simulated $G_{\max}$ without neutralization and perfect neutralization (**a**) and f_{stab} versus neutralization capacitance C_n (**b**) for a 25 μm pseudo-differential pair

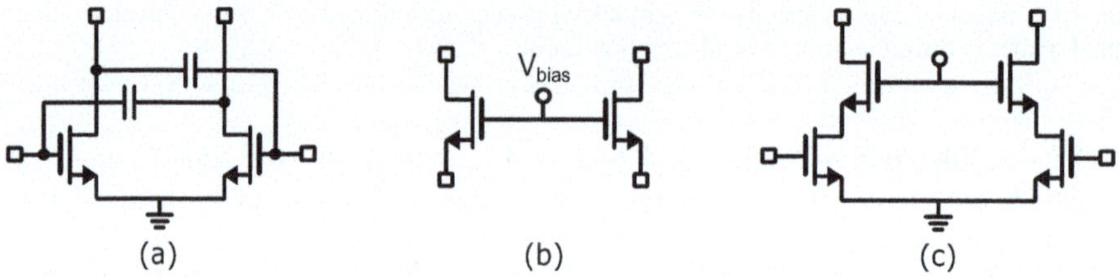

Fig. 4.21 Popular differential amplifier topologies at mm-Wave frequencies: neutralized common-source (**a**), common-gate (**b**), and cascode (**c**)

drop in the lines. Even more complicated systems require ad hoc analyses, like those described in Sect. 4.2, and techniques to ensure unconditional stability.

A more thorough analysis of capacitive neutralization and its effect on stability can be found in [76, 83].

4.3.4 Amplifier Topologies

Most of the discussion so far involved transistors in a common-source topology (meaning the source terminals are tied to ground), in either a single-ended or differential configuration. Particularly the differential common-source amplifier with capacitive neutralization, or neutralized pseudo-differential pair, shown in Fig. 4.21a, shows promising properties at high frequencies while being a relatively simple circuit. Simplicity is key to high frequencies since the performance is highly sensitive to device and layout parasitics, as demonstrated by expression 3.102 of $f_{\max}$. At lower frequencies, techniques such as gain boosting are employed to increase the gain of a single stage at the cost of a few extra transistors. However, the significant increase in parasitics from these extra elements and the more complicated layout nullify any intended improvement at higher frequencies. Therefore, at higher frequencies (>100 GHz), circuit topologies become more simplistic, but the design per device all the more challenging. Nevertheless, there are two other topologies that are sometimes used in mm-Wave design: the common-gate and cascode amplifier.

Firstly, the common-gate amplifier, shown in Fig. 4.21b, is often employed in the lower mm-Wave range at the input of an LNA. The common-gate has an intrinsically lower noise figure than the common-source, which makes it ideal for LNA implementations [76]. Furthermore, its low input impedance makes it easier to match to a 50 Ω input over a broad bandwidth [93]. However, the G_{max} is quite a bit lower than for a common-source, and its optimal noise match impedance is very dissimilar from its conjugate match impedance, resulting in an even larger power gain penalty if minimum noise matching is pursued. Finally, the common-gate is only conditionally stable, which requires cautious design and analysis to avoid potential instability.

Secondly, Fig. 4.21c shows a cascode amplifier topology where a common-gate is stacked on top of a common-source amplifier. This cascode configuration demonstrates a larger gain for the same current, mostly due to an increase in output resistance. Furthermore, the cascode topology enables the supply voltage to be scaled up to twice the nominal value, thereby providing higher output voltage swing and thus output power. To avoid gate-bulk breakdown issues, all devices must be placed within a triple well with the source tied to the bulk. It also shows higher input–output isolation, which theoretically is beneficial for stability. However, the increase in layout parasitics significantly limits the effectiveness of the cascode at higher frequencies and makes the amplifier more prone to internal instabilities. Indeed, the amplifier is very sensitive to oscillations through the gate node and at the output of the cascode device [94]. Moreover, some of the gain improvement is lost in the form of mismatch losses when the large output impedance has to be matched to a potentially much lower input impedance.

In the end, the neutralized pseudo-differential pair (NPDP) from Fig. 4.21a turns out to be the preferred topology at frequencies above 100 GHz since it offers a large power gain, low-noise figure, good simultaneous gain/noise match, and guaranteed unconditional stability (at least in DM). As almost all high-frequency blocks are implemented by means of this topology, the design of an NPDP will be described more extensively in the next section.

4.3.5 Implementation of an NPDP

As the NPDP is the most effective amplifying block above 100 GHz, the implementation of this core building block must be thoroughly optimized. Optimizing the NPDP involves optimal implementation of the neutralization capacitor as well as drawing the layout of the whole NPDP core.

Before implementing the neutralization capacitor, one must first choose the type of capacitor. One option is to implement this capacitor using the parasitic gate–drain capacitance of an MOS transistor combined with a large resistor to prevent any current from flowing through them [83]. This approach is very robust against

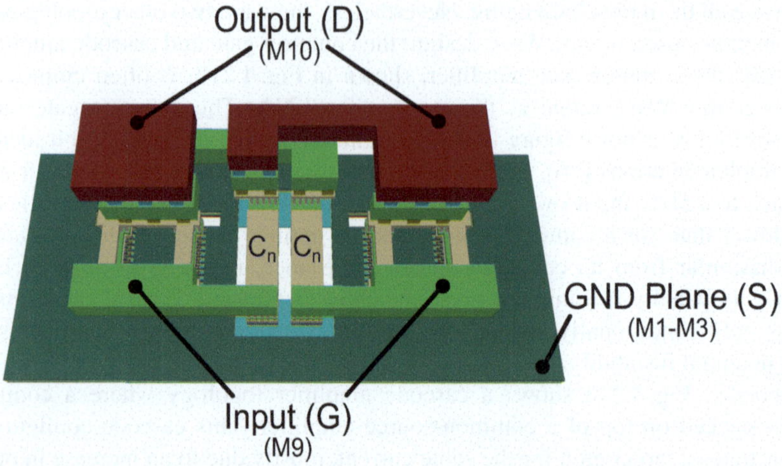

Fig. 4.22 Layout of a neutralized pseudo-differential pair with 25 μm transistors

PVT variations as the capacitor ideally experiences the same variation as the active device as they are of the same type. However, as such as MOS capacitor has a very low Q-factor at high frequencies, they introduce some loss and can significantly increase the noise of the NPDP. MOM capacitors, on the other hand, achieve very high Q-factors at these frequencies and therefore barely degrade the noise figure ($\approx$0.2 dB at 100 GHz). Furthermore, by implementing the MOM capacitor in as many layers as possible (in practice from M2 and up), its area is minimized thereby ensuring minimally short interconnects. Lastly, the value of C_n ideally equals the parasitic C_{gd} from the transistors, which corresponds to the largest frequency range where the NPDP is unconditionally stable, so where $\mu > 1$. The MOM capacitor should be accurately modeled using an EM simulator tool.

Next, for a given transistor size and designed capacitor, the layout of the complete NPDP must be drawn. Starting from the optimized layout of a single transistor (Fig. 3.23), the main goal is to draw the layout as compact as possible to minimize the interconnect parasitics. As shown in Fig. 4.22, the capacitors are placed in between the transistors and connected using the highest metal layers. Since the amplifier core will be connected via passive components situated in the top metals, the input and output are also routed in those metals to provide a well-modeled, low-loss interface. For large transistor widths, the transistor can be split up into multiple unit cells to better match the length of the capacitors, like in Fig. 4.22 where the 25 μm transistors consist of two 12.5 μm cells. Modeling these interconnects accurately is crucial to predict the performance of the NPDP. For small cells, a parasitic extraction tool can be used to fully extract the complete layout. However, for large transistor sizes the long interconnect can introduce significant inductance that degrades the neutralization performance and can even introduce oscillations.

4.3.6 Design of an NPDP

Finally, with the topology and layout strategy fixed and optimized, the NPDP core should be designed to meet the circuit specifications. Concretely, this means choosing the bias voltage and width of the transistors. As gain is very scarce above 100 GHz in CMOS, the NPDP is usually biased around the point that provides maximum $G_{\max}$ at the frequency of interest, while considering its current consumption and linearity performance. The size is then chosen depending on the desired output power. However, there are more intricate details that are worth discussing, beginning with the transistor size.

4.3.6.1 Transistor Size

Due to the large influence of layout parasitics, the transistor layout must be manually drawn for each size in order to accurately evaluate its performance. Note that this is not necessary if the transistor model provided by the foundry would include a highly optimized layout. In order to be able to quickly assess trade-offs between different sizes while designing, it is a good idea to compose an inventory of many different optimized transistor sizes. Table 4.1 shows a performance overview at a frequency of 140 GHz for an optimized NPDP with various transistor widths. The core transistor model includes everything up till metal 1 and is verified until 110 GHz by the foundry. A finger width of 630 nm is used, which is optimal in this 28 nm CMOS technology to achieve minimal gate resistance. Note that the width numbers are not exact but were rounded to make the table more easily readable. For each size, the transistor layout is fully drawn from metal 2 until metal 8 and modeled using a parasitic extraction tool. The neutralization capacitor is kept ideal, which is a decent approximation because its actual Q-factor would be very high, and sized for optimal DM stability. The bias is fixed at 500 mV, corresponding to a maximum $G_{\max}$ at 140 GHz for ultra-low V_T transistors. This model provides a quick way to evaluate a transistor size without having to fully draw and model each layout and predicts the eventual rigorously modeled performance of the NPDP very well. Of course, small changes are to be expected when all interconnects are included but are limited to several 0.1s of dBs.

The table reveals how the performance of an NPDP changes with increasing size, such as an increase in output power ($OP_{1\,\mathrm{dB}}$) and current consumption, decrease in $G_{\max}$ due to increased parasitics, and general decrease in equivalent parallel input/output resistance but increase in parasitic capacitance. The minimum noise figure NF_{min} was also added but barely changes with size. However, by comparing the complex conjugate of the optimal source impedance for minimal noise $Z^*_{S,\mathrm{NF}} \sim R_{S,\mathrm{NF}} \parallel C_{S,\mathrm{NF}}$ with the input impedance, the NF degradation can be estimated in case the input would be conjugately matched instead of noise matched.

Table 4.1 Performance overview of an NPDP in 28 nm CMOS with various transistor widths

Transistor width	20 μm	25 μm	30 μm	40 μm	45 μm	55 μm
Parallel cells	2	2	3	3	3	3
C_n	4.8 fF	6.2 fF	7.4 fF	9.4 fF	11.2 fF	13.2 fF
G_{max}	7.8 dB	7.8 dB	7.8 dB	7.6 dB	7.5 dB	7.3 dB
$OP_{1\,dB}$	5.6 dBm	6.2 dBm	7.1 dBm	7.9 dBm	8.7 dBm	9.1 dBm
I_{DC}	15 mA	19 mA	22 mA	27 mA	33 mA	38 mA
NF_{min}	4.4 dB	4.4 dB	4.4 dB	4.5 dB	4.5 dB	4.6 dB
$R_{S,NF} \parallel C_{S,NF}$	320‖8 fF	245‖10 fF	210‖12 fF	170‖15 fF	140‖19 fF	115‖20 fF
$R_{in} \parallel C_{in}$ (Q_{in})	520‖10 fF (4.4)	270‖12.5 fF (2.9)	240‖15 fF (3.1)	190‖18 fF (3)	160‖22 fF (3.1)	130‖24 fF (2.8)
$R_{out} \parallel C_{out}$ (Q_{out})	180‖10 fF (1.6)	160‖13 fF (1.8)	130‖16 fF (1.8)	110‖21 fF (2)	90‖23 fF (1.9)	85‖26 fF (1.9)

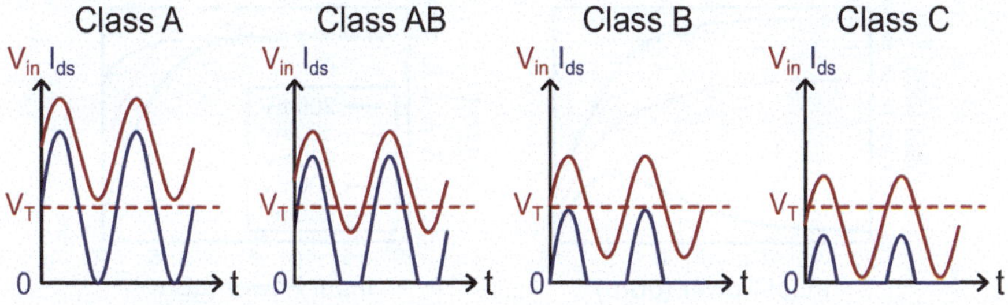

Fig. 4.23 Visualization of the input voltage and output current in time of an amplifier operating in classes A, AB, B, and C

4.3.6.2 Biasing

Next to sizing, the bias voltage of NPDP has a huge impact on the gain, linearity, and efficiency. By adjusting the bias voltage, the amount of DC current that the transistors conduct throughout the signal period can be controlled, thereby changing the efficiency, which is especially important for power amplifiers. At lower RF frequencies, this concept is pushed to the extreme by operating the PA as a switching amplifier driven by a rectangular signal, conducting no static current (like a CMOS inverter), but generating a highly nonlinear output with numerous harmonics. By providing specific harmonic terminations that filter out particular harmonics, the desired linearity can be achieved while benefiting from the superior efficiency [95]. At 100 GHz and above, however, generating a rectangular signal is very challenging as it requires higher order harmonics at frequencies where signal power is scarce. Furthermore, implementing high-Q passive resonators at these high harmonic frequencies is not practically possible in CMOS. Within the sub-THz range, PAs thus always behave relatively linearly with limited harmonics.

Nevertheless, even without significant harmonics, the performance of a PA can still be manipulated using the bias voltage. In classical PA theory, this results in four basic PA classes based on the biasing condition: A, AB, B, and C [60]. Depending on the bias voltage and input swing relative to the transistor threshold voltage, the transistor will conduct current during a certain fraction of the cycle, as illustrated in Fig. 4.23. However, the definition of the threshold voltage is very vaguely described in nanometer CMOS technologies, and in practice, the transistor will always conduct a small amount of (subthreshold) current, even if $V_{\text{in}} < V_T$. Furthermore, at very high frequencies, parasitics will always lead to some part of the input signal coupling to the output, making the class definitions even more vague. Therefore, practical PAs usually operate somewhere in class AB, where the value of the bias voltage determines whether the operation is closer to class A or class B. So even though classes can be useful for theoretical insight, it is more instructive to look at actual simulated behavior of an amplifier as a function of bias voltage to assess the relevant trade-offs.

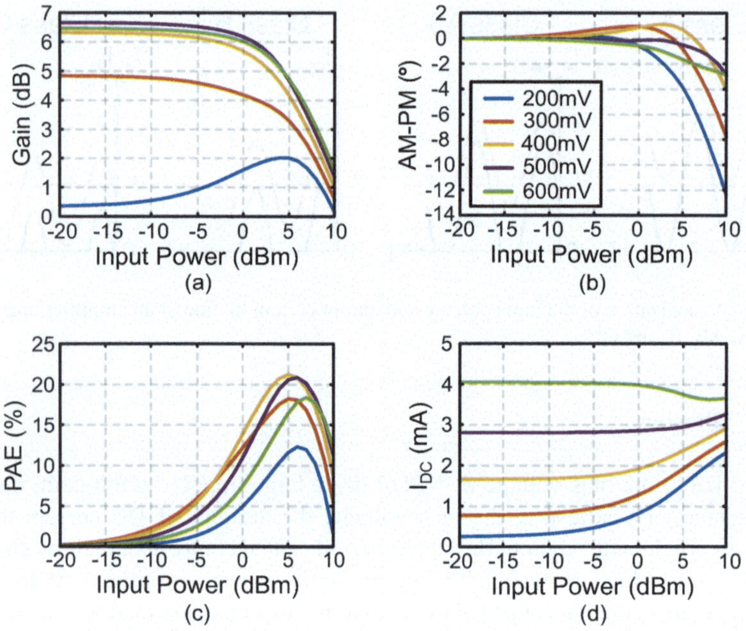

Fig. 4.24 Simulated gain (**a**), AM–PM (**b**), PAE (**c**), and DC current consumption (**d**) versus input power of a 40 μm width NPDP

Figure 4.24 shows the simulated gain, AM-PM, PAE, and DC current consumption versus input power of a 40 μm width NPDP. The plots show how the output power, linearity, and efficiency change with bias voltage. The more clear overview of several important amplifier metrics versus biasing is given in Fig. 4.25.

First of all, the simulated gain in Fig. 4.24a demonstrates clear gain compression for higher bias voltages, but gain expansion for very low voltages when the amplifier is operating near class B. While the gain expansive behavior can be useful in cascade with a gain compressing stage to improve overall gain linearity, the low power gain at that bias limits its use at very high frequencies. After all, due to the low gain more amplifying stages will be required, which negatively impacts the area and power consumption. As seen in Fig. 4.25, the $OP_{1\,\mathrm{dB}}$ is very large for small biases due to the gain expansion, then drops, and finally increases again when increasing the bias voltage.

Next to magnitude distortion (AM–AM compression/expansion), the phase distortion is quantified by the AM–PM distortion, signifying the change in phase response as the input power is increased, which was discussed in Sect. 2.3.3. This is mainly caused by the nonlinear input capacitance and transconductance of the NPDP [96]. As shown in Fig. 4.24b, the linearity of the phase response is highly dependent on the bias voltage but always decreases sharply when the amplifier is compressed at large input powers. Suppose the amplifier is operated at $OP_{1\,\mathrm{dB}}$,

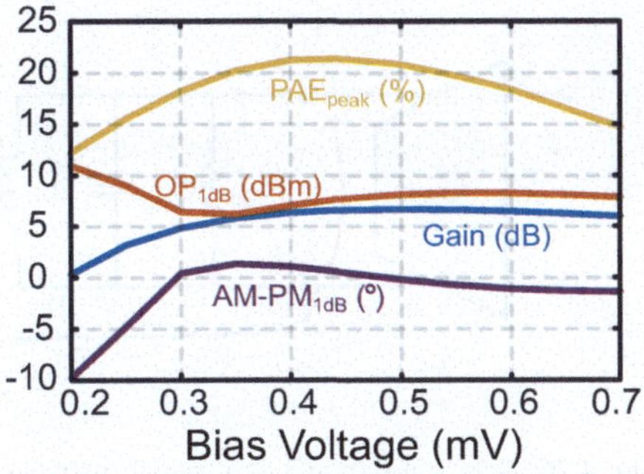

Fig. 4.25 Simulated gain, $OP_{1\,dB}$, peak PAE, and AM-PM at 1 dB compression versus bias voltage of a 40 µm width NPDP

and a convenient metric for AM–PM distortion is the AM–PM response at 1 dB compression, as plotted in Fig. 4.25. However, especially for complex modulated signals with numerous distinct power levels, it is important to consider the phase response over the complete range of input powers. For instance, while $AM-PM_{1\,dB}$ is very similar for 400 and 500 mV bias in Fig. 4.25, the response is clearly more linear over P_{in} at 500 mV bias when inspected in Fig. 4.24b. Furthermore, the output power at 1 dB compression is higher at 500 mV bias as well.

Finally, the efficiency can be evaluated by means of the plotted PAE and DC current consumption in Fig. 4.24c–d. The current clearly increases with increasing bias voltage but grows fast with larger input swings as the bias is lower. The peak efficiency (PAE), as also plotted in Fig. 4.25, achieves a maximum around 400–500 mV.

Combining all of the above, based on Fig. 4.25, any practical amplifier will be biased around 400–500 mV as the NPDP then provides close to optimal gain, AM–AM and AM–PM performance, and lastly, efficiency.

4.4 Transformer-Based Matching Networks

Having determined the optimal amplifier core, this section discusses the design of the transformer-based networks performing the impedance matching between the different amplifier blocks. For a given set of impedances (see Fig. 4.4), we want to find the matching network that can most closely approach the Bode–Fano limit (Eq. 4.13) and thus attain with minimal insertion loss for the specified bandwidth.

4.4.1 Why Transformer-Based Matching?

First of all, as discussed in Sect. 4.1.6, lumped-element-based networks can provide compact networks with low insertion loss at mm-Wave frequencies. Then a decision

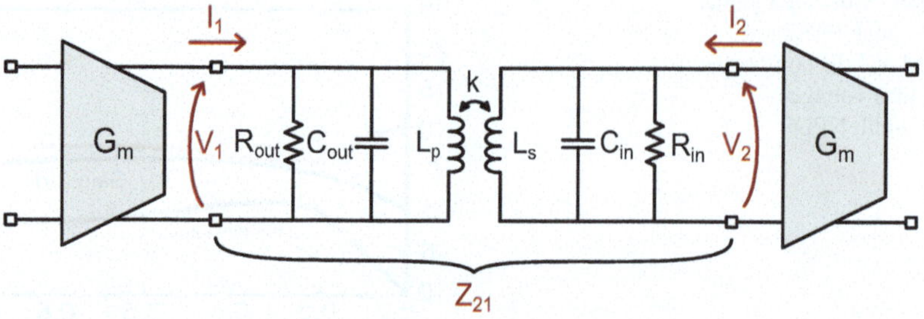

Fig. 4.26 Schematic model of a transformer-based matching network

must be made on the complexity of the network. High-order filters offer more control over the frequency response and can thus provide excellent matching performance over a wide frequency band with low in-band ripple. However, the high number of components generally increases the insertion loss of the network. Choosing the matching network topology thus entails a trade-off between bandwidth and insertion loss.

A magnetically coupled resonator network, or transformer-based network, has a 4th-order filter characteristic that can provide a broadband frequency response with minimal insertion loss. For a certain bandwidth, this network has the lowest ripple compared to other 4th-order filters [81]. Furthermore, transformers have a compact footprint and provide galvanically isolated AC coupling with convenient DC feeding via center tap connections. All of these properties make transformer-based networks the preferred topology to achieve impedance matching in this work.

4.4.2 *Frequency Response of a Transformer-Based Network*

To analyze the frequency response of the filter, the general diagram from Fig. 4.4 is converted into the schematic from Fig. 4.26. The transformer is modeled by lossless inductors L_p, L_s and a magnetic coupling coefficient k. Both amplifier stages are modeled as an ideal transconductance stage G_m with parallel input and output impedances R_{in}, C_{in} and R_{out}, C_{out}, respectively. Since a typical amplifier stage in CMOS functions as a transconductance amplifier, for instance the neutralized pseudo-differential pair from Sect. 4.3, the voltage V_2 at the input of the next stage must be maximized for a given output current I_1 out of the previous stage. In other words, maximum power transfer is achieved by maximizing the magnitude of the transimpedance Z_{21} of the filter network. Note that while the power gain, i.e., S_{21}, of a passive network is always $\leq$0 dB, it can still have a transimpedance gain >0 dBΩ.

By initially assuming $L_p = L_s = L$, $C_{\text{out}} = C_{\text{in}} = C$, and $R_{\text{out}} = R_{\text{in}} = R$, the transimpedance Z_{21} of this transformer network can be expressed as [81]

$$Z_{21} = \frac{-\omega_0^3\, k\, Q\, R\, s}{[Q\, s^2 + \omega_0\, s + Q(1+k)\, \omega_0^2]\, [Q\, s^2 + \omega_0\, s + Q(1-k)\, \omega_0^2]}\,, \tag{4.52}$$

where the natural resonance frequency ω_0 and quality factor Q of the resonator filter are given by

$$\omega_0 = \frac{1}{\sqrt{L(1-k^2)C}} \tag{4.53}$$

$$Q = \frac{R}{\omega_o L(1-k^2)} = \omega_0 RC\,. \tag{4.54}$$

The denominator of expression 4.52 indeed shows a 4th-order dependence on the frequency that yields two complex pole pairs. Assuming a high Q-factor, these poles correspond to two distinct resonant frequencies, ω_L and ω_H, that can be calculated as [81]

$$\omega_L = \frac{1}{\sqrt{L(1+|k|)C}} \tag{4.55}$$

$$\omega_H = \frac{1}{\sqrt{L(1-|k|)C}}\,. \tag{4.56}$$

To illustrate the transimpedance response of this filter, consider two symmetrical amplifier load situations at 150 GHz, represented by $R \parallel C = 250\,\Omega \| 15\,\text{fF}$ ($Q = 3.5$) and $250\,\Omega \| 7.5\,\text{fF}$ ($Q = 1.75$). Using the derived equations, the required inductance L is calculated from a center frequency of 150 GHz for several coupling coefficients k going from 0.2 to 0.6. Figure 4.27a shows the resulting transimpedance gain for the case where $Q = 3.5$, clearly showing the two predicted poles. As k increases, the distance between the poles grows and so does the bandwidth at the cost of more ripple in the response. However, for the lowest values of k, the peaks are not so clearly visible, and the response looks more like an RLC resonator tuned at a single frequency for $k = 0.2$. This effect is even more clear in Fig. 4.27b where a lower Q-factor results in a flat response even for the highest k values.

It turns out that the kQ product is a key parameter that determines the behavior of the transformer network. If $kQ < 1$, no resonant peaks can be distinguished, and increasing k leads to an increase in transimpedance, as observed in Fig. 4.27b. As kQ approaches a value of 1, the gain reaches its maximum level of $\sqrt{R_1 R_2}/2$ [81] and the bandwidth starts expanding without introducing ripple. Finally, for $kQ > 1$, increasing k further results in an even larger bandwidth, but at the expense of larger in-band ripple as the two resonant peaks start to show, as shown in Fig. 4.27a. Note that in case the input and output loads have different Q-factors, the equivalent Q of the network can be calculated as: $Q = \sqrt{Q_{\text{in}} Q_{\text{out}}}$.

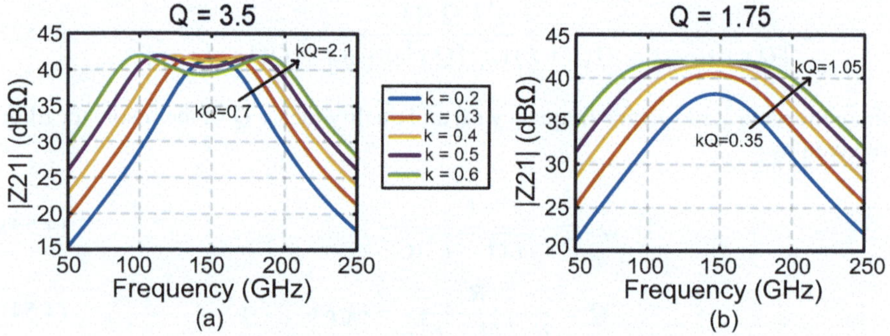

Fig. 4.27 Transimpedance frequency response of the transformer network at a center frequency of 150 GHz for various values of k with Q-factors of 3.5 (**a**) and 1.75 (**b**)

From studying this behavior, several important conclusions can be drawn regarding the design of broadband matching networks. The first conclusion is that low-Q loads are easier to match over a broad bandwidth than high-Q loads. Indeed, by comparing both graphs in Fig. 4.27, the same bandwidth can be achieved with much less ripple if the Q-factor of the load is smaller. This is in direct agreement with the Bode–Fano criterion discussed earlier (Eq. 4.13). Note, however, that the lower Q-factor of the load is a direct result of more resistive loss in the transconductor, meaning that this lower Q is often accompanied with a lower $G_{\max}$. Furthermore, the type of load dictates how to design the transformer. For low-Q loads, the magnetic coupling of the transformer must be maximized to boost the transimpedance and bandwidth of the network. For high-Q loads, a more weakly coupled transformer with a smaller k is desired to avoid significant in-band ripple while obtaining a wide response.

4.4.3 *Realistic Transformer Networks*

While the response for a fully symmetrical and lossless network has given some useful insight, the assumptions do not hold in realistic situations. First of all, in general the capacitive load at the input and output will not be identical, so $C_{\text{in}} \neq C_{\text{out}}$. This unbalanced capacitive load is easily dealt with by scaling the inductances L_p, L_s such that the LC product and therefore the complex poles remain as desired at the input and output of the network. If the Q-factors are still the same, i.e., R is scaled such that the RC product remains unchanged, the transimpedance has the same frequency response apart from a constant scaling factor [81]. However, more realistically, the Q-factors at the input and output are different. The response is then governed by the overall Q-factor $Q = \sqrt{Q_{\text{in}} Q_{\text{out}}}$. Of course, capacitance can always be added on either side to make the load more balanced. For instance, while interstage matching networks interface two relatively similar impedances arising

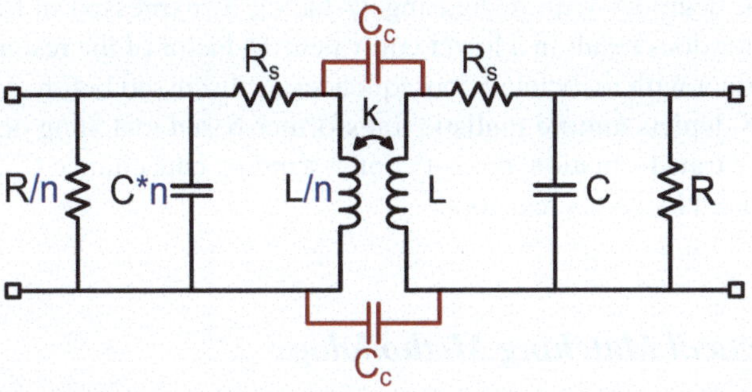

Fig. 4.28 Schematic of a more realistic transformer-based matching network including impedance transformation, parasitic inter-winding capacitance and inductor losses

from amplifier stages, sometimes one of the loads is almost purely real, for example, when matching to a 50 Ω source or load, or even inductive (caused by interconnect parasitics). In that case, a larger capacitance can simply be added to attain the desired resonance. This comes, however, at the cost of a larger network loss as the voltage over that added capacitance does not contribute to a larger voltage swing over the C_{in} of the next stage.

Secondly, most matching networks must perform an impedance transformation, i.e., $R_{in} \neq R_{out}$. Similar to an unbalanced capacitive load, the values of C and L must be scaled to maintain the same Q-factor and resonance frequency. For instance, if an impedance upconversion is required, i.e., $R_{in} < R_{out} = nR_{in}$, L_1 must be scaled by $1/n$ and C_1 by n, as illustrated in Fig. 4.28. Again, the frequency response remains identical, but a transformation of $1/n$ costs a factor of $1/\sqrt{n}$ in Z_{21}.

Next, as discussed in Sect. 3.3, transformers suffer from intra-winding capacitance within each coil as well as inter-winding capacitance C_c between the inductors, as depicted in Fig. 4.28. While the intra-winding capacitance can be absorbed in the filter load, the inter-winding capacitance C_c cannot and will thus affect the network behavior. In differential-mode, a nonzero C_c causes the lower pole to drop in frequency (for inverting transformers, i.e., $k < 0$, which is usually the case), leading to an increase in bandwidth but with larger ripple. In common-mode, the capacitance C_c provides a low-impedance path for the CM current to flow between the coils, which can cause imbalance issues if the input or output signal is not perfectly balanced or differential. When designing a transformer as a balun (which performs balanced-to-unbalanced signal conversion), CM coupling through this parasitic capacitance is especially problematic. Moreover, the added capacitance lowers the SRF of the transformer hence limiting the highest usable frequency.

Finally, as also mentioned in Sect. 3.3, physical inductors exhibit loss, which can be represented by a series resistance (that is generally frequency dependent). While inductors in at mm-Wave frequencies with acceptable Q-factors (>15) do not significantly affect the frequency response of the transformer, the insertion loss of the network degrades with decreasing Q-factor, as expressed in Eq. 3.123. The limited Q-factor does result in a lower equivalent Q-factor of the resonator and thus lower kQ product with associated consequences as discussed before.

Figure 4.28 depicts a more realistic transformer-based matching network with a $1/n$ impedance transformation, parasitic inter-winding capacitance C_c, and inductor losses represented by series resistors.

4.4.4 Practical Matching Methodology

The behavior of a 4th-order transformer network is now well understood as is the effect of various changes, desired or undesired, to the ideal network. In this section, a practical design methodology is discussed for realistic circuits. First, it is important to note that the equivalent representation of the input (source) and output (load) impedance is not accurate over a large range of frequencies. From inspecting expressions 4.11 and 4.12, the Z_{out} of a typical amplifier can be well modeled by a parallel RC network, but the Z_{in} resembles a series RC network. When designing broadband matching networks above 100 GHz, the variation of both impedances is significant over the band and must be well modeled to avoid inaccurate results. Furthermore, it is important to separate the source and load impedance from the filter components. For instance, if extra capacitance is added to manipulate one of the resonances in the network, it might improve the response but will also lead to some additional insertion loss. This extra loss is not modeled in the calculation of Z_{21} because all of the admittances at the source and load side are merged.

Therefore, an improved design methodology must focus on maximizing S_{21} as opposed to Z_{21}, as illustrated in Fig. 4.29. The source and load ports are characterized by the simulated output and input impedances over frequency from the input and output stages, respectively. The transformer-based matching network, including optional added capacitance elements at the input C_i or output C_o, is completely separated and can be easily characterized by a set of Z-parameters. Based on this schematic, the S_{21} of the network can be calculated as [97]

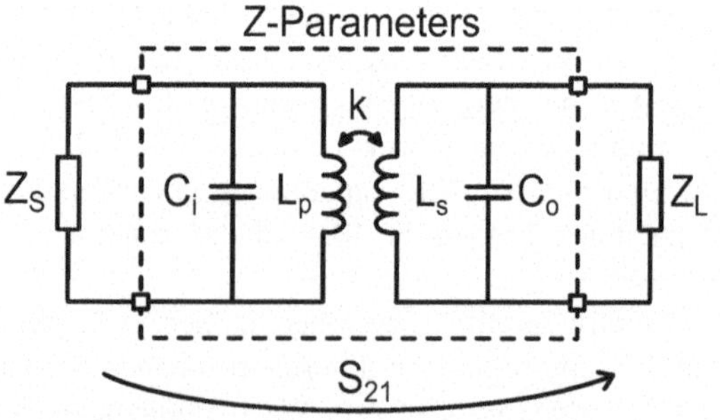

Fig. 4.29 Schematic for the design of a practical matching network

$$S_{21} = \frac{2Z_{21}\sqrt{R_S R_L}}{(Z_{11}+Z_S)(Z_{22}+Z_L) - Z_{12}Z_{21}}\,, \tag{4.57}$$

where $Z_S = Z_{out}$ and $Z_L = Z_{in}$. The S_{21} of the network quantifies exactly how much power from the source will be delivered to the load, thus measuring the actual insertion loss, in contrast to Z_{21} which describes the total voltage over all elements at the input of the next stage. Nevertheless, analyzing Z_{21} with parallel RC loads is still useful to gather theoretical insight. Even though S_{21} is the more interesting metric in this context, its expression is nonintuitive and is best calculated using circuit simulators.

When maximizing the magnitude of S_{21}, or equivalently minimizing the insertion loss, of the transformer matching network, one must make the distinction between mismatch loss and transformer loss. The mismatch loss denotes how much power is lost due to imperfect conjugate match, i.e., when the generalized $S_{11} \neq 0$ and $S_{22} \neq 0$. The transformer loss, on the other hand, is quantified by Eq. 3.123 and becomes larger with decreasing Q-factor or coupling factor k. By solely focusing on achieving a perfect conjugate match, a sub-optimal result might be reached due to an excessive transformer loss. Designing the optimal matching network thus entails finding a perfect balance between these two factors.

In practice, the network parameters can first be estimated based on the derived equations and approximated RC loads. Next, all parameters are be fine-tuned using a simulator tool to optimize S_{21} as in the setup from Fig. 4.29 with exact values of Z_S and Z_L over frequency. The optimization is performed with ideal (but lossy) inductors and capacitors, which are then all implemented and EM simulated. In case the performance of the implemented components differs from the ideal modeled elements, particularly the transformer, the network can be slightly retuned. To avoid many time-intensive design iterations, one must take the implementation limitations into account while optimizing all elements. For instance, the maximal attainable coupling factor in a transformer decreases as the diameters of the primary and secondary coils diverge. Also, for large inductance or capacitance values, the SRF of the transformer or capacitors might approach the band of interest, hence significantly influencing the frequency response. As shown in Fig. 3.34a, the effective inductance sharply rises close to the SRF, which makes it significantly more challenging to obtain a flat, broadband characteristic. As long as the optimized transformer has an attainable coupling coefficient with realistic inductances, the ideal and implemented transformer should have a very similar behavior as the more ideal model.

4.4.5 Transformer Implementation

Maximizing the performance of the transformer is obviously one of the big keys to success. After inspecting the input and output impedances of an NPDP at 140 GHz in Table 4.1, the loaded Q-factors all lie within a range of 2–3. Since the coupling

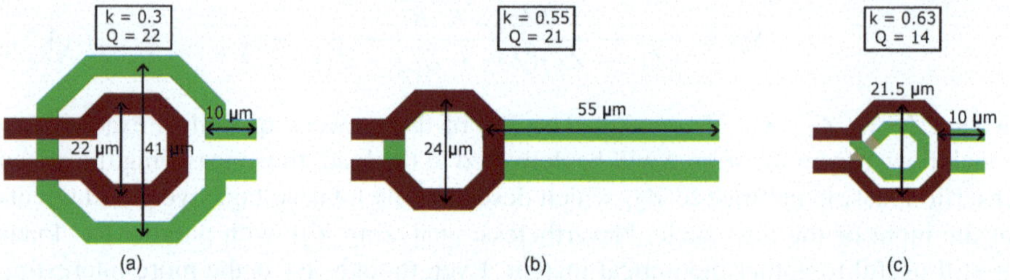

Fig. 4.30 Three different stacked transformer implementations with $L_p = 40\,\text{pH}$ and $L_s = 80\,\text{pH}$: one-to-one transformer with scaled diameters (**a**), one-to-one transformer with same diameters and additional leakage inductance (**b**), and one-to-two transformer (**c**)

factor of a transformer at that frequency is limited to around 0.6 and inductors can obtain Q-factors around 15–20, most of the designed matching networks will yield a response similar to Fig. 4.27b. The goal when implementing the transformers for these networks will thus be to maximize the coupling factor to achieve minimal insertion loss and maximal bandwidth.

Obtaining a high-k transformer at these frequencies is no easy feat, especially when a large difference between the primary L_p and secondary L_s inductance is required to perform an impedance transformation. Instead of acquiring the desired L_p/L_s simply by scaling the diameter of the windings, which results in very weakly coupled coils and hence low k coefficients, the diameters of both coils are made identical and as large as possible for maximum coupling, while the difference in inductance is obtained by adding additional leakage inductance in series where necessary. Figure 4.30a and b show both transformer types with $L_p = 40\,\text{pH}$ and $L_s = 80\,\text{pH}$, demonstrating that a coupling factor of 0.55 can be achieved using this strategy of maximal overlap as opposed to only 0.3 in the classic design. The leakage inductance can also be curled into a separate coil to save area compared to just a straight line [98].

Alternatively, the larger secondary inductance can be implemented by means of a two-turn inductor, shown in Fig. 4.30c, which drastically increases inductance for the same area and thus overlap. Indeed, an even higher k of 0.63 can be obtained this way, but at the cost of a smaller Q-factor caused by the lower SRF and thinner traces. Due to the high mutual inductance of the double turn, the inductor must be scaled down considerably, sometimes resulting in thinner traces, to achieve the relatively low inductances that are typically needed above 100 GHz. Using a two-turn inductor thus increases the intra-winding capacitance (lowering the SRF) and often needs thinner traces and additional vias (decreasing the Q-factor of the inductor). A two-turn transformer is therefore mostly used for very large impedance transformations with high necessary inductance values, when the desired coupling cannot be obtained using a one-to-one transformer with leakage inductors.

Finally, one of the big advantages of using transformers is the convenience of supplying DC signals, such as a bias voltage or supply voltage. As shown in Fig. 4.31, by connecting the DC signals through the center tap of the primary or

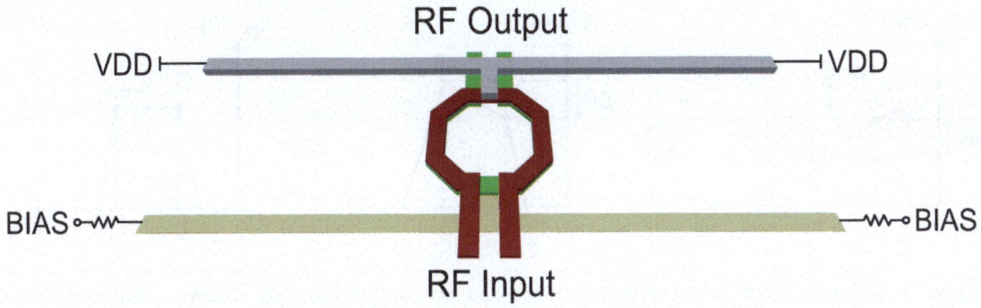

Fig. 4.31 A stacked one-to-one transformer providing DC supply and bias connections through the center taps of two windings

secondary winding, the differential-mode characteristic is unaffected, as the center tap behaves as a virtual ground in differential-mode, while being a convenient access point in common-mode. As discussed in Sect. 4.3, the gate bias is connected through large resistors to ensure CM stability. Notice that this center tap connection is not as symmetrical for a two-turn inductor, thus slightly complicating the routing compared to a single-turn winding.

4.4.6 Broadband Multistage Amplifier Design

Finally, several additional considerations should be borne in mind when designing a multistage amplifier. First and foremost, while the design methodology from Fig. 4.29 takes the actual impedance frequency characteristic into account, the gain response of the amplifying block is completely ignored until now. Yet, as demonstrated in Eq. 3.100 and shown in Fig. 4.20a, the maximum power gain is inversely proportional to the square of the frequency, resulting in a 6 dB gain per octave roll-off. While this roll-off is fundamentally caused by the RC networks at the input and output of a CMOS transistor, it is not compensated for by simply ensuring a conjugate match for these impedances. After all, if the amplifier is conjugately matched at all frequencies, its gain response is identical to G_{max}, including the gain roll-off. To compensate this roll-off, the matching network should produce a positive 6 dB/octave slope, resulting in a flat response when combined, as illustrated in Fig. 4.32. This brings about an important consequence for broadband amplifiers at higher frequencies. Namely, if a flat response is required, the maximum gain per stage is limited by the gain at the upper frequency of the band and the minimum insertion loss of the matching network at that higher frequency. Even if more gain is available at lower frequencies, that gain is sacrificed in the matching network to ensure a flat frequency characteristic.

Next, when several tuned amplifier stages are cascaded, the frequency responses of all stages are combined, which not only has a consequence for the gain, but

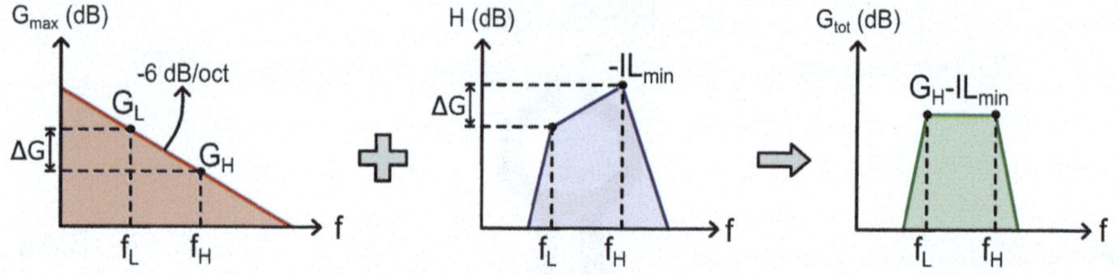

Fig. 4.32 Illustration of how the gain roll-off can be compensated to achieve a flat combined response

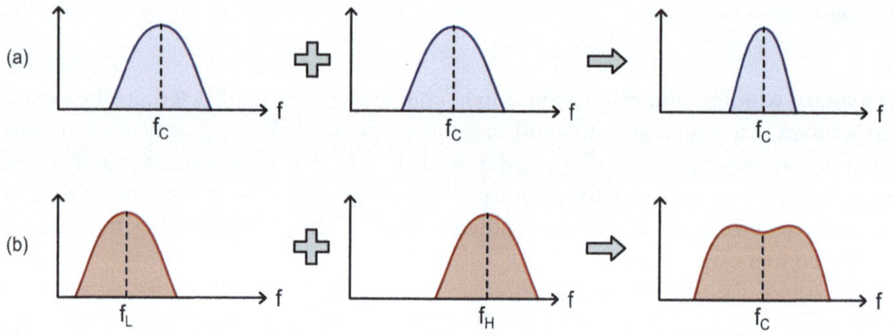

Fig. 4.33 Illustration of how cascading identical bandwidth-limited responses leads to bandwidth reduction (**a**) and how stagger tuning can increase global bandwidth at the cost of lower gain (**b**)

also the bandwidth of the overall system. For frequencies where the individual responses are exactly flat, the combined response will also be flat. However, in reality there is some (acceptable) drop-off near the edges of the band, which is why the bandwidth is often defined where the response drops 3 dB relative to its maximum. When multiple stages are cascaded, the relative drop-off accumulates, and hence the response of the overall system narrows, as illustrated in Fig. 4.33a. If N stages are all tuned at the same frequency, the total 3-dB bandwidth reduces according to Eq. 2.92. However, by tuning the stages at different frequencies, called stagger tuning, the overall bandwidth can be enhanced, as shown in Fig. 4.33b. However, the increased bandwidth caused by stagger tuning comes at a cost of lower total gain compared to single-frequency tuning. Moreover, the spacing between the distinct tuning frequencies determines the trade-off between bandwidth, ripple, and gain at the center frequency. Ideally, the spacing is chosen such that maximum bandwidth is obtained without any in-band ripple. It must be mentioned that stagger tuning makes the system particularly more sensitive to process variations, as slight shifts in the individual center frequencies have much more degrading consequences (in terms of bandwidth and ripple) on the overall response compared to single-frequency tuning.

When designing a multistage amplifier, it is hence important to include the combined frequency responses of all amplifying and matching network blocks. The overall S_{21} response can then be optimized by co-designing the various blocks, using techniques such as stagger tuning and explicitly compensating the frequency-dependent gain roll-off of the transistors. After having optimized each matching network element, they should be implemented step by step. Each time an implemented component is included in the simulation, and the other non-implemented elements can be re-optimized to compensate for any non-idealities the newly implemented component has brought. A good strategy is to start with the most difficult components (i.e., transformers which require a high k and winding diameter ratio) and finish with the components that should behave very close to its ideal model. For complex systems, this implementation and optimization procedure should be started from good initial values based on theory and performed with the relevant design trade-offs in mind, as opposed to mindlessly optimizing from scratch which will often lead to a sub-optimal result with a lot of wasted simulation time.

4.5 Overview of Amplifier-Based Circuit Blocks

Finally, to end the section on the basics of high-frequency circuit design, various amplifier-based circuit blocks with different specifications will be examined. For each block, the main design methodology is briefly discussed using the neutralized pseudo-differential pair and transformer-based matching network as primary building blocks.

4.5.1 Designing for Gain: Small-Signal Amplifiers

Small-signal amplifiers function as a general amplification block where power gain is of primary concern. Furthermore, the gain should be attained over a specified bandwidth while minimizing power consumption.

As noise is not important and only small-signal levels are expected, the amplifier can be optimized to deliver maximum small-signal power gain. The power gain per stage is determined by the maximum available power gain of the amplifying core (G_{max}) and the insertion loss of the matching network. The transistor size must therefore not only be chosen to maximize G_{max}, but as importantly to ensure low matching insertion loss. In general, the minimal achievable matching network loss becomes larger as the Q-factor of the impedances ($Q = \omega_0 R_p C_p$) increases and as the impedance transformation ($R_{\text{out}}/R_{\text{in}}$ or $R_{\text{in}}/R_{\text{out}}$) increases. For example, taking Table 4.1 as a reference, a transistor width of 25 μm might be a good option. After all, 20 μm width transistors have a higher Q input impedance, while 30 μm wide transistors have a similar Q-factor at their input and output but consume more current for the same G_{max}. The biasing is usually chosen for maximal G_{max} but can be decreased slightly to lower current consumption with minimal gain degradation.

4.5.2 Designing for Power: Power Amplifiers

Power amplifiers (PAs) are amplifiers specifically designed to deliver a large output power to a load and are therefore common as the final block in a transmitter. Additionally, as power generation has become more challenging at higher frequencies, LO generation above 100 GHz also requires dedicated power amplifiers to generate LO signals with sufficient power to drive subsequent blocks. Obviously, output power is the main specification of a PA. Depending on the type of signal to be amplified, different linearity requirements are associated with that power specification, as discussed in Sect. 2.3.3. For LO drivers, the LO signal is a continuous wave (CW) signal with a PAPR of 0 dB and is thus insensitive to nonlinearity. In that case, the maximum (saturated) output power must be optimized with no concern for linearity. For modulated signals, however, the linearity specifications become more stringent as the PAPR of the signal increases. For these signals, the output power at a certain power back-off is the metric to be optimized, for instance, the output power at 1-dB compression ($OP_{1\,\text{dB}}$).

Next to power, a minimal amount of gain is of course necessary since the large output power must be generated from a lower power input signal. At high frequencies, the available power gain is so low, especially when the amplifier starts to compress, that the PA usually consists of one or more driver stages that drive the main output stage. That way, the final stage can be fully optimized for output power instead of also having to deliver the required gain at the same time.

Furthermore, since PAs generate the largest signals, they typically dominate the power consumption in the whole transceiver. As such, the power efficiency is another crucial metric that needs to be optimized, which is preferably quantified using the PAE, as discussed before. In general, the efficiency will be maximal at saturation and decreases quickly as the PA operates at back-off. This indicates that PAs amplifying modulated signals will have much worse efficiencies than CW PAs. Improving the efficiency at back-off is a popular research topic that has inspired to invention of many more complex PA architectures.

Finally, bandwidth appears again as an important metric when working with modulated signals. As CW LO signals occupy no meaningful bandwidth, only little bandwidth is needed in the PA to support some tuning in the LO frequency. For modulated signal PAs, requiring a large bandwidth is yet another reason why their efficiency will generally be lower than CW signal PAs. After all, broadband matching networks typically have a high insertion loss than narrowband networks, as discussed in the previous section.

The design of PAs can be very intricate, as small improvements in output power and efficiency have rather large improvements on the performance of the overall system. While some general PA design aspects will be highlighted here, a more in-depth investigation into the design of PAs at mm-Wave frequencies can be found in previous works [78, 95, 98].

In order to maximize the power delivered to the load, the output stage must be able to generate maximal output power and simultaneously ensure minimal insertion loss in the output matching network. As discussed in Sect. 4.1, for maximal generation a large-signal power match is required as opposed to a small-signal conjugate match. However, at frequencies above 100 GHz, the ideal load impedance for maximal output power in a CMOS PA is very similar to the complex conjugate of its output impedance. This is mainly caused by the decrease in output resistance at higher frequencies due to the output parasitics [98]. So compared to lower mm-Wave frequencies, a conjugate match at the PA output leads to minimal degradation in output power. However, while the lower output resistance, corresponding to a more lossy and thus lower Q output impedance, is beneficial for output matching, it limits the obtainable output power and efficiency of PAs at such high frequencies. Nevertheless, next to maximizing the transistor output power, its output resistance ideally lies close to the load resistance to ensure low output matching insertion loss. From looking at Table 4.1, the largest width of 55 μm seems ideal, as it provides the highest $OP_{1\,\mathrm{dB}}$ with decent $G_{\max}$, while having an output resistance R_{out} relatively close to a 50 Ω load. For LO power amplifiers, the load might be the input of a mixer or modulator, and so a different size might provide a better Z_{out} for output matching. Even larger transistor sizes, though not included in the table, lead to only marginal improvements in output power with a significant degradation in power gain, due to considerable interconnect parasitics. Furthermore, the larger output capacitance will require a smaller inductance in the output matching network, which increases the insertion loss of the transformer (due to smaller magnetic coupling or thinner traces).

It must be noted that the gain reported as $G_{\max}$ in Table 4.1 is the available small-signal gain. The actual gain will be smaller as the amplifier stage operates more in compression. Taking compression into account (e.g., 1 dB compression), interconnect losses as well as matching network insertion loss, the power gain of the output stage will be smaller than 5 dB. Due to this low gain, the driver stage must not only provide high gain, but also moderately large output power. It is thus good practice to scale up the different stages in the whole power amplifier to gradually build up the signal power with maximal gain. To ensure the output stage receives enough input power to reach the desired compressed output power, the previous driver stage must not compress before the output stage. Concretely, if the output stage needs about 3 dBm of input power to reach 1 dB compression, the driver stage better be able to deliver 4–5 dBm including the matching losses with minimal compression. A driver with 40 μm transistor width will do the trick with some healthy margin. Extending this line of reasoning to the stage driving the driver stage, the amplifier can be further scaled down to save power. Designing a power amplifier chain thus entails a trade-off between output power (for fixed linearity specifications), robustness (i.e., power margin between stages), and power consumption. Being able to accurately predict the performance of each stage will therefore limit the needless over-design in power and efficiency for the whole PA.

Finally, biasing the stages above 100 GHz is quite straightforward. As discussed in Sect. 4.3, very low bias voltages yield insufficient gain, while a large bias results in a huge power consumption without any added benefit in gain. As summarized in Fig. 4.25, a bias voltage around 400–500 mV is ideal in this technology to provide adequate gain, close to optimal efficiency, and acceptable AM–AM and AM–PM linearity performance.

4.5.3 Designing for Noise: Low-Noise Amplifiers

A low-noise amplifier (LNA) is a crucial block at the input of a receiver to ensure a minimal receiver NF. As such, a key specification is the NF of the LNA, as this has the largest contribution to the overall receiver NF. However, as demonstrated in Eq. 2.78, the gain of the LNA must be sufficiently high in order to suppress the noise generated by subsequent blocks. Next, the specified characteristics must be attained over a specified bandwidth since the LNA is amplifying modulated signal. Finally, all of the above must be optimized while consuming the least amount of DC power.

At lower RF frequencies, linearity (e.g., input-referred $P_{1\,\mathrm{dB}}$) is another vital specification in order to support a large range of input powers and to be resilient against large interference signals. At frequencies above 100 GHz, however, the incoming signal will be quite limited in amplitude due to the large FSPL, as will any interfering signals be due to the directive nature of propagation at these frequencies. Linearity will therefore not be of particular concern in this work.

As exemplified by Eq. 4.10, the noise figure of a single amplifying stage is determined by the inherent minimal noise figure ($\mathrm{NF_{min}}$) as well as how much the provided source impedance differs from the optimal impedance for minimal noise. From Table 4.1, it is clear that $\mathrm{NF_{min}}$ practically identical between the stages and also does not vary significantly with bias. Therefore, minimizing the noise comes down to optimizing the noise match while simultaneously ensuring a conjugate input match for maximal power transfer. After all, the insertion loss at the input of the LNA is directly added to the total NF of the receiver. To this end, Table 4.1 also shows the complex conjugate of the optimal noise impedance and the input impedance of NPDPs with various sizes. The optimal size can be determined by looking at a high similarity between the aforementioned impedances, while giving a preference to small sizes to keep the current consumption to a minimum. An NPDP with a width of 25 μm satisfies our criteria, as a conjugate input match provides an NF of 4.8 dB, which is a degradation of only 0.4 dB compared to $\mathrm{NF_{min}}$.

To determine the optimal number of LNA stages, one must make a compromise between the NF of the whole LNA (more stages give more noise) and the gain of the LNA (more gain gives more suppression of subsequent noise sources). Using Eq. 2.78, a system-level analysis can reveal how many stages will lead to the lowest overall NF. Again, power consumption and bandwidth must be kept in mind, as more stages lead to an increase in current consumption and require more matching networks that limit the total bandwidth. Once the number of stages is determined, the design follows the same procedure as for a small-signal amplifier, since power gain is mainly the priority for subsequent stages.

4.5.4 Designing for Nonlinearity: Frequency Multipliers

A special case that will be briefly discussed is the frequency multiplier. Multipliers are mainly used to generate a high-frequency LO signal from a lower frequency reference. Additionally, multipliers can enable signal generation above $f_{\max}$, which is not feasible with regular amplifiers as they have no power gain at those frequencies. Multipliers employ an amplifier-like topology but are designed to maximize the nonlinear behavior of the transistor and hence the generation of harmonics.

As seen in Fig. 3.16, the transistor current follows a nonlinear characteristic as a function of input voltage. This nonlinear behavior can be maximally exploited by adequately biasing the signal depending on the input swing. For small input swings, the signal is best biased around the transition from the subthreshold into the strong-inversion region, so around the threshold voltage V_T. For large input swings, the signal should be biased deep in class A to extract the harmonics arising from the transition between saturation and triode operation.

Next to maximizing the generation of harmonic power, that power should be optimally extracted from the transistors via matching networks. One straightforward way to achieve this is by tuning the network to the desired harmonic frequency, maximizing harmonic power transfer to the load. However, next to optimizing transfer at the harmonic, the network should also provide the optimal load impedance to the multiplier at the fundamental frequency. After all, the multiplier requires a high-impedant load to generate a large output swing and thus harmonic-rich output. This is called harmonic matching, as the impedance is optimized at multiple frequencies. Additionally, other harmonics might indirectly influence the power generated at the desired harmonic [99, 100]. For instance, to maximize the 3rd harmonic output power of a frequency tripler, the output swing at the fundamental must be maximized, while the second harmonic output swing must be minimized. In general, designing a frequency multiplier requires intricate nonlinear circuit analysis and complex matching networks optimized at various harmonics [101]. In Chap. 6, the design of a 390 GHz frequency tripler is more extensively discussed.

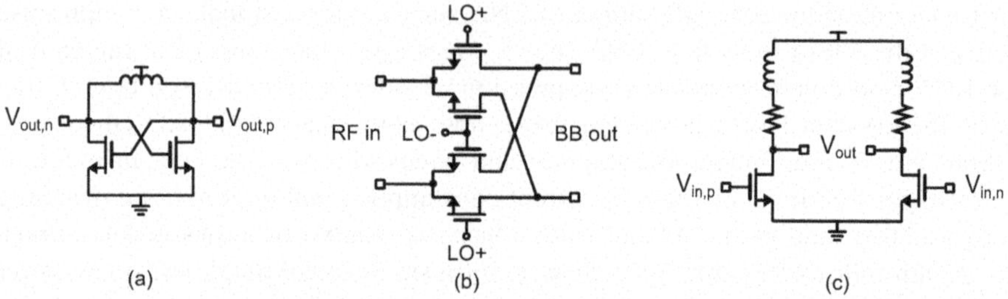

Fig. 4.34 Schematic of a VCO (**a**), passive mixer (**b**), and inductively peaked baseband amplifier (**c**)

4.5.5 Non-amplifier-Based Circuit Blocks

Finally, it must be noted that many other essential circuit blocks exist that do not directly rely on resonant amplifier-based topologies, which were the focus of this section.

A voltage-controlled oscillator (VCO), for example, can be implemented by cross-coupling two MOS transistors with an inductive load, as shown in Fig. 4.34a. Cross-coupling effectively creates a feedback loop that provides a negative resistance ($-2/g_m$). The inductive load resonates with the parasitic capacitance of the transistors, resulting in a low-loss positive feedback loop at the resonance frequency, triggering an oscillation. The oscillating signal can then be extracted and used as a standalone LO frequency or as part of a larger PLL system to generate a more stable LO. When designing a VCO, the transistor size and load inductance are mostly optimized for the desired oscillation resonance frequency and maximal g_m from the transistor pair.

Another example is a passive mixer, depicted in Fig. 4.34b, which downconverts a high-frequency RF input to a baseband output. The transistors are driven by a large swing LO signal to sharply turn on and off, operating in the triode region. Optimizing this mixer follows very different tactics than an amplifier and mainly focuses on finding the best transistor size to function as a low-loss, high-bandwidth switch.

The last example is shown in Fig. 4.34c. Despite employing an inductive load, this amplifier has a baseband characteristic instead of a resonant one. The inductor enables a bandwidth enhancement by compensating the drop of impedance caused by the output capacitance at higher frequencies.

The design of amplifier- and non-amplifier-based circuits is more extensively discussed for practical, concrete examples in Chaps. 6 and 7. Additional details can be found in the following excellent textbooks: [7, 60].

4.6 Conclusion

This chapter focused on the fundamentals of high-frequency circuit design. Starting from the basics of impedance matching, the core characteristics of resonators were discussed, and several fundamental trade-offs in impedance matching were uncovered. After a brief discussion on stability, the neutralized pseudo-differential pair was introduced as one of the most effective amplifying building blocks above 100 GHz. The implementation and design trade-offs of this block were extensively outlined. Then, transformer-based impedance matching was analytically investigated with an in-depth discussion of practical matching scenarios and implementation challenges. Finally, the design of several important amplifier-based circuit blocks was discussed, with the two described building blocks as a foundation.

Chapter 5
System-Level Considerations

The demand for higher data rates is continuously rising as is the necessity for energy-efficient systems. The high complexity of communication systems characterized by many degrees of freedom makes it rather challenging to identify the optimal ways toward improvement. This chapter attempts to reveal the most essential high-level trade-offs in mm-Wave links, using Shannon's capacity equation as a starting point. First, a qualitative discussion of the different system-level choices will be given. Then, the limitations in data rate and energy efficiency will be modeled in a quantitative manner, providing an overview of the primary performance bottlenecks in mm-Wave communication systems.

5.1 Qualitative Discussion on Link Capacity and Efficiency

The quest for higher data rate links has long been guided by the insights provided by Claude Shannon [13] who devised an elegant equation describing the capacity C (bit/s) of a channel given a certain bandwidth BW (Hz) and signal-to-noise ratio SNR. The capacity is the maximum data rate that can theoretically be achieved through that channel and was derived based on insights in information theory, as discussed in Sect. 2.1.1. Extending his expression to N individual channels yields the following equation for the capacity of an N-channel system:

$$C = N \cdot \mathrm{BW} \cdot \log_2(1 + \mathrm{SNR}), \tag{5.1}$$

which will be the starting point of the discussion in this chapter. The equation reveals three primary degrees of freedom to increase the capacity of a system: the number of channels N, BW, and SNR, as visualized in Fig. 5.1. The available SNR determines

C. D'heer, P. Reynaert, *THz and Sub-THz CMOS Electronics for High-Speed Telecommunication*, Analog Circuits and Signal Processing,
https://doi.org/10.1007/978-3-031-64439-9_5

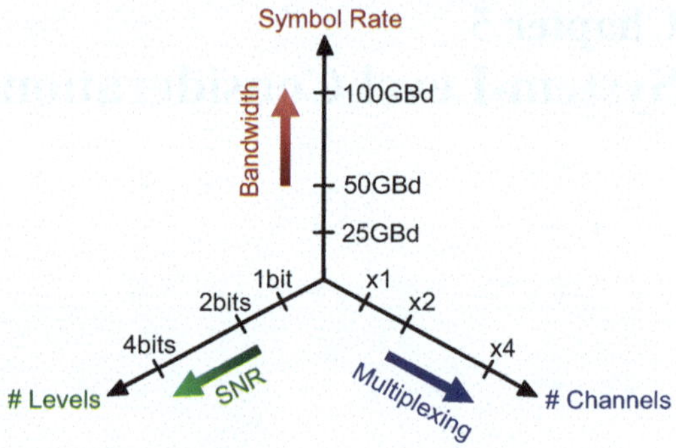

Fig. 5.1 Illustrating the three degrees of freedom to increase the capacity of a communication system

which spectral efficiency can be supported with acceptable BER performance. The absolute BW is mainly related to the carrier frequency of the link and sets an upper limit on the symbol rate. And lastly, the number of channels can be expanded by multiplexing in several dimensions, such as frequency, or polarization.

These three factors directly determine the maximal data rate that can be achieved, as well as the energy efficiency of the system. All three elements will now be discussed more in depth and assessed in the broad scheme of communication system design.

5.1.1 SNR and Modulation Scheme

A first way to boost data rate is by increasing the SNR, or rather, by increasing the available SNR and more effectively exploiting it by using higher order modulation schemes. This is a popular strategy at lower frequencies (<10 GHz) where SNR is relatively abundantly available, at least for reasonably short link distances. For instance, the latest IEEE 802.11be Wi-Fi 7 standard at 6 GHz is planning to implement modulations as dense as 4096-QAM [102].

However, further increasing the link SNR, for the same distance, proves troublesome due to several reasons. First of all, improvements in transmitter output power are stagnating due to technology limitations and are bounded by health regulations. The noise figure of receivers also hit a physical limit making grand reductions in noise infeasible. Moreover, pursuing innovations in output power or noise figure close to the technological limits comes at a large cost of power consumption, trading off link performance for efficiency.

Furthermore, next to providing the necessary SNR value, the associated linearity performance is also essential when increasing the modulation density. Indeed, as the modulation becomes more dense and thus more sensitive to system nonidealities, the linearity specifications become more stringent, resulting in an additional growing penalty in SNR. This trade-off between spectral efficiency and SNR/linearity requirements was extensively discussed in Sect. 2.3 and summarized in Table 2.4.

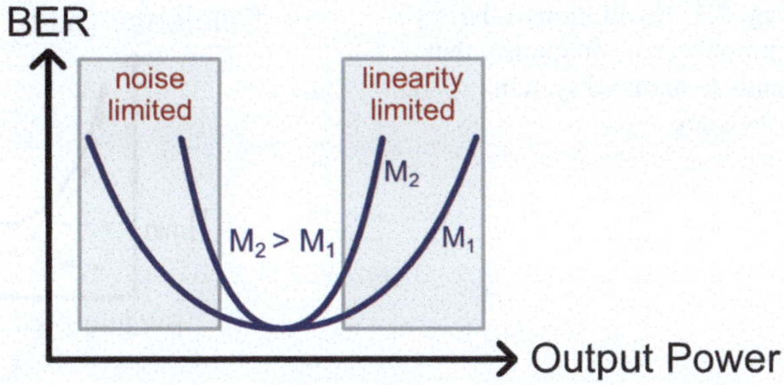

Fig. 5.2 Visualization of the dynamic range of the signal bounded by linearity and noise, showing a more restricted range as the modulation order increases

Figure 5.2 further illustrates how the dynamic range becomes smaller for denser modulations. As the modulation order increases, the maximum output power is more constrained by linearity, while the minimum output power is higher due the larger SNR requirement for the same BER.

Lastly, all of the aforementioned challenges (SNR and linearity) become considerably more extreme as the operation frequency increases. Not only does the raw SNR quickly degrade with increasing frequency, because of the free-space path loss and circuit performance, the required linearity conditions are more difficult to meet. Therefore, the achievable spectral efficiency steadily decreases as the system moves up in frequency.

However, even though the maximum achievable SNR seems quite fixed by the link distance, frequency, and technology, a popularly emerging technique has reignited the quest for higher SNR: phased arrays. Indeed, by increasing the number of antenna cells, the link budget can be significantly improved at the cost of extra area and efficiency. Due to a decrease in antenna size at higher frequencies, compact large-element high-frequency arrays become feasible. Of course, phased-array implementations come with their challenges and limitations, as discussed later in this chapter. Nevertheless, phased arrays can significantly boost the otherwise limited SNR at higher frequencies, enabling more efficient exploitation of the large available bandwidths at very high frequencies.

5.1.2 Bandwidth and Carrier Frequency

Discounting phased arrays, improving the SNR in single-cell systems is becoming quite difficult and leads to diminishing improvements in capacity due to their logarithmic relation. Therefore, a more effective capacity improvement can be attained by focusing on increasing the bandwidth. This search for higher bandwidths has led research to explore higher frequencies in search of large bandwidth channels. Especially the sub-THz frequency band (100–300 GHz) has received considerable interest in recent publications [103–108]. Note that increasing the bandwidth generally comes with an SNR penalty. After all, the integrated noise power increases as the signal band becomes larger. So even when increasing the

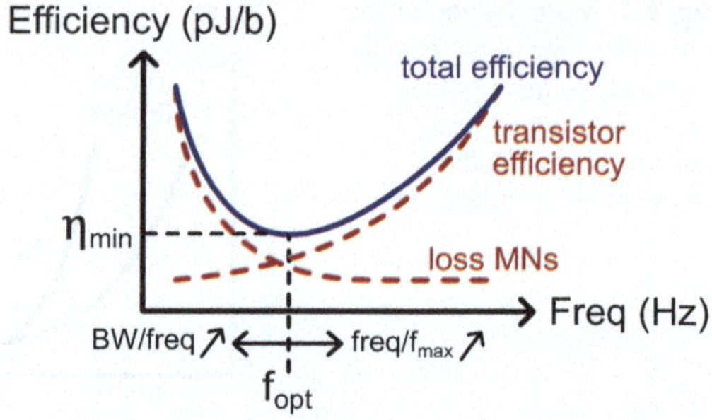

Fig. 5.3 Illustration of the optimal carrier frequency that leads to minimal system efficiency

bandwidth, the improvements in capacity eventually diminish due to a decrease in SNR. Nevertheless, improving the bandwidth still is a much more effective way to attain ultra-high data rates rather than just focusing on SNR increments.

However, the improvements in absolute BW at higher frequencies come at a steep cost of lower SNR and energy efficiency as the frequency approaches the limitations of the technology (e.g., f_{max}). But, at the same time, the same absolute bandwidth can be obtained with more efficient, narrowband circuits as the fractional BW lowers at higher carriers. Indeed, as discussed in Sect. 4.1.6, broadband matching inevitably requires more lossy, low-Q networks. For a specified absolute BW, an optimal frequency exists that leads to an optimal system efficiency, as illustrated conceptually by Fig. 5.3. The loss of matching networks increases with fractional BW (BW/freq), while the efficiency of transistors degrades with freq/f_{max}, leading to an optimal frequency that depends on the required data rate and link distance. While low data rates are more efficiently realized at lower microwave frequencies, higher data rates above tens of Gb/s require a carrier in the mm-Wave range to avoid excessively large fractional BWs. As a general rule of thumb, it is a good idea to operate below $f_{max}/2$ to ensure decent transistor performance and system efficiency. Furthermore, longer link distances will favor smaller frequencies due to their more favorable link budget, as long as the bandwidth specification can be met.

5.1.3 Number of Channels and Multiplexing

Finally, the channel can be split up into various sub-channels and multiplexed together, with various potential benefits as a consequence. A first obvious benefit is the added throughput, as the data rates of the various sub-channels are combined. For example, by operating at two orthogonal polarizations, two data streams can be independently propagated at the same frequency without interfering. However, dedicated dual-polarization antennas are required to combine (TX) and separate (RX) the two polarizations, which can introduce cross-polarization coupling and distort the data streams [109].

A second reason to split up the channel in several sub-channels is to reduce the distortion and improve efficiency. For instance, by splitting a large frequency band into two separate bands, the bandwidth of each individual channel is reduced by half [30, 110]. This can lead to a flatter frequency response with less ISI as well as more efficient (narrowband) circuit design in each band. Furthermore, the effects of dispersion become more pronounced with increasing signal BW, as will be discussed further. Combining two bands with smaller bandwidths will thus enable higher data rates than one single band with double the bandwidth, if the performance is limited by dispersion.

Combining and separating different channels, either in frequency or polarization, is not trivial, especially at high data rates. The assessment of the advantages and disadvantages of multiplexing multiple channels thus requires a careful evaluation of each specific case, as the potential benefits need to be weighed against the accompanying challenges on an individual basis. For instance, splitting an existing band in half simplifies each individual channel but roughly doubles the amount of circuit blocks. Such trade-offs will not be investigated in this chapter as the main focus will be on single-channel links. However, most of the conclusions will also apply to multichannel systems of course.

5.1.4 System-Level Design

The discussed system-level design choices highly depend on the specifications that are given. The most important specifications that will be considered are: data rate, link distance, energy efficiency, and type of channel (wireless or guided). Additional secondary specifications may include: cost, footprint, latency, BER... In many cases, the specification is not a fixed number, but rather a metric that should be optimized. For instance, a better efficiency for the same data rate and distance is of course always desirable. Obviously, depending on what metric is prioritized, the system design will differ.

While optimizing the link capacity is an interesting and useful enterprise, it will most certainly not necessarily lead to the most efficient system. Indeed, as conceptually visualized in Fig. 5.4, initial improvements in data rate come with negligible increases in power consumption and hence lead to a lower energy

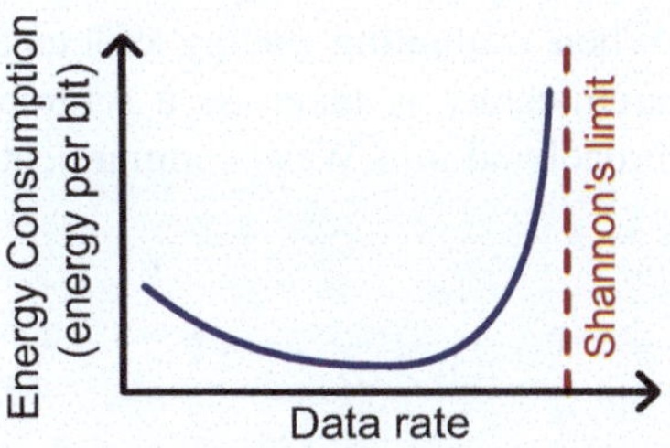

Fig. 5.4 Conceptual energy consumption versus system performance

consumption. However, an excessive amount of power needs to be devoted in order to achieve data rates close to the Shannon limit, with an increase in energy consumption (poorer efficiency) as a consequence. Shannon's capacity is, after all, a theoretical limit that needs complicated coding, advanced modulation schemes, and ideal circuits to be obtained. Since perfect circuits do not exist, power consuming calibration and compensation mechanisms will need to be employed, such as equalization, pulse shaping, predistortion, etc. Data rate must therefore always be co-optimized with energy efficiency.

5.1.5 A Note on Architectures

When the important system-level decisions are made (frequency, modulation, channels) based on the necessary specifications (distance, data rate, BER…), the next step is to determine the optimal architecture to implement the system. The architecture details which circuit blocks are required to perform the necessary operations, such as amplification, modulation, demodulation, LO generation, etc. While the general high-level trade-offs were outlined, more specific architectural and circuit-level details are needed to numerically optimize these choices and estimate the actual energy efficiency. For example, ultra-high bandwidth channels will be significantly easier to realize in a direct-conversion architecture compared to a heterodyne one, as the intermediate frequency (IF) stages will prove to be a significant bandwidth bottleneck. Another example is the influence of the required demodulation method on the efficiency of the system. Simple modulations (OOK, FSK) can be decoded using incoherent methods, which will result in a much more efficient system than coherent QAM demodulators that require IQ LO generation with carrier recovery.

While an in-depth discussion of all possible architectures falls outside the scope of this book, most of them are discussed throughout the book. An overview of the most common modulation and demodulation architectures for the different modulation schemes was detailed in Chap. 2. High-speed transmitter architectures above f_{max} are outlined in Chap. 6. Finally, an extensive discussion of heterodyne and homodyne architectures is given in Chap. 7 with a comparison to the newly introduced direct-digital system. More details on transceiver architectures can be found in [7, 60, 111]. For the rest of this chapter, the architecture will be kept very general as it usually is not important for the high-level performance. When evaluating energy efficiency, however, a more concrete direct-conversion architecture is taken as a reference since it is the most suitable framework for broadband mm-Wave communication.

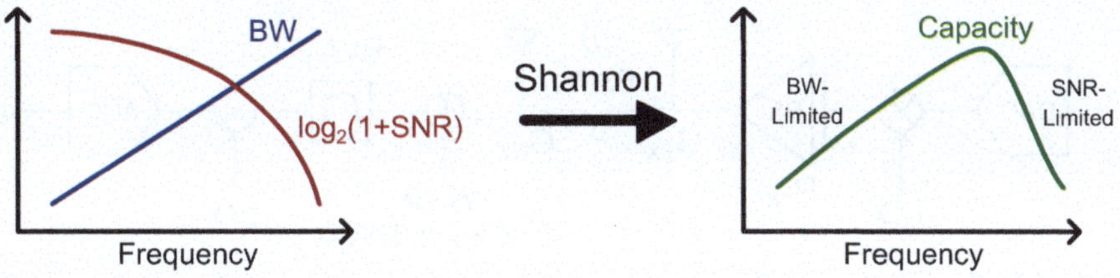

Fig. 5.5 Visual representation of Shannon's capacity equation over frequency

5.1.6 Let Us Get Numerical!

Choosing the optimal carrier frequency and modulation scheme is a multidimensional problem, in the end mainly coming down to a trade-off between system bandwidth and signal-to-noise ratio. At lower frequencies, the dynamic range of the system enables the use of very dense modulation schemes, but the achievable data rate is limited by the available channel bandwidth. At higher frequencies, a lower SNR limits the density of the modulation, but large absolute bandwidths can be exploited to obtain very high speeds despite the moderate spectral efficiency. Figure 5.5 qualitatively shows the link capacity versus frequency, showing a distinction between a BW-limited and SNR-limited regime for low and high frequencies, respectively. In terms of energy efficiency, higher frequencies enable lower fractional bandwidths at the cost of lower transistor efficiencies, implying the existence of a global minimum, as was visualized in Fig. 5.3.

In order to quantitatively determine the optimal frequency and modulation scheme, and provide insight into the fundamental limitations of high-speed systems, the various components of a generic link should be numerically modeled over frequency and plugged into the all-encompassing equation of Mister Shannon. To estimate the energy efficiency, the power consumption of each block will be modeled as a function of the properties of the system. Of course, there are other factors influencing the exact choice of the center frequency, such as regulatory spectrum allocation and atmospheric absorption. These are not considered in this analysis but were briefly discussed earlier in the book in Sect. 1.3.

5.2 Shannon's Link Capacity over Frequency

The first step in our quantitative model is to describe Shannon's equation, for a single-channel system, over frequency:

$$C = \text{BW} \cdot \log_2(1 + \text{SNR}). \tag{5.2}$$

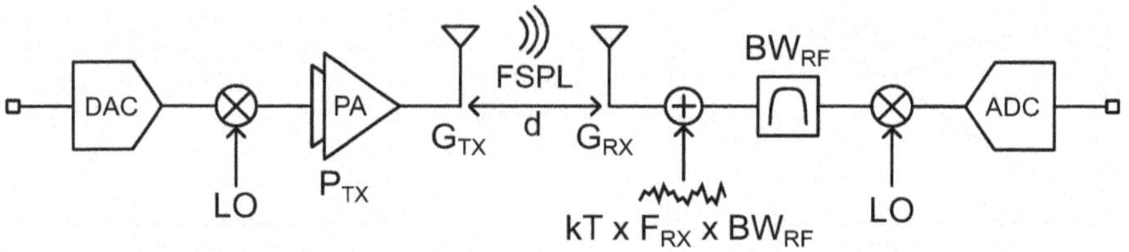

Fig. 5.6 A general framework for a communication system

To achieve this, the two degrees of freedom, BW and SNR, must be characterized over frequency. For this purpose, a framework is needed that models a general communication link.

5.2.1 General Link Model

Our starting point is a general framework to represent any communication system, as shown in Fig. 5.6, consisting of a transmitter (TX), a wireless channel, and a receiver (RX). To determine the link capacity, the BW and SNR of the system need to be modeled over frequency. As discussed in Sect. 2.3.5, the system SNR can be calculated from the link budget, based on Friis' transmission equation:

$$\mathrm{SNR} = \frac{P_{\mathrm{signal}}}{P_{\mathrm{noise}}} = \frac{P_{\mathrm{TX}} \cdot G_{\mathrm{TX}} \cdot G_{\mathrm{RX}} \cdot \frac{1}{\mathrm{FSPL}}}{kT \cdot F_{\mathrm{RX}} \cdot \mathrm{BW}_{\mathrm{RF}}}, \tag{5.3}$$

where P_{TX} is the TX output power, G_{TX},G_{RX} the RX/RX antenna gains, F_{RF} the RX noise figure, and the free-space path loss (FSPL) can be calculated from the link distance d and center frequency f_0:

$$\mathrm{FSPL} = \left(\frac{4\pi d}{\lambda_0}\right)^2 = \left(\frac{4\pi\, d\, f_0}{c}\right)^2. \tag{5.4}$$

The concrete values of each block heavily depend on the adopted technology. As a high-speed communication system highly benefits from large-scale integration and low-cost, high-yield production is crucial for its long-term economic viability, the following analysis will mainly focus on CMOS technologies when determining concrete values. However, the interested reader can follow the analysis along similar lines whilst plugging in the numbers for his/her favorite front-end technology where appropriate.

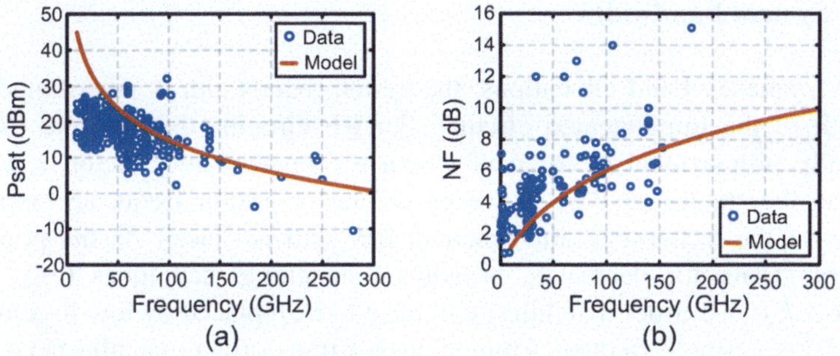

Fig. 5.7 Performance of state-of-the-art mm-Wave CMOS transmitters (output power) [21] (**a**) and receivers (noise figure) [22] (**b**)

5.2.2 *Performance over Frequency*

5.2.2.1 Transmitter Output Power

As initial model, the TX output power is chosen equal to the saturated output power (P_{sat}) of the power amplifier (PA). While G_{max} is inversely proportional to the square of the frequency (Eq. 3.100), the P_{sat} degrades approximately inversely proportional to the frequency cubed:

$$P_{sat} \propto f_0^{-3} \,. \tag{5.5}$$

A reference value of 15 dBm was taken at 100 GHz, inspired by the latest PA survey from ETH [21]. This fitted model shows to be an acceptable representation of the state-of-the-art CMOS transmitters, as shown in Fig. 5.7a. Even at frequencies above f_{max}, this model accurately estimates the maximum achievable output power, despite needing a nonlinear output stage, such as a multiplier.

5.2.2.2 Receiver Noise Figure

The noise figure of the RX is roughly estimated based on data from published low-noise amplifiers (LNAs) collected by ETH [22]. The RX noise figure F_{RX} can be described over frequency as follows:

$$F_{RX} \approx 1 + K \cdot f_0 \,, \tag{5.6}$$

where K is a proportionality constant determined by a reference noise figure of 6 dB at 100 GHz, as based on the LNA survey. Figure 5.7b compares the model with state-of-the-art LNAs from the literature.

5.2.2.3 System Bandwidth

Ignoring frequency band allocations, the system bandwidth is determined by the bandwidth of the implemented circuits. For RF circuits, the absolute (RF) BW is generally proportional to frequency with a proportionality factor α, which is defined as the fractional BW. However, as the operation frequency approaches the limits of the technology, the practical BW will be lower. As the power gain of a single transistor decreases, broadband matching techniques (e.g., stagger tuning, low-k transformer matching) will need to be replaced by low-loss matching techniques (e.g., single-frequency tuning, high-k transformer matching) to minimize the sacrifice of amplifier gain for bandwidth. To model this, the total system bandwidth is described as a 1st order filter with the transition frequency f_T as the corner frequency:

$$\mathrm{BW}_{\mathrm{RF}} = \frac{\alpha \cdot f_0}{\left|1 + j \cdot \frac{f_0}{f_T}\right|} , \tag{5.7}$$

where an f_T of 300 GHz is chosen to represent a standard CMOS technology together with a fractional BW α of 20 %.

5.2.2.4 Antenna Gain

This analysis will mainly focus on single-cell systems to expose the most important limitations of the circuits. Later in the chapter, a phased-array system will be discussed as well. For now, a single TX and RX antenna is considered, both omnidirectional, so $G_{\mathrm{TX}} = G_{\mathrm{RX}} = 1$, with a constant gain over frequency.

5.2.3 Evaluating Shannon's Capacity

Figure 5.8 depicts an overview of the different system parameters modeled over frequency. Plugging in all of the frequency-dependent parameters into the Shannon's equation enables us to evaluate the capacity of the channel over frequency. Figure 5.9 shows the calculated Shannon capacity for different link distances in the single-cell case. As expected from intuition, shorter link distances lead to a larger available system SNR and as a result a larger maximum link capacity peaking at a higher optimum frequency.

While these concrete curves give some information already, the ideal nature of the model limits the profoundness of insight we can acquire. For instance, Shannon's capacity expression assumes ideal coding, ideal constellation shaping, optimal modulation, and does not include any implementation limitations. In the next iteration of this model, a specific modulation scheme will be chosen with uniform constellation shaping and no error correcting coding.

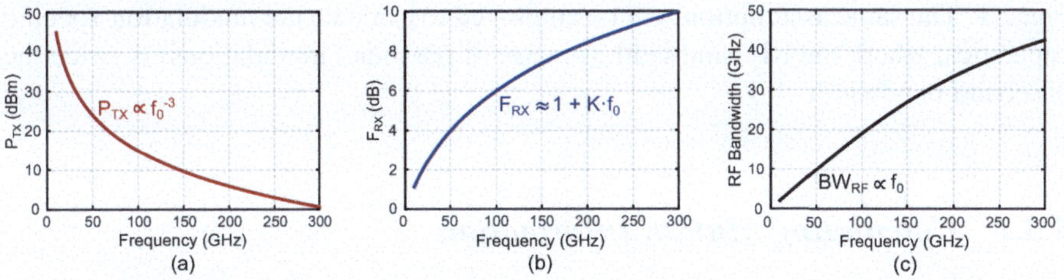

Fig. 5.8 Modeled system parameters versus frequency: TX output power (**a**), RX noise figure (**b**), and system RF BW (**c**)

Fig. 5.9 Shannon link capacity versus frequency for various link distances in the single-cell case

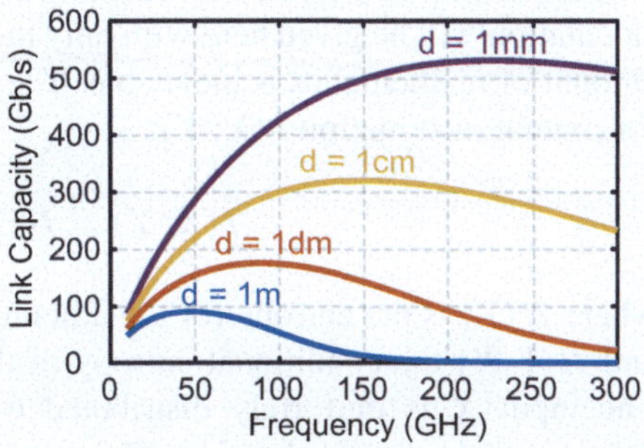

5.3 Modulation-Specific Link Capacity

In order to calculate the link capacity for a specific modulation, it is useful to divide the capacity C (in bits/s) into symbol rate D (in symbols/s) and spectral capacity C_S (in bits/symbol) [112]:

$$C = D \cdot C_S \,. \tag{5.8}$$

Firstly, the spectral capacity is described in information theory as the mutual information between input (X) and output (Y), also denoted as $I(X;Y)$ [34]. It quantifies (in bits per symbol) the reduction in uncertainty of a certain variable (e.g., the input symbols X) due to the knowledge of the other variable (e.g., the output symbols Y). The mutual information is bound by the spectral efficiency of the modulation and diminishes as more noise corrupts the output symbols with lower SNR values. In Shannon's capacity expression, the maximum mutual information is equal to $1/2 \cdot \log_2(1 + \text{SNR})$. This mutual information should be re-evaluated for each modulation as a function of system SNR.

Secondly, the symbol rate determines the number of transmitted symbols per unit of time and is tightly linked to the available system bandwidth and applied pulse shape. In Shannon's expression, a Sinc pulse shape is assumed, and thus a system with baseband bandwidth B can support a symbol rate of $2B$, as was illustrated in

Fig. 2.1. The same assumption will be employed to evaluate the modulation-specific capacities, where the RF bandwidth (in case of passband modulations) is twice the baseband bandwidth.

5.3.1 Calculating Mutual Information

The fundamentals of information theory can be found in [34], and the calculation of mutual information for a Gaussian channel is extensively documented in [36]. Hence, in order to preserve the readability of this section, a concise overview of the calculation will be given here with only the essential equations. However, a minimal amount of mathematics is inevitable. The starting point is the following expression for mutual information $I(X;Y)$:

$$I(X;Y) = H(Y) - H(Y|X), \tag{5.9}$$

where $H(Y)$ is the entropy (or the amount of information) of the output Y (in bits) and $H(Y|X)$ the conditional entropy of the output Y when the input X is known. The input X is uniformly distributed over all possible discrete input symbols, determined by the modulation. The output Y is continuous in nature and consists of the transmitted input symbols X combined with additive white Gaussian noise (AWGN).

To calculate the conditional entropy $H(Y|X)$, we need to know the output distribution corresponding to a given transmitted input symbol, denoted by conditional density function $f(y|x)$. This is fully determined by the normal distribution of the AWGN added to each input symbol:

$$f(y|x) = \frac{1}{\sqrt{\pi N_0}} e^{-(y-x)^2/N_0}\ , \tag{5.10}$$

where $N_0/2$ is the total noise power. Due to the fact that the transmitted input symbol does not affect the conditional entropy (it only translates the distribution in symbol space), $H(Y|X)$ can be evaluated as follows [36]:

$$H(Y|X) = -\int_{-\infty}^{\infty} f(y|x) \cdot \log_2 f(y|x)\, dy = \frac{1}{2} \log_2 \pi e N_0\ . \tag{5.11}$$

To calculate the entropy of the output $H(Y)$, we use the general expression of (differential) entropy:

$$H(Y) = -\int_{\mathbb{R}} f(y) \cdot \log_2 f(y)\, dy\ , \tag{5.12}$$

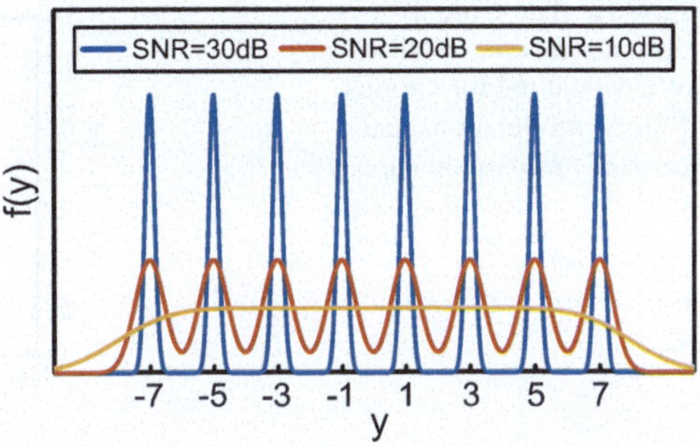

Fig. 5.10 The output density function $f(y)$ for an 8-PAM signal with different SNR values

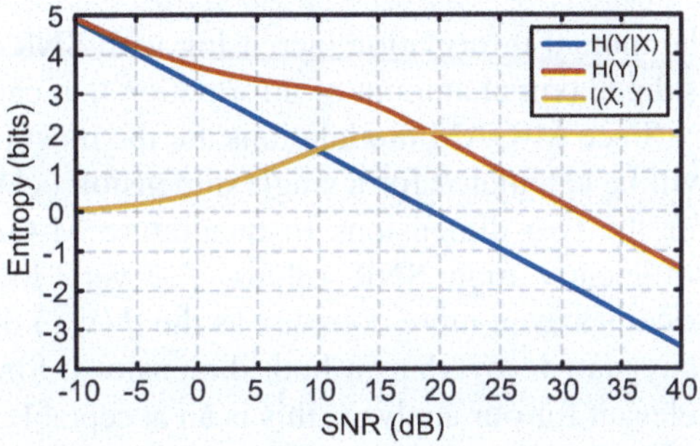

Fig. 5.11 Conditional entropy $H(Y|X)$, entropy $H(Y)$, and mutual information $I(X;Y)$ versus SNR of a PAM-4 signal

where the output density function $f(y)$ can be written as a function of the conditional density function $f(y|x)$ and the input distribution $p(x)$, which is assumed uniform over all M possible input symbols:

$$f(y) = f(y|x) \cdot p(x) = \frac{1}{M} \sum_{x} f(y|x) . \tag{5.13}$$

Finding an analytical expression for $f(y)$ is difficult, so evaluating $H(Y)$ is accomplished via numerical integration. Figure 5.10 visualizes $f(y)$ for an equidistant 8-PAM signal with different SNR values. As the SNR drops, the 8 output levels become less distinct until noise makes the different levels indistinguishable.

Combining the expressions of $H(Y)$ and $H(Y|X)$, we can now evaluate the mutual information $I(X;Y)$ versus SNR for different modulation schemes. Figure 5.11 shows how the different entropies vary over SNR for a PAM-4 signal. For very low SNR values, both the conditional entropy and output entropy are large and equal, as they are fully determined by the noise. As a result, the mutual information is close to zero as the output does not contain a lot of information about the input, but mostly noise. As the SNR increases, the conditional entropy drops because it is fully determined by the decreasing noise power. The output entropy decreases comparably less since it also contains the information from the input signal that progressively becomes more visible at the output. This effect leads to

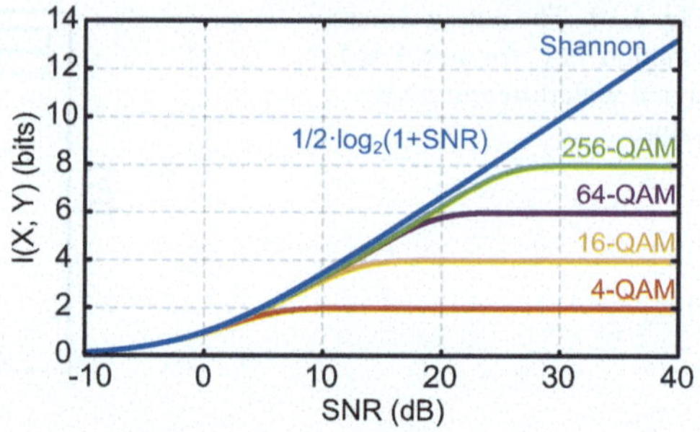

Fig. 5.12 The mutual information versus SNR of an AWGN channel for various M-QAM modulations and Shannon's maximum capacity

the mutual information increasing with SNR and finally saturating at 2 bits, which is the maximal amount of information that can be carried by a PAM-4 signal.

Since M-QAM modulations are the most popular in wireless networks, $I(X; Y)$ will be calculated for a single dimension ($\sqrt{M}$-PAM) and then doubled to account for the two dimensions in quadrature M-QAM modulations, which is valid for sufficiently high SNR values. For very low SNRs, the two-dimensional $f(y)$ becomes even more complex as the $f(y|x)$ distribution from neighboring symbols may start to overlap in both dimensions. Since this very low SNR regime is less relevant for our analysis, this is an acceptable simplification.

Figure 5.12 shows the mutual information (or spectral capacity) versus SNR of an AWGN channel for different M-QAM modulations. The spectral capacity saturates to the maximum transmitted bits per symbol as the channel improves. The mutual information from Shannon's maximum capacity expression is also added as a reference and continues to grow as SNR increases since it is modulation-agnostic.

5.3.2 *Evaluating Modulation-Specific Capacities*

Now we can combine the calculated spectral capacity (mutual information) with the determined symbol rate to work out the total link capacity for each of the modulations. Figure 5.13 shows the capacity versus frequency for different modulations and Shannon's case for a 0.01 m, 0.1 m, and 1 m single-cell link.

It is clear that at lower frequencies, where more SNR is available, the denser modulations perform better, while higher frequencies prefer more simple modulation schemes. Although this plot already uncovers some underlying trends, the results still lead to unrealistic conclusions. For instance, reaching close to 160 Gb/s using 256-QAM at 120 GHz seems quite challenging and will probably be impossible due to certain implementation impairments. In order to get the full, realistic picture, system implementation nonidealities need to be included in the model.

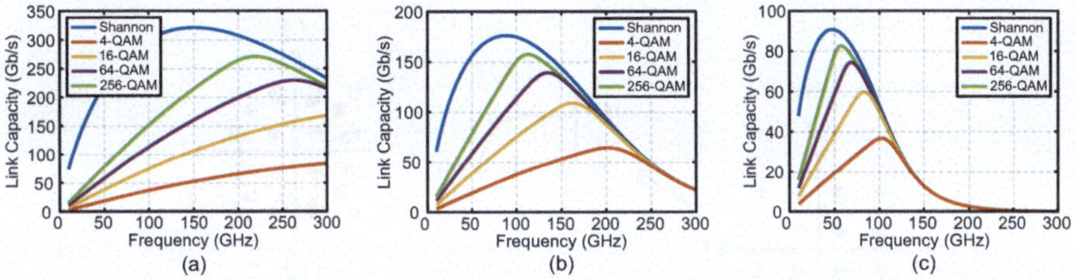

Fig. 5.13 Channel capacity versus frequency for various M-QAM modulations and Shannon's capacity for a 0.01 m (**a**), 0.1 m (**b**), and 1 m (**c**) single-cell wireless link

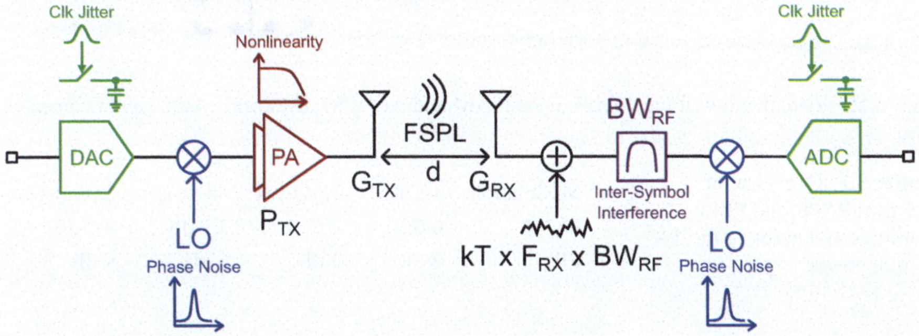

Fig. 5.14 A general framework for a communication system with the most critical impairments

5.4 Adding System Impairments

Figure 5.14 shows the same generic model of a communication link as before, but with the most limiting implementation impairments highlighted in color. First of all, the transmitter output power will be limited at higher order modulations by the linearity of the PA. Secondly, at higher carrier frequencies the phase noise of the LO will greatly affect the obtainable signal purity in large constellations. Next, clock jitter will add uncertainty to the sampled signal in data converters, especially for high bandwidth signals. And finally, high symbol rate systems will suffer from inter-symbol interference due to the limited available bandwidth introducing a power penalty and thus limiting the usable symbol rate. Each of these impairments will now be discussed and modeled.

5.4.1 PA Nonlinearity

There are two effects which degrade the TX output power as the modulation density increases, which was outlined previously in Sect. 2.3.3. First of all, to correctly determine the system SNR, the average signal power needs to be considered instead

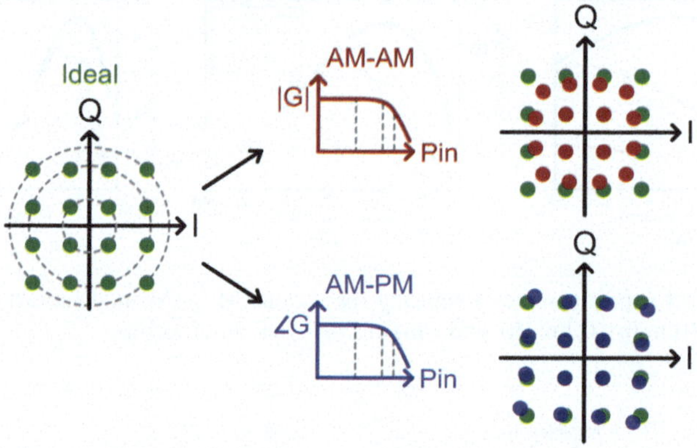

Fig. 5.15 Visualization of the effects of AM–AM and AM–PM distortion on the signal constellation

Table 5.1 Overview of selected PAPR and PBO numbers to determine the TX output power

Modulation	4-QAM	16-QAM	64-QAM	256-QAM
PAPR	0 dB	2.6 dB	3.7 dB	4.2 dB
PBO	0 dB	3 dB	6 dB	9 dB

of the peak signal power. Hence, the peak output power of the PA needs to be subtracted by the peak-to-average power ratio (PAPR) to obtain the average PA output power. As the PAPR increases for more spectrally efficient modulation schemes, the average signal power will be lower. Secondly, operating the PA in saturation degrades the constellation of high-order modulations due to AM–AM distortion and AM–PM distortion, as illustrated in Fig. 5.15.

In order to limit signal degradation caused by these distortions, the PA must work in a power back-off (PBO), reducing the peak output power with respect to its P_{sat}. Again, denser modulations are more sensitive to these distortions, thus requiring a larger PBO and further exacerbating the decrease in output power (and efficiency as will be discussed later). Combining these effects, the average TX output power can be calculated as follows:

$$P_{\text{TX}} = P_{\text{sat}} - \text{PAPR} - \text{PBO}\,, \qquad (5.14)$$

where P_{sat} has been modeled before over frequency, and Table 5.1 gives an overview of the PAPR and PBO numbers employed in this analysis. The PAPR values are calculated assuming unfiltered waveforms, and the PBO numbers are roughly estimated from the literature but in reality highly depend on the exact gain characteristic of the PA over input power. More exact expressions quantifying the amount of AM–AM and AM–PM distortion can be found [55] but are only valid for a specific magnitude and phase behavior. The numbers from Table 5.1 aim to be generally applicable to most PAs.

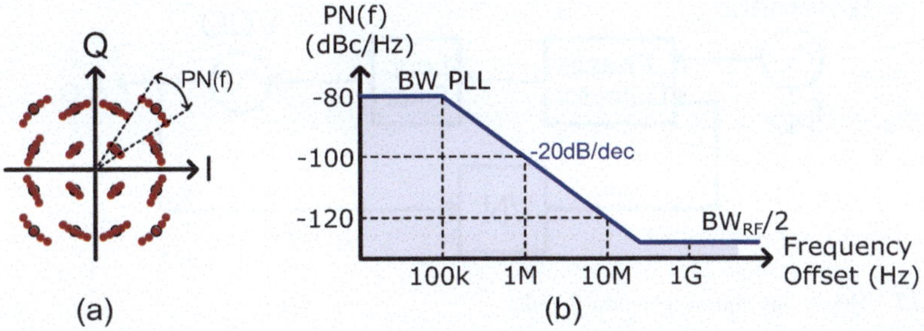

Fig. 5.16 Stochastic rotation of the constellation points in the presence of phase noise (**a**) and modeled phase noise profile versus frequency offset for a 60 GHz LO signal (**b**)

Note that the 3 dB increase in PBO every time the number of constellation points quadruples is no coincidence. Indeed, for a cubic nonlinearity, see Sect. 2.3.3, increasing the back-off by 3 dB leads to a 6 dB improvement in signal quality (EVM), which compensates halving the distances between constellation points when switching to a four-times more dense QAM modulation [113].

5.4.2 LO Phase Noise

Phase noise (PN) represents random fluctuations in the frequency (and thus also phase) of the LO signal, which is expressed as a broadening of the LO impulse in the frequency domain. As briefly mentioned in Sect. 2.3.2, this uncertainty in the LO phase leads to a stochastic rotation of the constellation points which degrades the signal SNR, as visualized in Fig. 5.16a. For generic VCO, the phase noise PN at a given offset frequency Δf can be expressed as

$$\mathrm{PN}(\Delta f) = \frac{2FkT}{P}\left(\frac{f_0}{2Q\,\Delta f}\right)^2, \tag{5.15}$$

where F is the noise factor of the VCO, k the Boltzmann constant, T the temperature, P the power of the oscillating signal, and Q the Q-factor of the VCO tank. Known as Leeson's equation [114], it shows that the phase noise at a fixed offset degrades with the square of frequency.

Since the phase noise of a free-running VCO is quite bad, the VCO is usually embedded within a phase-locked loop (PLL), which can be generally represented by a block diagram like in Fig. 5.17. The main goal of the PLL is to generate a stable high-frequency LO signal based on a lower frequency reference with very good phase noise. An in-depth discussion of PLLs is outside the scope of this work and is described in [7, 60]. Due to the PLL operation, part of the phase noise at small frequency offsets is filtered by the loop, drastically improving the phase noise

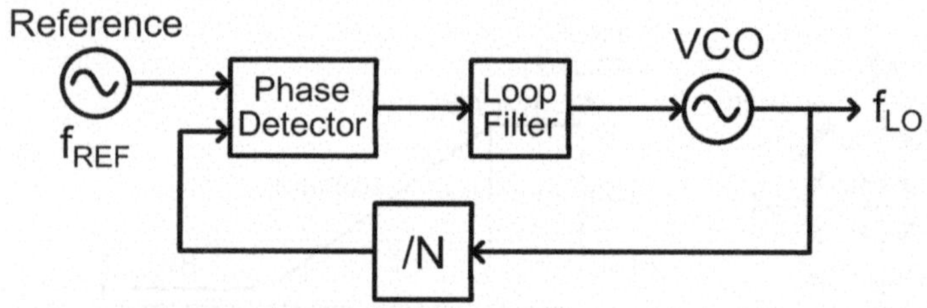

Fig. 5.17 Block diagram of a standard PLL

of the high-frequency LO signal. Due to this operation, the phase noise, typically quantified by the power spectral density of the LO signal versus frequency offset with respect to the main LO frequency, has a characteristic as shown in Fig. 5.16b. At small offsets, the phase noise is dominated by the phase noise of the reference, multiplied by the division factor of the loop, and the noise of PLL components (such as the phase detector and charge pump). At higher offsets outside of the loop BW (BW PLL), the phase noise is determined by the VCO, characterized by the typical −20 dB/dec slope predicted by Leeson's equation. At very high offsets, the noise floor is set by the amplitude noise floor, e.g., by LO buffers.

For a given phase noise profile, the RMS phase error and SNR of the LO signal can be calculated as follows:

$$\frac{1}{\mathrm{SNR_{LO}}} = \phi_{\mathrm{rms}}^2 = 2\int_0^{\mathrm{BW_{RF}}/2} \mathrm{PN}(f)\,df\ . \qquad (5.16)$$

A representative phase noise profile for a 60 GHz LO was constructed based on the state-of-the-art mm-Wave VCOs and PLLs [115–119] and is shown in Fig. 5.16b. To determine the PN at other frequencies f_0, the created phase noise profile was shifted in magnitude by a factor of "20 log(f_0/60GHz)," as phase noise generally scales with f_0^2. The phase noise is also doubled to account for the effects of the TX as well as the RX LO signal. Ultimately, the chosen phase noise model yields an RMS phase error of 6° at 100 GHz and equivalent RMS jitter of 168 fs (independent of frequency), where the RMS jitter can be calculated from the RMS phase error as follows:

$$t_{jit,\mathrm{rms}} = \frac{\phi_{\mathrm{rms}}}{2\pi f_0}\ . \qquad (5.17)$$

5.4.3 *DAC/ADC Aperture Jitter*

In high-speed data converters (DACs and ADCs), aperture jitter is arguably the most critical factor limiting the accuracy of data converters at high (>5 GS/s) sampling

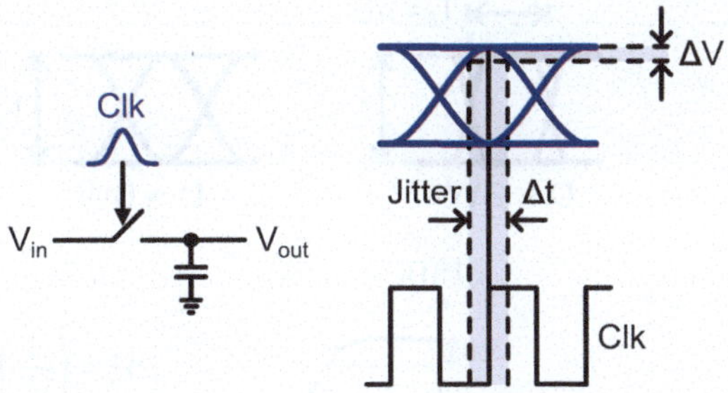

Fig. 5.18 Visualization of how jitter in the sampling clock leads to a voltage error in the sampled output

speeds [45]. Uncertainty in the edge of the sampling clock leads to uncertainty in the sampled output, as visualized in Fig. 5.18.

For a given RMS jitter, the SNR degradation caused by this sampling uncertainty can be characterized as follows [43]:

$$\mathrm{SNR}_{jit} = \frac{1}{(2\pi f_{\mathrm{in}} t_{jit})^2}\,, \tag{5.18}$$

where the input frequency f_{in} is determined by the bandwidth of the baseband modulated signal (in case a direct-conversion architecture is adopted). However, this equation assumes a sinusoidal input signal, which is not necessarily the case. The amount of voltage error will be lower if the pulse has a flatter shape around the sampling instant, and this equation thus leads to a slightly pessimistic estimation of the error caused by jitter. However, since the pulse will be filtered by the limited bandwidth of the system, see Fig. 2.48, this sinusoidal approximation is accurate enough. Based on the literature, an RMS jitter of 150 fs is assumed, and the input frequency is equated to half of the system's RF bandwidth. The total SNR degradation is finally doubled to account for the TX-side DACs' and RX-side ADCs' aperture jitter.

5.4.4 Inter-symbol Interference

The last impairment is inter-symbol interference (ISI) and is the main limiting factor of symbol rate at high speeds. ISI is a phenomenon where signal filtering causes the symbol waveform to spread out and spill over into the following symbols causing interference and consequently signal distortion, as discussed previously in Sect. 2.3.4. Assuming a distortionless channel, most of the undesired filtering is caused by the TX and RX circuits which have a limited bandwidth and thus act as a low-pass filter. A larger symbol rate D, resulting in a larger occupied signal bandwidth, will result in more pronounced signal distortion for the same circuit bandwidth BW and thus a more severe reduction of the vertical eye height (V_E), as illustrated in Fig. 5.19. As the vertical eye height decreases, the effective SNR

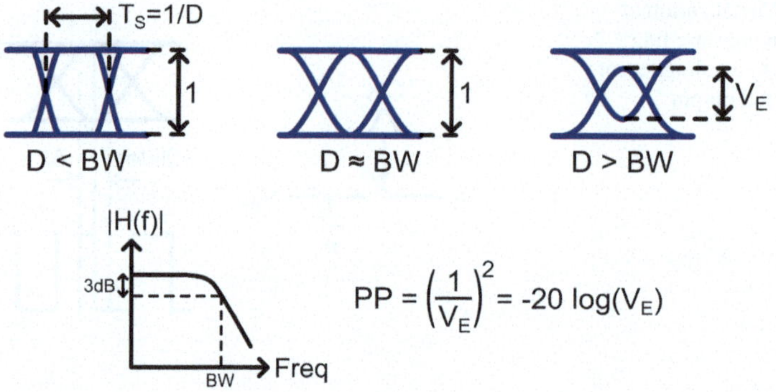

Fig. 5.19 Conceptual overview of ISI and associated power penalty

drops and thus an increase in signal power would be necessary to obtain the same SNR as before (without ISI). This decrease in signal power due to ISI is called the power penalty (PP) [120] and is visualized in Fig. 5.19 for a binary signal. Equalization schemes (e.g., CTLE, FFE, DFE, etc.) are very popular to partially compensate the predictable effects of ISI. By absorbing their effects into the global system description, the following model is still applicable.

5.4.4.1 1st Order Filter

As initial example, consider a simple 1st order (Butterworth) filter system with baseband, binary signaling. The transfer function is given by the following expression:

$$H_1(f) = \frac{1}{1 + s/(2\pi f_c)}, \tag{5.19}$$

where f_c is the cut-off frequency (3-dB BW) and s the complex Laplace variable. In order to calculate the degradation in vertical eye height, the time-domain response is required. The normalized step response has an exponential behavior which settles closer to 1 the longer you wait (relative to the time constant of the system τ):

$$V_{\text{step},1}(t) = 1 - e^{-2\pi f_c t} = 1 - e^{-t/\tau} \,. \tag{5.20}$$

The filtered vertical eye height is then determined by the worst-case output in case of a binary 1 (i.e., the step response at $t = T_s$ = symbol period) subtracted by the worst-case output in case of a binary 0 (i.e., 1 minus the step response at $t = T_s$), leading to a worst-case filtered vertical eye height of

$$V_{E,1} = 2 \cdot V_{\text{step},1}(t = T_s) - 1 = 1 - 2 \cdot e^{-2\pi f_c T_s} \,. \tag{5.21}$$

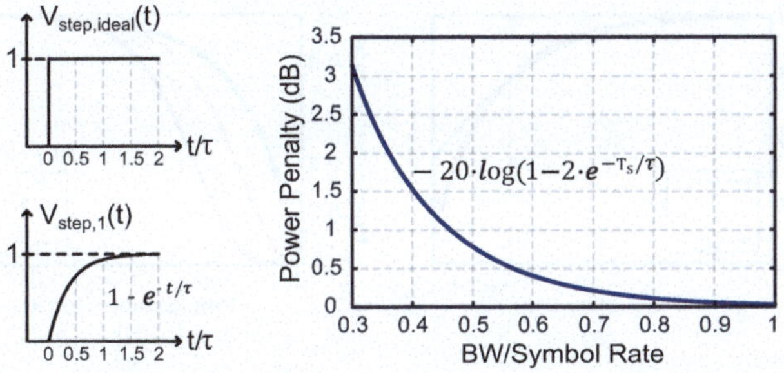

Fig. 5.20 Ideal and filtered step response by a 1st order filter, and associated power penalty versus BW

Since the response is normalized, the power penalty due to ISI can then be evaluated by squaring the ratio of the eye height with and without ISI, which yields in decibel units:

$$\mathrm{PP_{ISI}} = -20 \cdot \log(V_E) \ . \tag{5.22}$$

The ideal and filtered step response, along with the calculated power penalty versus bandwidth, is depicted in Fig. 5.20.

5.4.4.2 Multistage Filter

Now we would like to model the ISI power penalty for a practical, multistage RF system. While a realistic RF system can have a quite complex frequency behavior (e.g., frequency translations, stagger tuning, asymmetries, etc.), the essence can be captured using a more simple high-order frequency response. The frequency response of a generic RF transceiver is mostly determined by a cascade of passband RLC resonators (matching networks) and baseband RC low-pass filters. To model the ISI, the RF transceiver will be represented as an equivalent baseband system and modeled as a high-order low-pass filter, constructed as a cascade of identical 1st order low-pass filters, where the filter order and cut-off frequency are determined to approximate the RF system response. Again, this simplified response does not aim to accurately represent every RF transceiver implementation, but rather to model the effects of ISI caused by limited bandwidth and provide an analytical expression to do so.

The previous calculation of the power penalty can be extended to higher order filters, constructed as a cascade of identical 1st order filters with cut-off frequency f_c, where the transfer function of the Nth order filter is given by

$$H_N(f) = H_1(f)^N = \left[\frac{1}{1 + s/(2\pi f_c)}\right]^N \ . \tag{5.23}$$

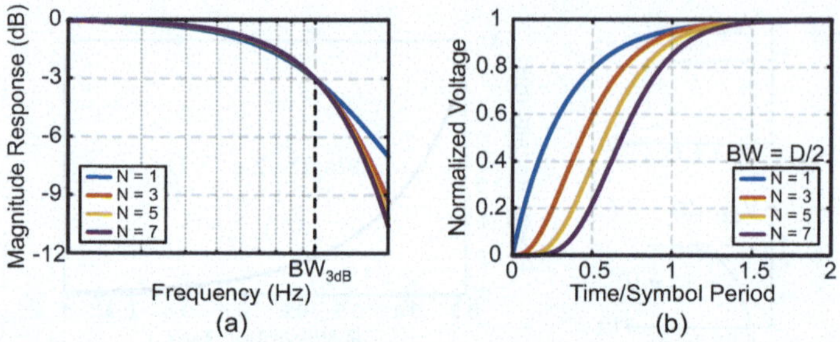

Fig. 5.21 Frequency response of the Nth order filter with identical 3dB bandwidths (**a**) and step response of the Nth order filter with identical bandwidths equal to half the symbol rate (**b**)

The 3-dB cut-off frequency of this Nth order filter can be calculated as

$$f_{c,N} = f_c\sqrt{2^{1/N} - 1}\,. \tag{5.24}$$

It can be shown[1] that the step response of this Nth order filter can be expressed as

$$\begin{aligned} V_{\text{step},N}(t) &= 1 - e^{-t/\tau} \cdot \left[1 + t/\tau + \frac{(t/\tau)^2}{2} + \ldots + \frac{(t/\tau)^{N-1}}{(N-1)!}\right] \\ &= 1 - e^{-t/\tau} \cdot \sum_{n=1}^{N} \frac{(t/\tau)^{n-1}}{(n-1)!}\,, \end{aligned} \tag{5.25}$$

where the time constant τ is again determined by the corner frequency of the 1st order filter f_c. Figure 5.21 shows the frequency response and step response of this filter for different orders N but identical overall bandwidth $f_{c,N}$ equal to half the symbol rate $D/2$, demonstrating a slower settling behavior as the filter order increases.

Using the calculated step response, the vertical eye height after filtering can be evaluated in a similar fashion as in the case of a 1st order filter:

$$V_{E,N} = 2 \cdot V_{\text{step},N}(T_s) - 1 = 1 - 2 \cdot e^{-T_s/\tau} \cdot \sum_{n=1}^{N} \frac{(T_s/\tau)^{n-1}}{(n-1)!}\,. \tag{5.26}$$

Based on this expression, the power penalty due to ISI can be calculated for different filter orders and varying filter bandwidths, as displayed in Fig. 5.22. The power penalty grows as the filter order increases and as the global bandwidth decreases relative to the symbol rate.

[1] The derivation is rather trivial and left as an exercise to the reader (hint: refresh your Laplace transforms!).

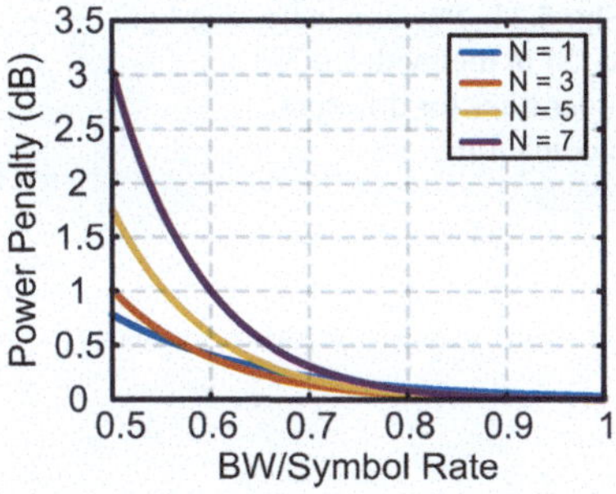

Fig. 5.22 Power penalty due to ISI for different filter orders in case of binary signaling

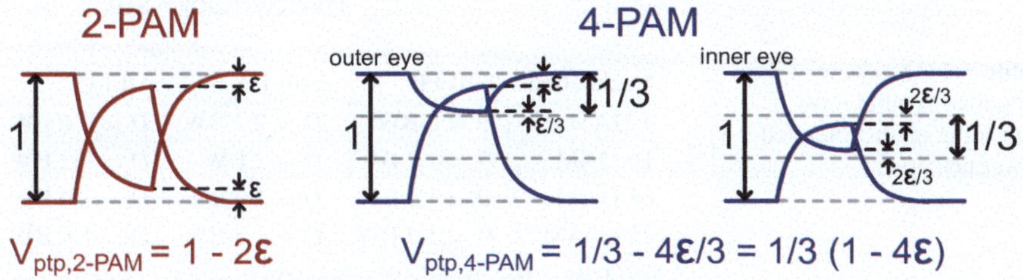

Fig. 5.23 Illustration of the decrease in vertical eye height for PAM-2 and PAM-4 signaling

5.4.4.3 Power Penalty for Multilevel Signaling

Until now, only binary signaling was considered when evaluating the power penalty. However, multilevel signals are more sensitive to ISI than binary signals, suffering a worse power penalty for the same settling error. For example, for the same ISI, the relative decrease in vertical eye height will be twice as large (or 3 dB higher) for a PAM-4 signal than for a binary PAM-2 signal, as illustrated in Fig. 5.23.

In general, the vertical eye degradation increases by a factor of $\log_2(M)$ for a M-PAM signal compared to a binary signal. This increased sensitivity to BW limitations will result in a smaller tolerated symbol rate for the same system BW as the number of signal levels increases. Note that for M-QAM modulations, the number of signal levels in each dimension is equal to $\sqrt{M}$. So the final vertical eye height expression in case of an Nth order filter with M-QAM modulation is

$$V_{E,N,\text{M-QAM}} = 1 - \sqrt{M} \cdot e^{-T_s/\tau} \cdot \sum_{n=1}^{N} \frac{(T_s/\tau)^{n-1}}{(n-1)!} \quad . \tag{5.27}$$

Finally, based on all of these expressions, the power penalty due to ISI can be evaluated for a given system order, bandwidth, and modulation scheme. A realistic system model is first determined based on the behavior of actual implemented circuits. Given a certain 3-dB bandwidth, the filter order of the equivalent model is chosen based on the steepness of the response, in this case by also matching the 6-dB bandwidth. Following this procedure, a 3rd order filter, with specified

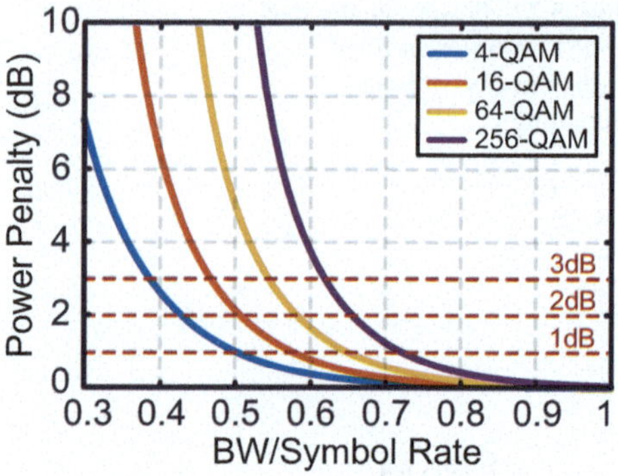

Fig. 5.24 Power penalty due to ISI in the modeled 3rd order filter for different modulations

Table 5.2 Overview of the supported symbol rates depending on the tolerated power penalty

Modulation	1 dB PP	2 dB PP	3 dB PP
4-QAM	$D = 2$ BW	$D = 2.3$ BW	$D = 2.6$ BW
16-QAM	$D = 1.7$ BW	$D = 2$ BW	$D = 2.2$ BW
64-QAM	$D = 1.5$ BW	$D = 1.7$ BW	$D = 1.9$ BW
256-QAM	$D = 1.4$ BW	$D = 1.5$ BW	$D = 1.6$ BW

NB: BW = Baseband BW = RF BW/2

3-dB bandwidth for the global system, emerged as an accurate model within the 6-dB bandwidth. While this filter order might seem rather low for an intricate RF transceiver and might not accurately model the behavior past the 6-dB BW point, that part of the frequency behavior has a negligible effect on ISI as the response within the signal bandwidth is the most essential.

Figure 5.24 shows the power penalty due to ISI in the modeled system with a given 3-dB bandwidth in the case of different modulations. It can be observed that the power penalty worsens for smaller bandwidths and in case of more dense modulations. Assuming a certain power penalty can be tolerated, the allowed symbol rate can be determined relative to the available system bandwidth. Table 5.2 gives an overview of the supported normalized symbol rate for various power penalties. Note that the (equivalent baseband) bandwidth numbers used in these ISI analyses are half of the RF bandwidth values used in the link capacity expressions.

For the remainder of this analysis, an accepted power penalty of 1 dB is assumed (and subtracted from the system SNR), which fixes the symbol rate as a function of BW for each modulation according to the second column in Table 5.2.

5.4.5 *Other Transceiver Impairments*

Although the discussed impairments are the most prominent ones in high-speed mm-Wave links, there are other effects that may possibly degrade the performance. Since most of these effects can be calibrated out or completely avoided at design time, these do not form fundamental limitations in high-speed systems and are

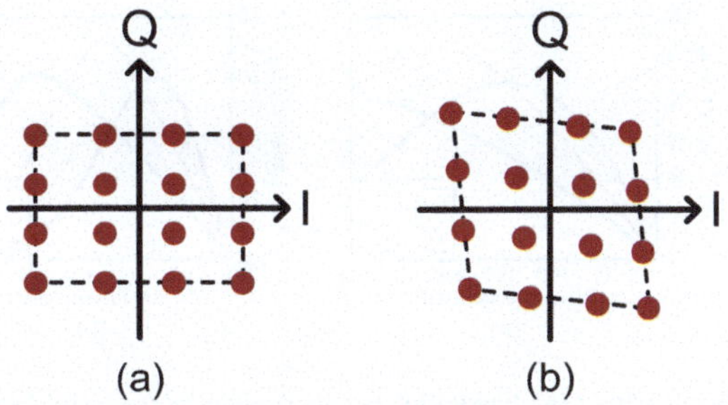

Fig. 5.25 Visualization of IQ magnitude (**a**) and phase (**b**) imbalance on a 16-QAM constellation

thus not included in the model. One example is a possible imbalance between the generated quadrature I and Q LO signals. As visualized in Fig. 5.25, a magnitude and phase imbalance can lead to substantial constellation distortion and coupling between the I and Q channels. As discussed in Chap. 7, this type of distortion can be sufficiently minimized. Another example is the presence of spurious LO tones at the sub-harmonics of the LO frequency. This can cause severe performance degradation, as extensively discussed in Chap. 7. But again, these can be mostly avoided with good design and are thus not included in this analysis. More advanced phenomena, such as frequency-dependent IQ coupling, may be extremely relevant for very broadband systems, as again discussed in Chap. 7, but are too complex to model in a generic way and are only relevant in specific cases.

Another aspect excluded from the model is interference. As discussed in Sect. 1.3, the frequency spectrum is shared between many different users and applications, possibly leading to large interference signals within the same band or a neighboring band. These signals can desensitize the receiver and make correct demodulation of the desired signal significantly more difficult [60]. To cope with this scenario, the receiver must be able to handle very large signals at its input and thus exhibit much better linearity performance. The interference signal can then be separated from the desired signal by means of specialized analog or digital techniques. The influence of interference is not included in the model as it depends heavily on the specific interferer scenario and is generally less relevant at higher frequencies, as mentioned in Sect. 1.3.

5.4.6 Evaluating Impaired Link Capacities

All of the modeled impairments can now be applied to the general expression for link capacity:

$$C = D\,(\mathrm{BW}, M) \cdot C_S\,(\mathrm{SNR}, M) \quad . \tag{5.28}$$

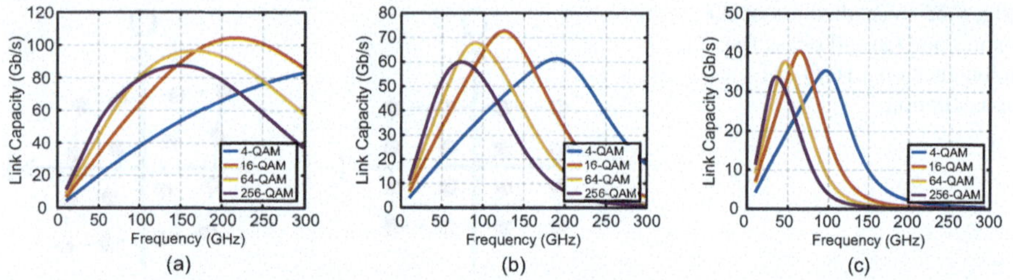

Fig. 5.26 Link capacities versus frequency for a single-cell link of 0.01 m (**a**), 0.1 m (**b**), and 1 m (**c**) distance, including implementation impairments

The symbol rate D is chosen relative to the available BW and modulation for the tolerated power penalty of 1 dB, as discussed in the previous section. The exact numbers can be found in Table 5.2.

The spectral capacity C_S, or mutual information as discussed before, can be calculated based on the available SNR for a specific (M-QAM) modulation. Compared to the previous section, the SNR is now impacted by the SNR degradation caused by the PA nonlinearity, LO phase noise, and aperture jitter of the data converters. The resulting SNR can be calculated as follows:

$$\mathrm{SNR_{tot}} = \left[\frac{1}{\mathrm{SNR_{AWGN}}} + \frac{1}{\mathrm{SNR_{LO}}} + \frac{1}{\mathrm{SNR}_{jit}}\right]^{-1}, \tag{5.29}$$

where $\mathrm{SNR_{LO}}$ and SNR_{jit} have been expressed previously. $\mathrm{SNR_{AWGN}}$ is computed by the Friis equation with adapted modulation-dependent P_{TX} determined by the PA nonlinearity and decreased by the accepted ISI-induced power penalty of 1 dB.

Figure 5.26 shows the evaluated link capacities versus frequency, including implementation impairments, for a single-cell link of 0.01 m, 0.1 m, and 1 m distance. Generally, the region where the capacity increases with frequency is bandwidth limited, while a declining capacity suggests the SNR is the limiting factor. For these distances, 16-QAM emerges as the optimal modulation scheme, with the optimal frequency dropping as the link distances increase.

It is also insightful to quantify how the capacity changes at a fixed frequency while changing the link distance, as shown on Fig. 5.27 for a carrier of 70 GHz and 140 GHz. For short distances, the link is mainly bandwidth and phase noise limited, while noise (AWGN) becomes the bottleneck for longer distances. Furthermore, a link operating at a higher frequency achieves higher data rates but degrades more quickly as the distance increases. For the shortest distance, 256-QAM performs the best at 70 GHz, while 64-QAM is optimal at 140 GHz, indicating the more stringent phase noise specification is not met at such high frequency. Of course, as suggested in Fig. 5.26, when the frequency can be optimized at each distance, 16-QAM will come out on top for nearly every link distance.

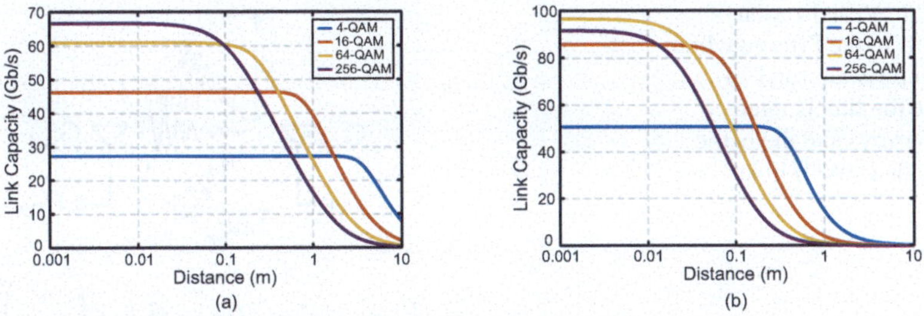

Fig. 5.27 Link capacities versus distance for a single-cell link at 70 GHz (**a**) and 140 GHz (**b**)

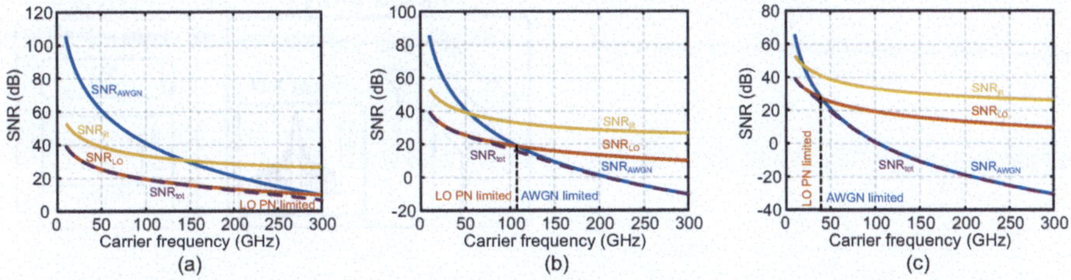

Fig. 5.28 Overview of the different SNR degradations and total SNR of a 0.01 m (**a**), 0.1 m (**b**), and 1 m (**c**) single-cell link, revealing an LO PN limited and AWGN limited regime

To get a more visual understanding which impairment has the biggest impact on the performance, each of the three SNR degradation terms can be plotted separately over frequency, as depicted in Fig. 5.28 for a 16-QAM link over all three link distances. We can see that the SNR is mainly limited by the phase noise of the LO at lower frequencies, while AWGN (including the PP due to ISI) limits the link's SNR at higher frequencies. The exact crossover frequency will vary with other modulations and link distances. Indeed, as the distance increases, the frequency at which AWGN starts limiting the SNR becomes progressively smaller, as demonstrated in Fig. 5.28.

5.5 Extending the Range of Single-Cell Wireless Systems

As can be deducted from Fig. 5.26, the capacity of single-cell wireless links drops down quickly as the communication range increases up to only 1 m. Due to the limited available SNR, the large absolute bandwidth at sub-THz frequencies cannot be adequately exploited. The reason for this low SNR at high frequencies is the enormous free-space path loss. As already discussed in Sect. 1.3, the term FSPL is quite misleading because it is not a real "loss" in the sense that power is dissipated. Rather, the large FSPL is the result of the spreading of the electromagnetic energy and scaling of the antenna aperture with frequency. After all, as the electromagnetic

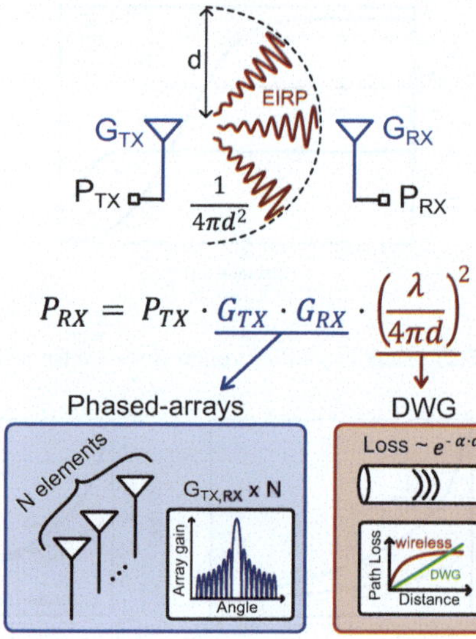

Fig. 5.29 Visualization of the FSPL in the Friis equation and two possible ways to increase the link budget: phased-arrays and dielectric waveguide (DWG) links

wave radiates outward from its (isotropic) source, its energy is spread across a sphere whose area (A_s) is expanding with the square of the distance. Moreover, only part of the radiated energy is captured by the antenna, whose aperture (or effective area A_e) scales with the square of the wavelength. Indeed, the received power can be calculated by dividing the EIRP over the area of a sphere and multiplying by the effective area of the receiving antenna, effectively deriving Friis' transmission equation:

$$
\begin{aligned}
P_{\mathrm{RX}} &= \mathrm{EIRP} \cdot \frac{1}{A_s} \cdot A_e \\
&= (P_{\mathrm{TX}} \cdot G_{\mathrm{TX}}) \cdot \left(\frac{1}{4\pi d^2}\right) \cdot \left(\frac{G_{\mathrm{RX}}\,\lambda^2}{4\pi}\right) \\
&= P_{\mathrm{TX}} \cdot G_{\mathrm{TX}} \cdot G_{\mathrm{RX}} \cdot \left(\frac{\lambda}{4\pi d}\right)^2 .
\end{aligned}
\tag{5.30}
$$

To extend the range of communication systems in the mm-wave and sub-THz range, two solutions are possible by both focusing on different parts of the Friis equation, as visualized in Fig. 5.29. First, phased arrays can be implemented to increase the directivity, and thus gain, of both antennas, thereby boosting the EIRP of the TX and enlarging the effective area of the RX antenna to capture more incoming power. Second, guided RF transmission over a dielectric waveguide (DWG) can reduce the spreading loss by confining the power within the waveguide. Both options and their effects on the achievable link capacity will now be discussed.

5.5.1 *Phased-Array Wireless Links*

Replacing a single antenna by an array of antennas has long been used to increase the directivity and thus extend the range of wireless links. Combining the signals from an antenna array in free space creates an interference pattern which can constructively interfere in the desired direction while destructively interfering in all other directions, effectively creating a directive beam. By modifying the delay of each antenna element, the beam can be steered electronically in free space.

5.5.1.1 Benefits and Challenges of Phased Arrays

Recent research has focused on creating phased arrays with more and more antenna elements at mm-wave frequencies [121, 122]. The motivation is a theoretical N^3 improvement in SNR of an N-element array compared to a single element system, while the total area and power consumption only scales by N. On the TX-side, the total output power and antenna directivity both grow by N, yielding an effective isotropic radiated power (EIRP) improvement of N^2. On the RX-side, the SNR improves as N by receiving signal power from an N times larger aperture, or equivalently having N times larger antenna directivity. Large-scale phased arrays can thus possibly offset the considerable FSPL and lead to broadband high-frequency links over large distances.

However, expanding the number of elements in a phased array becomes increasingly more challenging, especially at higher frequencies, and implementation nonidealities partially offset the previously mentioned gains in EIRP and SNR [123]. While the antenna area scales down with λ^2, the size of the ICs does not. Therefore, as the operation frequency increases, it becomes more challenging to fit the ICs within the area occupied by the N antenna elements. Besides fitting the IC in the package, other implementation challenges impair the phased-array performance as the number of elements grows: RF distribution losses to/from antenna elements, LO routing and phase coherence, element-to-element matching, limited amplitude and phase resolution in the VGAs and phase shifters, thermal management, etc., all elaborately discussed in [123]. Furthermore, most published phased arrays replace the tunable delay by phase shifters, which is a valid approximation for narrowband signals. However, as the modulation bandwidth grows, more challenging true time-delay circuits will need to be implemented with an increasing delay range in case of more antenna elements.

In order to observe the effects of phased arrays on the link capacity without coping with the complex implementation intricacies described above, a simple yet adequately realistic model should be arranged. First of all, only the number of antenna elements will be scaled, not the number of TX/RX circuits. As a result, the total antenna gain will be scaled, but the total TX output power remains constant regardless of the number of elements. While not realistic, it takes into account the reduction in EIRP/SNR improvement of ideal arrays caused by all of the

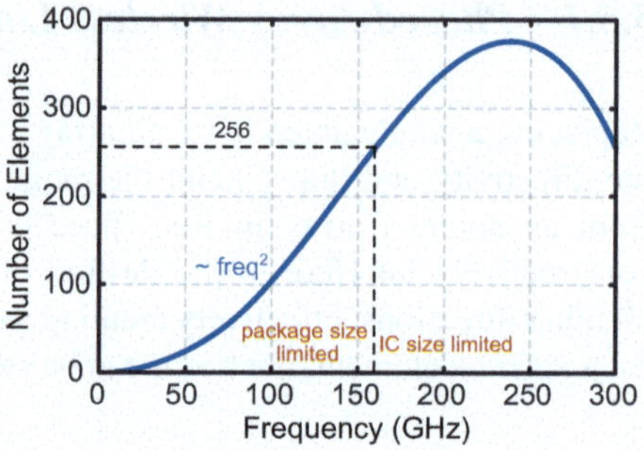

Fig. 5.30 Overview of the modeled number of antenna elements versus frequency

impairments mentioned before. Secondly, to determine the number of elements, a constant aperture is assumed. This means that generally the number of elements increases with the square of the frequency. However, as discussed before, the number of elements should saturate for very high frequencies due to the difficulty of fitting the ICs into the antenna package.

The phased-array survey in [123] along with the most recent research is used as inspiration to choose concrete values. Based on Bell Labs' latest phased array [124], up to 256 elements can be integrated at frequencies up to D-band (110–170 GHz), while the number of demonstrated elements in the literature at higher frequencies is significantly lower. In this analysis, we start with one antenna element at 10 GHz, scaling up the number of elements according to the square of frequency up to 256 elements at 160 GHz. Above 160 GHz, the increase in the number of elements saturates and eventually drops down as 300 GHz is reached. The resulting modeled number of elements versus frequency is shown in Fig. 5.30.

5.5.1.2 Evaluating Phased-Array Link Capacities

Plugging in the derived phased-array antenna gains G_{TX}, G_{RX} into the latest link model (with impairments) produces new capacity curves versus frequency, shown in Fig. 5.31 for distances of 1 m, 10 m, and 100 m. Note that evaluating the phased-array model for shorter distances would yield unrealistic results, as the TX and RX should be located in the far field of each other's antenna arrays.[2] The obtained results show significantly higher data rates for much larger distances, as can be expected. For relatively short distances, 16-QAM emerges once again as the optimal modulation scheme, where 4-QAM takes over for very long distances due to the limited SNR. To further increase the communication range, the number of elements should be further increased, with the many associated implementation challenges as discussed before.

---

[2] To comply with the far-field requirement for all relevant frequencies, the minimal distance is about 0.3 m, determined by the 350 cells at 200 GHz.

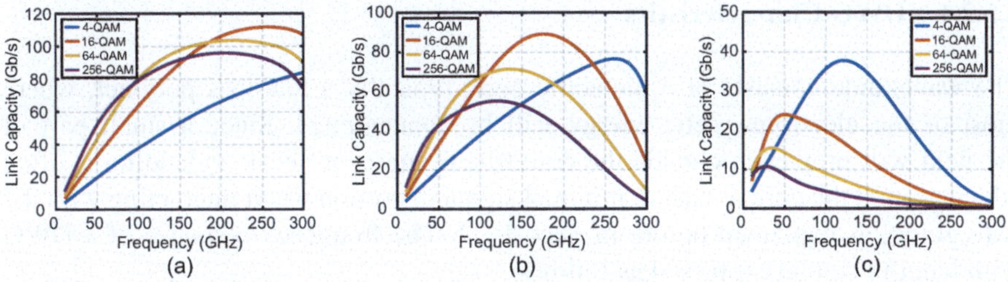

Fig. 5.31 Link capacities versus frequency for a phased-array link of 1 m (**a**), 10 m (**b**), and 100 m (**c**) distance, including implementation impairments

Fig. 5.32 Conceptual energy consumption comparison of the three main interconnect technologies: copper, DWG, and optical interconnects, and their optimal working regions

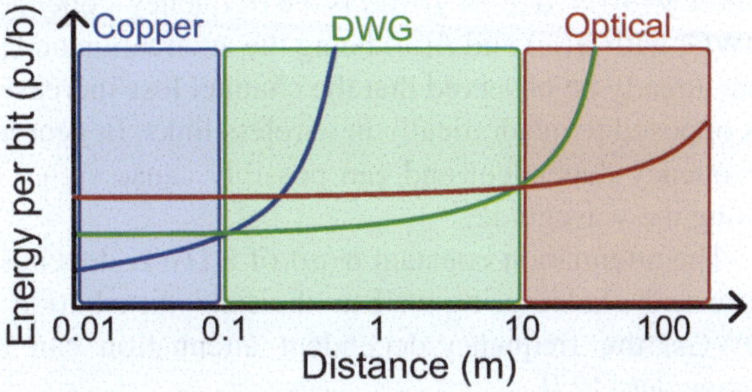

5.5.2 *Dielectric Waveguide Links*

As introduced in Chap. 1, high-speed DWG links have gained a lot of popularity in the last decade, combining high bandwidth mm-Wave transceiver chips with a directive, low-loss DWG channel [125, 126]. Compared to wireless links, a DWG link does not offer the same mobility but exhibits much lower channel loss compared to the FSPL enabling ultra-high data rates over short-to-medium distances. Copper wireline links already achieve high data rates over a relatively short distance, but copper cables are heavy, prone to electromagnetic interference, and their large and strongly frequency-dependent channel loss requires complex, power-hungry equalization techniques. Optical communication, on the other hand, benefits from low-loss, broadband optical fibers to support an enormous data throughput over long distances but requires very precise fiber alignment and costly optical/electrical signal conversion, rendering this solution too expensive for shorter distances. Mm-Wave DWG links can provide a high-speed, energy-efficient, and robust interconnect for short-to-medium distances (0.1–10 m) filling the gap between copper and optical interconnects, as visualized conceptually in Fig. 5.32.

5.5.2.1 DWG Characteristics

The waveguide consists of a dielectric core, typically a low-loss polymer, where most of the electromagnetic energy will be concentrated. Since a small part of the field will propagate outside the core (the evanescent field), a cladding is often added around the core to ensure minimal signal distortion when interacting with the waveguide, as explained before in Sect. 3.1.2. The frequency response of a DWG with length l can be expressed as follows:

$$H(\omega) = e^{-\gamma(\omega)l} = e^{-\alpha(\omega)l} \cdot e^{-j\beta(\omega)l} \quad , \tag{5.31}$$

where $\gamma(\omega) = \alpha(\omega) + j\beta(\omega)$ is the frequency-dependent propagation constant of the DWG, with $\alpha(\omega)$ and $\beta(\omega)$ being the attenuation and phase constant, respectively. It can already be observed that the channel loss increases exponentially with distance, as opposed to quadratically in wireless links. In general, however, both constants are frequency-dependent and can possibly cause signal distortions while propagating along the waveguide.

The attenuation constant $\alpha(\omega)$ of a DWG depends on the materials used for the core and cladding, as well as the field distribution in both regions. For circular DWGs, the frequency-dependent attenuation can be calculated by Meyer and Schneider [127]

$$\alpha(\omega) = \frac{\pi}{\lambda_0} \left(\sqrt{\epsilon_{r,1}}\, R_1 \tan \delta_1 + \sqrt{\epsilon_{r,2}}\, R_2 \tan \delta_2 \right) \quad , \tag{5.32}$$

where $\epsilon_{r,1}$,$\epsilon_{r,2}$ are the relative permittivities and tan δ_1,tan δ_2 the material loss factors of the core and cladding, respectively. R_1 and R_2 are geometric loss factors which can be calculated based on the ratio between the power loss and average power flow in the core and cladding. Since the DWG attenuation is roughly proportional to frequency, this model assumes a loss of 5 dB/m at 140 GHz, which is scaled linearly (in dB) to other frequencies, as shown in Fig. 5.33. As the attenuation generally varies within the signal bandwidth, the pulse shape will be distorted causing ISI similar to the previous ISI analysis caused by a limited system

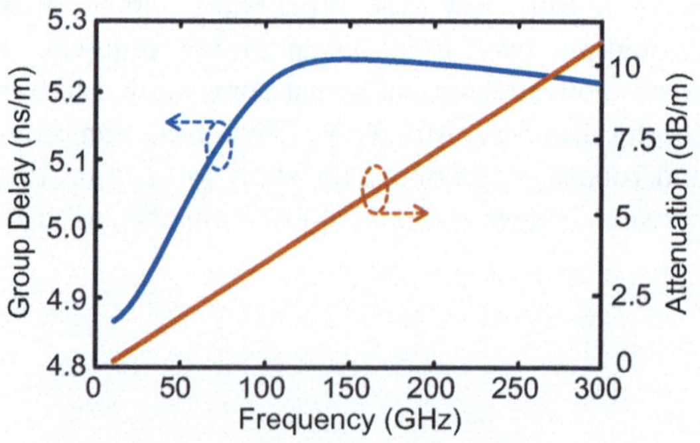

Fig. 5.33 Typical attenuation and group delay of a DWG optimized for 140 GHz operation

bandwidth. However, since the frequency variation of the attenuation constant is typically fairly limited within the signal band and its effect much less detrimental than dispersion [127], the attenuation will be assumed constant over the bandwidth (but can be included in the global system BW). Nevertheless, the loss of a DWG still has a large influence on the SNR of the link and should not be discarded during the optimization of the waveguide.

The phase constant $\beta(\omega)$ of a DWG ideally changes linearly over frequency, which would result in identical propagation delays for all spectral components of the signal thus conserving the pulse shape. However, due to a change in the field distribution in the DWG as the frequency varies, the effective dielectric constant becomes frequency-dependent causing the group delay to vary as well:

$$\beta(\omega) = \frac{\omega}{c}\sqrt{\epsilon_{r,\text{eff}}(\omega)}\ , \tag{5.33}$$

where c is the speed of light in vacuum. As a result, the distinct frequency components of the signal undergo different propagation delays causing pulse distortion. This phenomenon is called chromatic dispersion and is the main cause of dispersion in single-mode, single-polarization fibers as discussed in Sect. 3.1.2. Figure 5.33 shows the group delay versus frequency for a typical DWG, clearly showing an optimal band where group delay variation is minimal. In order to describe the dispersion in the DWG, the phase constant (per unit length) is often expressed as a 2nd order Taylor series around the center frequency:

$$\beta(\omega) \approx \beta_0 + \beta_1\omega + \frac{\beta_2}{2}\omega^2\ , \tag{5.34}$$

from which the group delay variation in a frequency band can be calculated:

$$\Delta\tau_g = \frac{d\tau_g(\omega)}{d\omega}\Delta\omega = \frac{d^2\beta(\omega)}{d\omega^2}L\ \Delta\omega = \beta_2 \cdot L \cdot \Delta\omega\ , \tag{5.35}$$

where $\tau_g(\omega)$ is the frequency-dependent group delay of a DWG of length L.

5.5.2.2 Dispersion Modeling

Dispersion is best analyzed by performing time-domain simulations based on realistic DWG characteristics, either simulated or measured. However, for the purposes of obtaining general conclusions in dispersion-limited DWG links, dispersion should be modeled in a more general manner. To quantify the limitations that dispersion imposes on the system, a popular condition is to set the minimal symbol period greater than the pulse broadening along the waveguide [61]:

$$T_s = \frac{1}{D} > \Delta\tau_g = \beta_2 \cdot L \cdot \Delta\omega\ . \tag{5.36}$$

While this expression serves as a decent rough estimate, it fails to model the actual broadening effect and the impact on multilevel signals. Rather, the effect of pulse broadening can be modeled on a Gaussian pulse [128] and converted into a power penalty [129] similar to how ISI was modeled in a bandwidth-limited system. For this analysis, a Gaussian pulse shape $f(t)$ is assumed with the same energy and peak amplitude of a rectangular pulse with duration $T_s = 1/D$:

$$f(t) = \frac{1}{\sqrt{2\pi}\sigma_0} \cdot e^{-\frac{t^2}{2\sigma_0^2}} = A_0 \cdot e^{-\frac{t^2}{2\sigma_0^2}} , \tag{5.37}$$

where the variance is calculated by

$$\sigma_0 = \frac{T_s}{\sqrt{2\pi}} = \frac{1}{\sqrt{2\pi}\, D} . \tag{5.38}$$

The filtered pulse (broadened by dispersion) can be expressed as [128]

$$\begin{aligned} g(t) &= \frac{1}{\sqrt{1 + \left(\frac{\sigma_D}{\sigma_0}\right)^2}} \cdot \frac{1}{\sqrt{2\pi}\sigma_0} \cdot e^{-\frac{t^2}{2(\sigma_0^2+\sigma_D^2)}} \\ &= A_0' \cdot e^{-\frac{t^2}{2(\sigma_0^2+\sigma_D^2)}} , \end{aligned} \tag{5.39}$$

where the dispersion variance is calculated as follows:

$$\sigma_D^2 = (\beta_2 \cdot L \cdot \sigma_\omega)^2 = (2.55 \cdot \beta_2 \cdot L \cdot D)^2 . \tag{5.40}$$

Figure 5.34a shows the original pulse and the pulse distorted by dispersion. Figure 5.34b visualizes how the broadening evolves as the signal propagates through

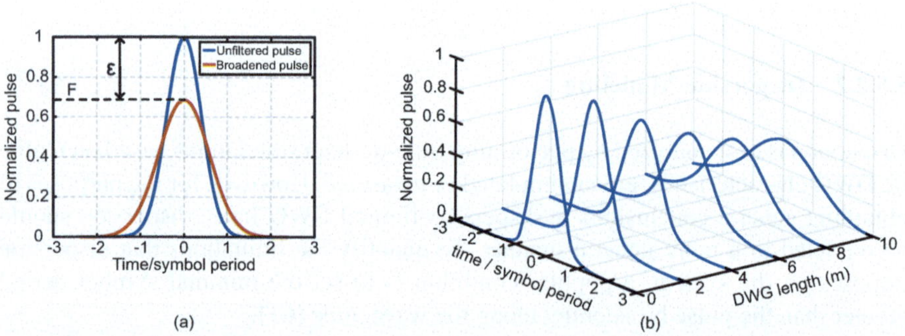

Fig. 5.34 Unfiltered pulse and pulse broadened by DWG dispersion (**a**) and visualization of the broadening of the pulse shape as the signal propagates through the DWG (**b**)

the DWG. The pulse is not only clearly broadened, the amplitude has significantly decreased as the total energy of the pulse has remained constant. The scaled amplitude (F) and vertical eye height (V_E) for an M-QAM modulated signal can be expressed as follows:

$$\frac{A_0'}{A_0} = F = \frac{1}{\sqrt{1+\left(\frac{\sigma_D}{\sigma_0}\right)^2}} = 1 - \epsilon \tag{5.41}$$

$$V_E = 1 - \sqrt{M} \cdot \epsilon = \sqrt{M} \cdot (F - 1) + 1 \ , \tag{5.42}$$

from which the power penalty can be calculated similar to ISI caused by limited bandwidth. All of the given equations can be summarized into a convenient equation which yields the maximum supported symbol rate given a tolerated power penalty PP_D (in dB), dispersion coefficient β_2, DWG length L, and M-QAM modulation:

$$D = \left[\left(\frac{10^{-\frac{\text{PP}_D}{20}} - 1}{\sqrt{M}} + 1\right)^{-2} - 1\right]^{1/4} \cdot \left[\beta_2 \cdot L \cdot 2.55 \cdot \sqrt{2\pi}\right]^{-1/2} \ . \tag{5.43}$$

In order to include the effects of dispersion in the model, concrete values must be plugged into this equation. The exact group delay variation within the bandwidth depends on which part of the group delay curve the link is operating, as is clear from Fig. 5.33. Based on DWG characteristics used in realistic links in the literature [52, 104, 106, 107, 130], a group delay variation per unit BW and unit length of 5 ps/GHz/m is a reasonable assumption. Furthermore, a dispersion-induced power penalty of 3 dB will be tolerated in this analysis, which corresponds to the results obtained from time-domain simulations and DWG link measurements.

5.5.2.3 Evaluating DWG Link Capacities

The maximum supported symbol rate in the link is now determined by the ISI caused by the limited system BW as well as the ISI caused by dispersion, as visualized in Fig. 5.35 for a 1 m DWG link using 4-QAM modulation.

At low frequencies, the symbol rate is confined by the limited bandwidth, while the symbol rate saturates at higher frequencies to the limit imposed by dispersion (which is independent of the operating frequency if the group delay behavior remains identical). After replacing the free-space path loss by the DWG loss for the calculation of the system SNR, the capacity of a DWG link can now be calculated using the same model as the wireless link. Figure 5.36 shows the link capacities for a 1 m, 5 m, and 10 m DWG link.

These capacity curves show three different regions of operation. At very low frequencies, the capacity is limited by the available system BW, indicated by a steep

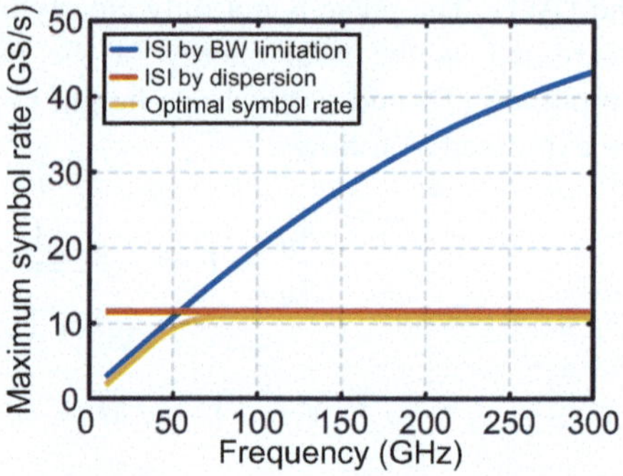

Fig. 5.35 Overview of the symbol rate limitations caused by limited BW, dispersion, and resulting global optimum symbol rate

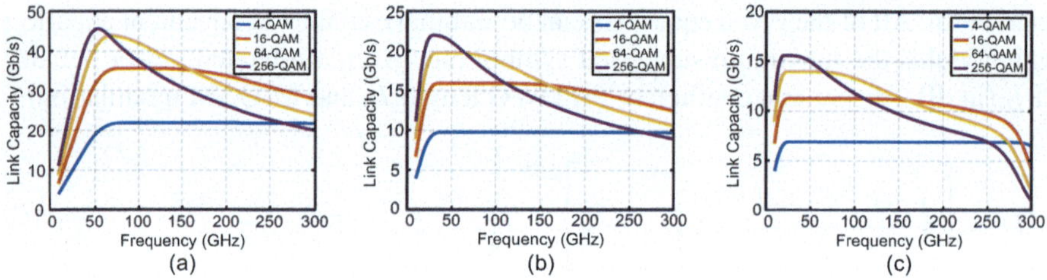

Fig. 5.36 Link capacities for a 1 m (**a**), 5 m (**b**), and 10 m (**c**) DWG link, including all modeled impairments

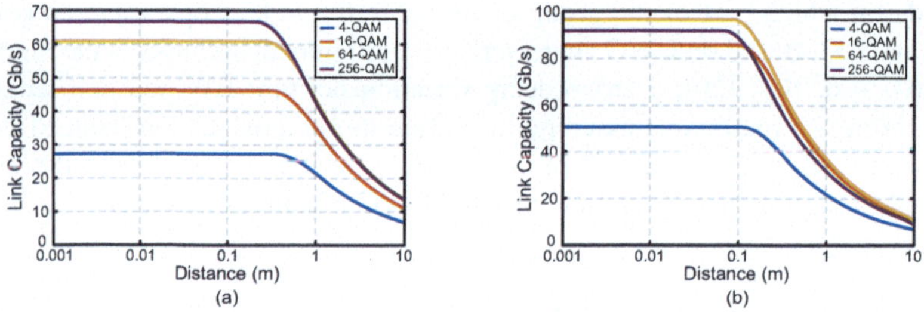

Fig. 5.37 Link capacities versus distance for a DWG link at 70 GHz (**a**) and 140 GHz (**b**)

increase in achievable data rate as frequency increases. The link capacity saturates and remains constant when the dispersion limit is reached. Finally, the capacity decreases again at higher frequencies when insufficient SNR causes the spectral efficiency C_S to decline toward zero.

An alternative viewpoint is given in Fig. 5.37, where the capacity is plotted versus DWG length for a link operating at 70 and 140 GHz. Where the link is mainly limited by BW and phase noise at short lengths, dispersion leads to a drop in capacity for longer lengths. The dispersion-dominated region can be identified by the slope proportional to the square of DWG length, as suggested by Eq. 5.43. Similar to the plots for wireless links, 256-QAM is optimal at 70 GHz, but at

140 GHz 64-QAM provides the best performance due to the higher phase noise. The main takeaway is that, with the current model values, dispersion is the main data rate bottleneck in DWG links longer than 1 m, as confirmed by the literature [52, 104, 127, 130].

5.6 Modeling Power Consumption and Energy Efficiency

Achieving high data rates is obviously highly desirable and is one of the key driving forces of technological and economical growth. Energy efficiency, on the other hand, is often ignored in academic research and considered less important. However, the exponentially growing energy consumption of global communication infrastructure calls for more attention toward efficiency. The last part of this chapter thus attempts to model the power consumption and energy efficiency of high-speed mm-Wave links to uncover important trends and reveal the most crucial efficiency bottlenecks as the world moves toward higher speed communication.

This section requires more nuance on the type of architecture, as the power consumption depends on the choice and specification of the circuit blocks in the system. As discussed before, the direct-conversion architecture is the most suitable for broadband mm-Wave communication. Compared to a heterodyne architecture, the data converters work at baseband and are hence less complex and consume less power.

The first step is to model the power consumption of each important block as a function of the system specifications: carrier frequency, link distance, modulation, and data rate. For this, the general framework is subdivided into three parts, as shown in Fig. 5.38, each consisting of their most essential building blocks: the RF front-end, LO generation, and data converters. Only single-cell (wireless and DWG) links will be considered as phased-array systems require a more complex model which is outside the scope of this book. For instance, the power consumption of the LO distribution blocks heavily depends on the exact phased-array implementation and physical layout. The power consumption modeling for each part will now be discussed.

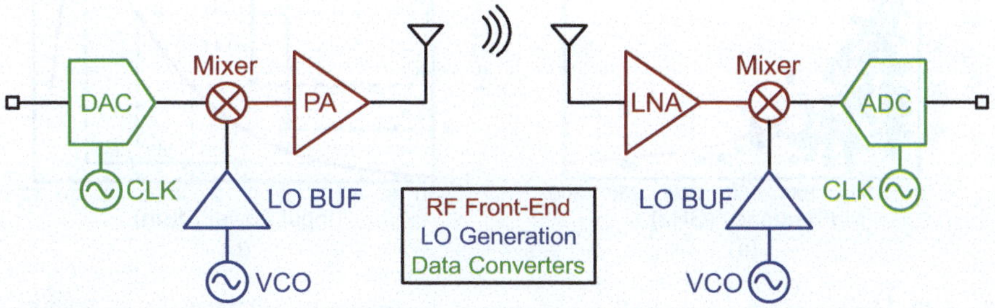

Fig. 5.38 Overview of the power consumption modeling framework

5.6.1 RF Font-End

The RF front-end is comprised of all the high-frequency blocks responsible for amplification, upconversion/modulation and downconversion/demodulation, excluding the LO generation. While this may contain many different blocks, only the most essential in terms of performance and power consumption are considered: the power amplifier, low-noise amplifier, and mixer.

5.6.1.1 Power Amplifier

The PA usually dominates the power consumption of the system, at least when large output powers are desired. The power consumption is modeled based on its output power and efficiency (PAE) as follows:

$$P_{\mathrm{DC}} = \frac{P_{\mathrm{out}} - P_{\mathrm{in}}}{\mathrm{PAE}} = \frac{P_{\mathrm{in}}(G-1)}{\mathrm{PAE}} \approx \frac{P_{\mathrm{out}}}{\mathrm{PAE}}, \tag{5.44}$$

where the approximation ($G \gg 1$) is valid in case of moderately high power gains. For instance, if the gain is larger than 20 dB, the approximation gives an error lower than 1% on the resulting P_{DC}. The PAE is modeled to degrade with the square of frequency, like $G_{\max}$, and was fitted to data from published state-of-the-art PAs [21]. Using a reference PAE of 15% at 100 GHz, the resulting PAE model is shown in Fig. 5.39a along with the individual data points. Note that, above $f_{\max}$, a PA with positive gain is not possible and a multiplier-last TX would be required, see Chap. 6. Instead of using efficiency numbers of published multipliers, the worse efficiency of above $f_{\max}$ transmitters is adequately modeled by the poor PAE from Fig. 5.39a.

An important aspect of a PA amplifying high-order modulated signals is the required power back-off to satisfy the linearity requirements. Not only does this lead to a lower peak output power, it also degrades the PAE compared to its maximum

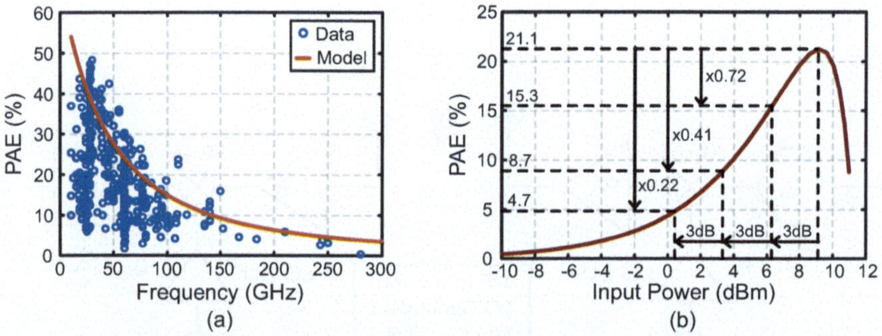

Fig. 5.39 Maximum efficiency ($\mathrm{PAE}_{\max}$) of state-of-the-art mm-Wave CMOS power amplifiers [21] (**a**) and simulated PAE of a 40 μm NPDP versus output power (**b**)

value at saturation, which was modeled in Fig. 5.39a. Based on the large-signal simulation performed in Sect. 4.3 for a 40 μm NPDP biased in class AB (bias of 400 mV), the degradation in PAE versus PBO can be estimated. As illustrated by the simulated PAE vs. output power in Fig. 5.39b, the PAE degradation when operating in a 3 dB, 6 dB, and 9 dB PBO is approximately equal to 28%, 59%, and 78%, respectively.

The power consumption of the PA should then be calculated based on the average output power and average efficiency:

$$P_{DC} = \frac{P_{out,avg}}{PAE_{avg}}, \tag{5.45}$$

where the average efficiency is calculated on the exact PAE characteristics versus output power and the statistical distribution (i.e., its probability density function or PDF), of the transmitted symbols (with corresponding output power):

$$\mathrm{PAE}_{avg} = \int \mathrm{PAE}(P_{out}) \cdot \mathrm{PDF}(P_{out})\, dP_{out}\,. \tag{5.46}$$

The $P_{out,avg}$ can easily be calculated based on P_{sat}, the required PBO, and PAPR of the modulation, as was done in the previous sections. To rigorously calculate PAE_{avg}, one must know the exact PAE characteristic as a function of P_{out}, which changes with each PA design, as well as the symbol distribution per modulation. However, for this analysis, an approximation will be made by calculating the power consumption based on the modulation-specific peak output power (P_{sat}—PBO in dB units) and the PAE at that PBO level:

$$P_{DC} = \frac{P_{out,peak}}{\mathrm{PAE}_{peak}}, \tag{5.47}$$

where the PAE_{peak}-numbers for several power back-offs can be deducted from Fig. 5.39b. Figure 5.40 gives an overview of the peak PAE numbers over frequency for the different modulations corresponding to different power back-offs, effectively combining Fig. 5.39a and b. The exact P_{DC} numbers of course depend on the concrete power level of the PA, which varies based on the situation.

Note that despite a lower PAE_{peak}, the P_{DC} of the PA for a more dense modulation can be lower than for a less dense one if the $P_{out,peak}$ is significantly lower (due to a large necessary PBO). Of course, in that case, the link with the more dense modulation has significantly lower available SNR. It is, however, not fair to compare the power consumption for two links with incomparable SNR performance. To make a more realistic comparison, the output power for each modulation will be scaled according to its SNR specifications, as will be explained later on. For instance, since 4-QAM requires lower SNR for the same BER than 64-QAM, the 4-QAM PA is able to scale down its output power while operating in saturation, thereby reducing its PA power consumption without any relative degradation in performance.

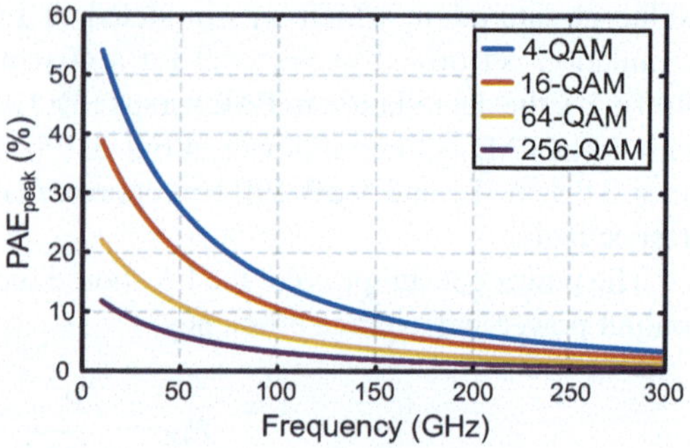

Fig. 5.40 Modeled peak PAE over frequency for various modulations and thus power back-offs

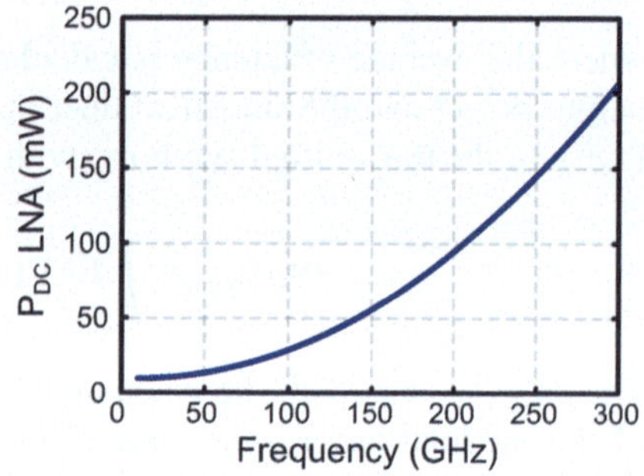

Fig. 5.41 Power consumption of a 20 dB gain low-noise amplifier versus frequency

5.6.1.2 Low-Noise Amplifier

The LNA can also consume a significant amount of power, especially at higher frequencies. It can be modeled like a small-signal amplifier where the power consumption is determined by the necessary gain with an efficiency degrading like G_{max} with freq^{-2}. As the frequency doubles, for instance, the gain per stages divides by four, which requires four times as many stages with four times the power consumption to achieve the same gain. A fixed gain of 20 dB will be assumed, independent of frequency, with a P_{DC} of 50 mW taken as a reference value at 140 GHz. Figure 5.41 shows the resulting LNA power consumption over frequency.

5.6.1.3 Mixer

The mixer blocks are assumed to be passive and thus consume zero power. The LO power required to drive the mixers, which can be very significant, is included in the power of the LO generation block, discussed next. In case one of the mixers would have an active implementation, with a small power consumption, less gain stages would be needed as well as lower LO driving power, eventually equating to about the same global power consumption.

The total power consumption of the RF front-end is thus calculated as

$$P_{\mathrm{RF}} = P_{PA} + P_{\mathrm{LNA}} \,. \tag{5.48}$$

5.6.2 LO Generation

The LO generation is highly important for good system performance, and its power consumption is becoming increasingly important at higher carrier frequencies. LO generation is split up into two parts: the generation of the actual oscillating signal by the VCO/PLL and the power generation by LO buffer amplifiers. Even though QAM modulations are examined in this analysis, the generation of the IQ phases (e.g., by a quadrature VCO or passive quadrature hybrid) is not explicitly considered since it can be done in a reasonably efficient way that does not heavily influence the power consumption.

5.6.2.1 Voltage-Controlled Oscillator

While a VCO generates the oscillating signal, it is usually embedded within a PLL to achieve better phase noise performance, as discussed before. Instead of modeling the power of each block in the PLL separately, the general trends of the total PLL consumption can be captured within the VCO model. Remember that the phase noise of a VCO can be expressed as follows:

$$\mathrm{PN}(f) = \frac{2FkT}{P_{\mathrm{VCO}}} \left(\frac{f_0}{2Q\,f} \right)^2 . \tag{5.49}$$

Using the previous expressions for the total RMS phase error, SNR of the LO signal, and jitter, they can be related to the power consumption of the VCO as follows:

$$\frac{1}{\mathrm{SNR}_{\mathrm{LO}}(f_0)} = (2\pi f_0 t_{jit})^2 = \phi_{\mathrm{rms}}^2 = \int \mathrm{PN}(f)\, df = \frac{K^2 f_0^2}{P_{\mathrm{VCO}}} \,, \tag{5.50}$$

where K is a constant that combines the non-changing parameters. This finally leads to an expression of the VCO power as a function of the required jitter, or LO SNR and frequency:

$$P_{\mathrm{VCO}} = \frac{K_1}{t_{jit}^2} = K_2 \cdot \mathrm{SNR}_{\mathrm{LO}}(f_0) \cdot f_0^2 \,, \tag{5.51}$$

where K_1 and K_2 are constants again. The power consumption is mainly determined by the required phase noise performance and thus increases for more dense modulations as they require lower phase noise. Note that the VCO P_{DC} increases with f_0^2 for the same phase noise ($\mathrm{SNR}_{\mathrm{LO}}$) but is fully determined by the required time-domain jitter, independent of frequency. However, the same jitter performance will have a more degrading effect on the system as the operation frequency increases or as the employed modulation scheme becomes more dense.

While these expressions describe the important trends, the absolute values, or constants K_1,K_2, still need to be determined based on the literature. By using PLL power consumption values as a reference, the extra consumption of the extra blocks in the PLL are accounted for, while preserving the trends described by the VCO. Based on the state-of-the-art mm-Wave PLLs, a P_{DC} of 25 mW is chosen as the reference power for a 60 GHz PLL with 168 fs jitter. This number is then scaled according to the required jitter and frequency specification.

Note that this model also includes all necessary multipliers that would be needed to generate the LO. For example, even if a 300 GHz would be generated by means of a 150 GHz VCO with a frequency doubler, this analysis models it as a 300 GHz VCO/PLL. What option would lead to the best phase noise or lowest P_{DC} is irrelevant for this model, as long as the power consumption increases realistically with frequency and phase noise performance. Furthermore, the power is doubled to account for the PLL power in the TX and RX.

5.6.2.2 LO Buffers

LO buffer amplifiers are needed to drive the mixers with a high (ideally rail-to-rail) voltage swing. While this is straightforward at lower frequencies, at high frequencies significant power must be generated by the buffer which is typically designed as a full-blown power amplifier. Assuming the mixer provides a capacitive load ($Z_L = 1/2\pi f_0 C_L$) to the LO buffer that decreases with frequency, the power that must be delivered to the mixer load for a rail-to-rail (V_{DD}) voltage swing can be written as

$$P_L = V \cdot I = V_{\mathrm{DD}} \cdot \frac{V_{\mathrm{DD}}}{Z_L} = V_{\mathrm{DD}}^2\, 2\pi f_0 C_L\,, \tag{5.52}$$

showing that the required LO power increases approximately linearly with frequency. Doubling the frequency thus requires about 3 dB more LO power, which leads to a larger LO buffer power consumption. A reference LO power of 5 dBm is taken at 140 GHz.

Note that this PA can work fully in saturation, since the LO is a CW signal. This means that the P_{DC} is calculated using $P_{\mathrm{out}} = P_{\mathrm{sat}}$ and $\mathrm{PAE} = \mathrm{PAE}_{\mathrm{max}}$ instead of the values at power back-off, leading to considerably higher efficiencies compared to the modulated PA. Figure 5.42a shows the required LO power into the mixer versus frequency, while the power consumption of a single LO buffer is plotted on Fig. 5.42b, using the previously modeled PAE values. For frequencies close to f_{max}, this power consumption represents the power of a low-efficiency multiplier chain, which also partly modeled in the VCO power. Also, this P_{DC} is again doubled to account for the LO buffers at TX and RX.

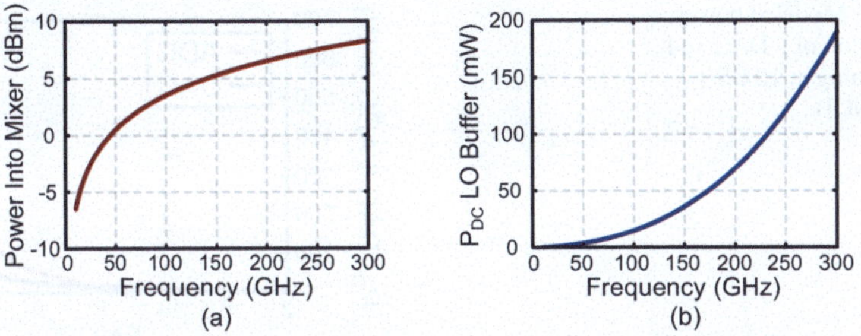

Fig. 5.42 Required power delivered to the mixers (**a**) and power consumption of the LO buffers versus frequency (**b**)

The total power consumed by the LO generation block is hence

$$P_{\mathrm{LO}} = 2\,P_{\mathrm{VCO}} + 2\,P_{\mathrm{BUF}}\ . \tag{5.53}$$

Note that LO synchronization, although essential in a high-speed communication system, is not included in the presented model. While LO synchronization is traditionally performed in the digital domain, analog approaches are on the rise, which can potentially do the job in a more energy-efficient way. The power consumed by digital and analog carrier recovery blocks depends heavily on the employed digital algorithms or analog circuit blocks, respectively, and would require quite a complicated model to predict accurately. Therefore, despite being a crucial component of communication systems, LO synchronization is not included in the model, which could have a small impact on the accuracy of the results. We can predict, however, that including this part would negatively impact more dense modulation schemes as the complexity of synchronization grows quickly as the number of modulation states increases.

5.6.3 Data Converters

5.6.3.1 DACs and ADCs

In a direct-conversion architecture, two DACs and two ADCs are needed in the TX and RX, respectively. Their resolution, expressed in effective number of bits (ENOB), is determined by the modulation scheme, and the sample rate is fixed by the symbol rate. Murmann provided a good energy model for ADCs in [131], leading to the following fitting function of the ADC power consumption:

$$P_{\mathrm{ADC}} = \left[k_1 \cdot \mathrm{ENOB} + k_2 \cdot 4^{\mathrm{ENOB}}\right] \cdot f_s\ , \tag{5.54}$$

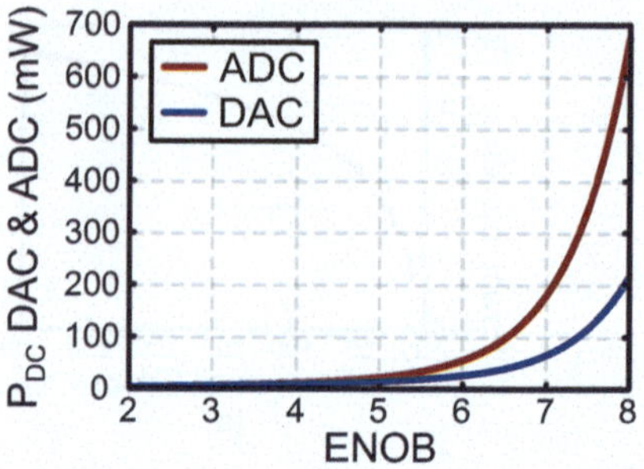

Fig. 5.43 Modeled power consumption of a DAC and ADC running at 10 GS/s versus ENOB

which accurately models low- as well as high-resolution converters. Furthermore, using ENOB as a metric for resolution as opposed to the number of bits yields a much more realistic power estimation. By fitting this model on state-of-the-art high-speed ADCs (e.g., [132–134]), k_1 and k_2 have been set at 250 fJ and 1 fJ, respectively.

A similar model can be used to estimate the power consumption of the DACs, but with different fitting parameters based on state-of-the-art high-speed DACs (e.g., [135–137]). Indeed, compared to ADCs the power consumption in DACs increases much less rapidly with increasing ENOB, leading to a k_1 and k_2 of 250 fJ and 300 aJ, respectively.

Setting the sample rate equal to the symbol rate is a reasonable assumption for high-speed direct-conversion systems, as oversampling is mostly done at lower speeds. The necessary ENOB is more challenging to determine, as it depends on the applied signal processing operations, such as equalization and pulse shaping, and the type of carrier and clock recovery circuits, as is discussed more in depth in Chap. 7. Based on the literature, an ENOB equal to the resolution of the baseband modulation plus an additional three bits is assumed. For instance, for 16-QAM, the baseband PAM-4 signals have a resolution of 2 bits and thus require data converters with an ENOB of 5 bits in this model. Figure 5.43 shows the power consumption of the modeled DAC and ADC running at 10 GS/s for increasing ENOB values.

5.6.3.2 Clock Generation

The generation of the clock signal used by the DACs and ADCs can also add significant power to the system, especially if low clock jitter is required. This clock generation is modeled similar to the VCO model, without the buffer, with a different jitter specification. Due to the lower frequency, the clock generation is typically much less power consuming than the LO generation. Nevertheless, its power is added (twice) to that of the DACs and ADCs.

5.6.3.3 Baseband Amplifiers

One important type of block missing from the model is the baseband amplifier. Apart from RF amplification (in the PA, LNA, LO buffers), baseband amplification is mostly needed to amplify the output of the downconversion mixers before it is digitized by the ADC. Its power consumption again heavily depends on the necessary bandwidth (determined by the symbol rate) and linearity specifications (determined by the modulation scheme). However, if these amplifiers are designed somewhat efficiently, they should only consume a fraction of the DC power consumption of the system. Since the model mostly focuses on the high-frequency RF blocks, the baseband amplifiers are ignored for this analysis but can be added to the model if desired.

In total, the data converter block consumes the following:

$$P_{conv} = 2\,P_{\mathrm{DAC}} + 2\,P_{\mathrm{ADC}} + 2\,P_{\mathrm{CLK}}\,. \tag{5.55}$$

5.6.4 *Energy Efficiency*

Knowing the power consumption of each block, the energy efficiency can be calculated based on the data rate as the link efficiency is typically quantified in energy per bit:

$$\eta = \frac{P_{\mathrm{DC}}}{R}\,. \tag{5.56}$$

Note that a "better" efficiency refers to a lower energy consumption per bit, which can be confusing at times. Furthermore, making a fair and insightful efficiency comparison is less straightforward than simply calculating the link capacity.

5.6.4.1 Modeling Methodology

First of all, we want to compare the efficiency across frequency and modulations at a fixed distance and while supporting a specific data rate, as opposed to at the maximum capacity. The theoretical capacity is evaluated by means of the mutual information determined by the SNR in the link. However, in a practical link, and hence in the efficiency model, the data rate is determined by the spectral efficiency of the modulation (maximum mutual information) and the symbol rate. The required total SNR can be deduced for a certain BER (see Table 2.4) and the symbol rate determines the bandwidth. Note that the degradation of link performance is now captured directly into a specified SNR value, instead of through the mutual information, which is a theoretical concept anyways.

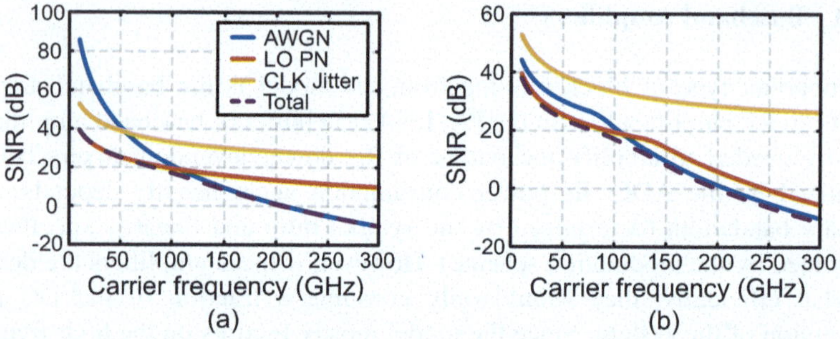

Fig. 5.44 Overview of the different SNR specifications without (**a**) and with (**b**) power consumption optimization while maintaining equal total link SNR

Furthermore, when maximizing the capacity, the performance of each block was always optimized to identify the bottlenecks of the system. For instance, the LO phase noise was always minimized and the SNR of the link maximized. However, when phase noise is the bottleneck, lowering the PA output power or increasing the LNA noise figure can significantly reduce the power consumption of the system without affecting the link performance. Since the PA is typically more power consuming than the LNA, reducing the PA output power will result in a more significant energy reduction than increasing the NF of the LNA would. Therefore, in each situation (frequency, data rate, distance) the LO phase noise and PA output power specifications are relaxed to a lower point with minimal performance degradation. Concretely, as illustrated by Eq. 5.29, when phase noise is the bottleneck, $\mathrm{SNR_{AWGN}}$ is lowered until it is a certain factor higher than $\mathrm{SNR_{LO}}$ to avoid any meaningful decrease in $\mathrm{SNR_{tot}}$. In case the system is noise limited, $\mathrm{SNR_{LO}}$ is lowered closer to $\mathrm{SNR_{AWGN}}$. SNR_{jit} remains unaffected as the aperture jitter of the data converters does not usually form the bottleneck, see Fig. 5.28, nor does the clock generation for these converters consume a lot of power compared to the other blocks. Figure 5.44 shows the various SNR impairment components with and without this optimization, demonstrating a negligible decrease in overall SNR whilst minimizing the individual specification to optimize power consumption.

From the specified data rate and modulation scheme, the necessary symbol rate and required SNR can be calculated. A BER of 10^{-3} is used as a reference in this model. Then, for each frequency and link distance, the required BW, PA output power, and LO phase noise are determined. From these specifications, the power consumption of each block can be determined based on the discussed model. And finally, the efficiency is calculated from the obtained power consumption and data rate. These calculations can be performed over a range of frequencies, data rates, and distances to gain valuable insights. Note that if the specification is higher than what is achievable by the technology, i.e., the desired data rate is higher than the link capacity, no P_{DC} number is reported as the link is not practically realizable. For example, the bandwidth cannot be higher than the specified ISI limit, and the PA output power is constrained by the values discussed before.

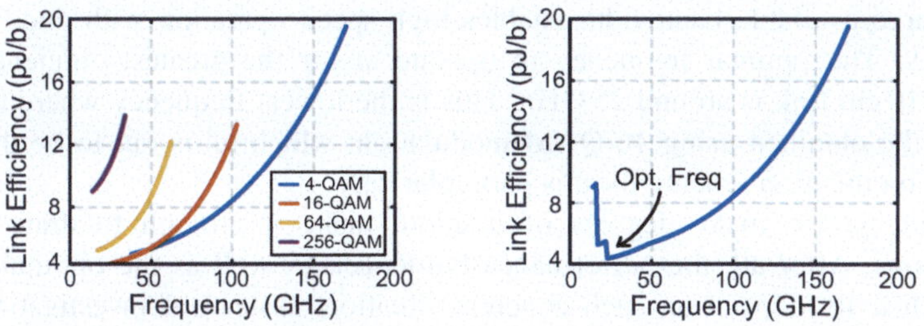

Fig. 5.45 Link efficiency versus frequency for a 15 Gb/s 10 cm link for different modulations (**a**) and minimum efficiency over modulation (**b**)

5.6.4.2 Link Efficiency Example

As an initial example, Fig. 5.45a plots the link efficiency versus frequency for a 15 Gb/s link at 10 cm distance for the various modulation schemes. Figure 5.45b shows the minimum energy efficiency taken over the difference modulation. Several observations can be made from these figures.

For each modulation, an efficiency value is only given in case the concrete link is deemed realistic, i.e., data rate is achievable at the given distance and frequency. At the lower frequency end, the data rate is limited by the available bandwidth. As the spectral efficiency increases, the minimum frequency at which the data rate can be obtained decreases. At higher frequencies, the link performance is ultimately constrained by the available SNR, either limited by AWGN or phase noise. Obviously, the highest frequency that can support the specified data rate decreases with modulation density, as the available SNR drops (lower PA output power and higher RX noise figure), while the SNR requirements increase (higher P_{sig} and lower PN needed) for the same BER.

Compared to the capacities of a 10 cm link, as plotted in Fig. 5.26b, the frequency range with a valid efficiency is smaller than the range where the data rate is smaller than the capacity, particularly at the high-frequency boundary. Indeed, as the SNR drops below its necessary value for a certain BER, it becomes progressively more difficult to get close to the theoretical capacity, as was illustrated in Fig. 5.4. To keep this efficiency analysis somewhat realistic, we are effectively only considering the first part of the capacity curve of Fig. 5.26b, where the data rate increases linearly with frequency (due to an increase in bandwidth) until the SNR is too low to achieve the desired BER. In order to push the data rate higher, more complex blocks are required to boost the SNR, such as error correction or predistortion, which consume considerable power but are not included in this model. Nevertheless, many conclusions can be drawn from this model.

From inspecting Fig. 5.45, one can conclude that the minimum achievable efficiency at lower frequencies is determined by the high-order modulation schemes, as their high spectral efficiency is required to produce the specified data rate given the low available bandwidth. At higher frequencies, the simpler schemes take over as the large available bandwidth enables high-speed operation with good energy efficiency. The optimal frequency to operate at for the greatest efficiency of a 15 Gb/s 10 cm link is around 25 GHz. This is the lowest frequency where the data rate can be obtained using 16-QAM modulation, which turns out to be the most efficient modulation scheme for this particular case.

Obviously, the many degrees of freedom make it difficult to draw overall conclusions. After all, the performance bottleneck as well as the optimal design choices heavily depend on each concrete situation. Instead of investigating each case and sweeping every possible parameter, the last section will discuss several concrete cases in order to obtain general conclusions in mm-Wave communication system design.

5.7 Efficiency Case Studies

This section investigates several interesting case studies to deduct overall trends and deeper insights in high-speed communication link design. Single-cell wireless and guided DWG links will be analyzed over data rate as well as link distance. Phased-array systems are not investigated here since a more complete model is required to realistically assess such links, e.g., modeling the effects of phase/gain resolution, high-frequency antenna interconnects, etc.

5.7.1 Case Study 1: 25Gb/s Wireless Link vs. Distance

In this first example, the data rate is fixed at 25 Gb/s, while the link distance is swept. The individual efficiency curves for the different modulations over distance are shown in Fig. 5.46a. At each distance, the minimum value (over frequency) is picked, which is usually the lowest frequency at which the data rate can be supported. To illustrate this, the efficiency versus frequency is shown for various distances in Fig. 5.47.

The optimal frequency to operate at as a function of distance can be determined by taking the minimum over all modulations and is plotted in Fig. 5.46b. Three regions can be identified and explained using the plotted figures. For short distances, 16-QAM is the most efficient modulation, see Fig. 5.47a, and thus determines the optimal frequency (just below 40 GHz). When the distance crosses about 1.1 m, 4-QAM takes over as the most efficient modulation, see zoomed-in portion of Fig. 5.46a, and changes the optimal frequency to a higher value, since 4-QAM needs a higher bandwidth to achieve 25 Gb/s (see Fig. 5.47b). Finally, at distances above 3 m, 4-QAM is not able to support 25 Gb/s anymore, as observed from Fig. 5.47c. This shifts the optimal frequency back down as 16-QAM once again becomes the optimal modulation. This can be explained as follows: Since 4-QAM needs a higher bandwidth than 16-QAM, it must operate at a higher frequency where the SNR is

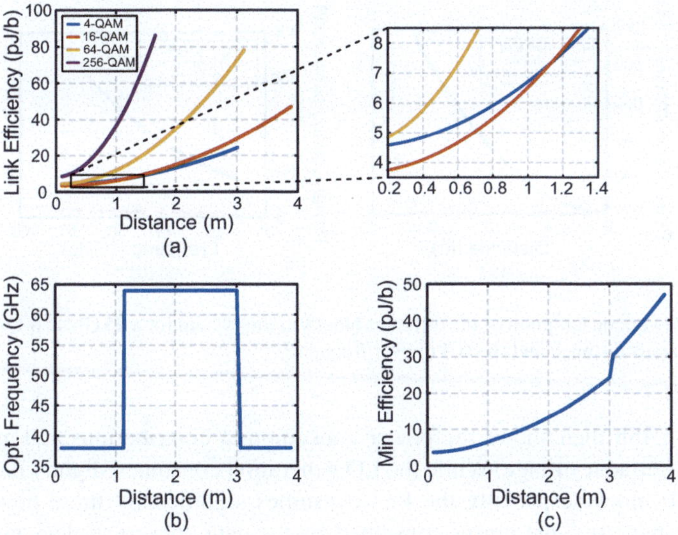

Fig. 5.46 The link efficiency of a 25 Gb/s link over distance for the different modulations (**a**), the optimal frequency (**b**), and corresponding minimum efficiency over modulation (**c**)

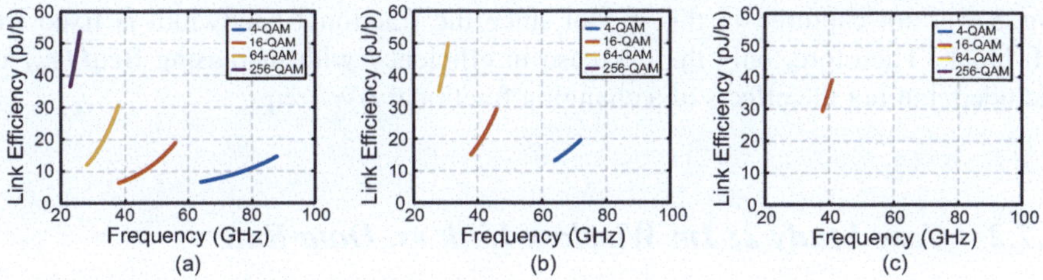

Fig. 5.47 The link efficiency of a 25 Gb/s link over frequency for the various modulations at 1 m (**a**), 2 m (**b**), and 3 m (**c**)

insufficient at 3 m distance, while 16-QAM has sufficient SNR at a lower frequency. Figure 5.46c then shows how the minimum efficiency, at the optimal frequency, changes over distance. While the transition at 1.1 m is smooth (4-QAM barely takes over from 16-QAM), it is discontinuous at 3 m as 4-QAM suddenly

To gain more insight into what blocks are dominating the power consumption, Fig. 5.48 breaks down the power in two situations, both for a 25 Gb/s 16-QAM link. Figure 5.48a plots the power consumption of the three blocks versus distance when operating at 40 GHz, which is close to the optimum frequency for 16-QAM, over the range of distances where 25 Gb/s is supported. The consumption of the RF block increases quickly with distance, since the required PA output power rises, while the LO generation and data converter blocks consume the same. The LO generation consumes the least, as the phase noise spec is quite mild given the small data rate, while the RF block dominates the power consumption from about 0.8 m onward.

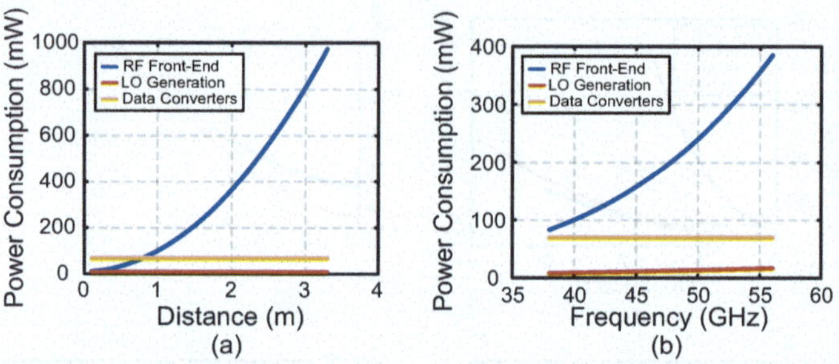

Fig. 5.48 Power consumption of the different blocks in the system for a 25 Gb/s 16-QAM link (**a**) at 40 GHz versus distance and (**b**) at 1 m over frequency

Figure 5.48b then shows the power consumption contributions at 1 m over the supported frequency range. While the LO generation consumes slightly more power and the RF block (especially the PA) consumes significantly more power as the frequency increases, the power consumed by the data converters does not change, as this solely depends on the resolution (modulation) and sample rate (symbol rate). The optimal frequency is hence the lowest frequency where the specified data rate is supported. The minimum point in the efficiency curve versus frequency from Fig. 5.4 is not captured in this model since the fractional bandwidth is fixed for RF block. Therefore, only the decrease in efficiency with increasing freq/f_{max} is included, but not the effects of a changing fractional BW/freq.

5.7.2 *Case Study 2: 1m Wireless Link vs. Data Rate*

The next case study keeps the link distance fixed at 1 m while increasing the data rate. Figure 5.49a displays the link efficiency versus data rate, demonstrating that 16-QAM provides the best efficiency across all data rates, only just better than 4-QAM. Compared to Fig. 5.26c, the maximum supported data rates indeed are slightly lower than the capacities for a 1 m link. In Fig. 5.49b, the different power consumption contributions are visualized for a 1 m 16-QAM link operating at 30 GHz. For small data rates, the consumption of the RF front-end is the most significant, while the data converters consume more and more power as the data rate rises, eventually dominating the total consumption at higher rates. Of course, since the frequency has to rise with increasing data rates, the consumption of the RF front-end will also increase but is not visible on the plot since the frequency is kept constant there.

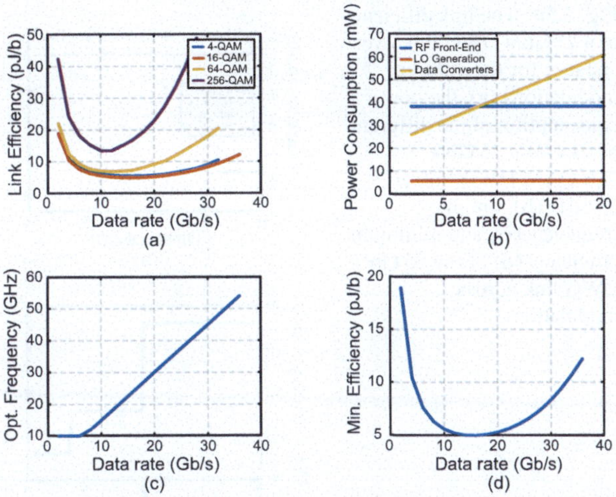

Fig. 5.49 The link efficiency of a 1 m link over data rate for the different modulations (**a**), the power consumption of the different blocks in the system for a 1 m 16-QAM link operating at 30 GHz (**b**), optimal frequency (**c**), and corresponding minimum efficiency (**d**) of the 1 m link across modulations

The optimal frequency and corresponding minimum energy efficiency are plotted on Fig. 5.49c and d. The optimal frequency is always as low as permitted by the bandwidth requirement, thus monotonically increasing with data rate once the minimal frequency (10 GHz in this analysis) provides insufficient bandwidth. Figure 5.49d shows how the global link efficiency first improves with increasing link rate and then increases again for larger speeds. This is reminiscent of the predicted curve from Fig. 5.4 but in this case demonstrates that the increase in power consumption (dominated by the data converters) is relatively larger than the gain in data rate, thus degrading the overall efficiency.

5.7.3 Case Study 3: 25Gb/s DWG Link vs. Distance

Now a 25 Gb/s DWG link will be analyzed over DWG length. To avoid, tiny ripples in the results, the transition between BW-limited and dispersion-induced ISI, as shown in Fig. 5.35, will be made more sharp. This way, a small increase in distance and thus dispersion will not influence the bandwidth if the link is already dispersion limited. Indeed, as shown in Fig. 5.50a, the energy efficiency of the 25 Gb/s DWG link changes negligibly as the DWG length or distance is increased. Not only does the data rate stay constant, since dispersion is the main limit in this case, the power consumption also does not vary significantly, as depicted in Fig. 5.50b for the 256-QAM link at 23 GHz. After all, since the link is dispersion limited, increasing the SNR (and thus PA output power and P_{DC}) will not improve the performance. Figure 5.50c and d displays the optimal frequency and minimum efficiency over distance, showing sudden jumps each time the link switches to a more dense modulation, with a degradation in efficiency as a consequence.

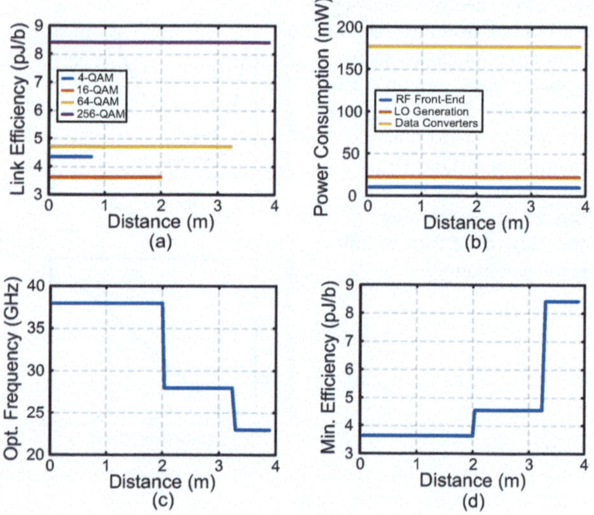

Fig. 5.50 The link efficiency of a 25 Gb/s DWG link over distance for the different modulations (**a**), the power consumption of the different blocks for a 25 Gb/s 256-QAM DWG link at 23 GHz (**b**), optimal frequency (**c**), and minimum efficiency (**d**) of the 25 Gb/s DWG link across modulations

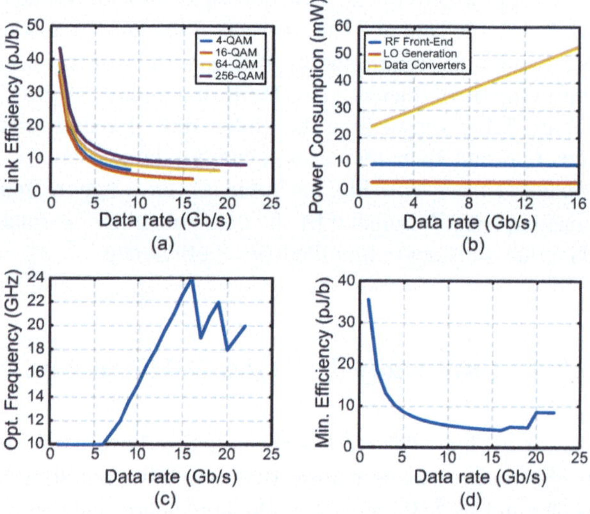

Fig. 5.51 The link efficiency of a 5 m DWG link over data rate for the different modulations (**a**), the power consumption of the different blocks for a 5 m 16-QAM DWG link at 24 GHz (**b**), optimal frequency (**c**), and minimum efficiency (**d**) of the 5 m DWG link across modulations

Indeed, since dispersion is the bottleneck, a high spectral efficiency is key to limit the required bandwidth. For short distances, however, 16-QAM remains the most efficient modulation.

5.7.4 Case Study 4: 5m DWG Link vs. Data Rate

Finally, the data rate in a 5 m DWG link is varied. Figure 5.51a shows how 16-QAM is the most efficient modulation again with the data converters dominating the power consumption, see Fig. 5.51b, since the link is dispersion limited and hence no large SNR is required. Figure 5.51c and d displays the change in optimal frequency every time the link switches to a more dense modulation, with small jumps in efficiency resulting from that. The minimum efficiency again shows a minimum, meaning that increasing the data rate beyond comes at the cost of a disproportionate increase in power consumption.

5.8 Conclusion

This chapter started with a qualitative discussion of the most important system-level considerations in high-speed communication links. For the quantitative analysis, the maximum data rate was first estimated using Shannon's capacity equation for a generic wireless link model. Next, by calculating the theoretical mutual information, modulation-specific capacity numbers could be evaluated. To make the data rates more realistic, implementation impairments were added to the model based on practical communication systems. Additionally, phased-array and DWG links were briefly discussed as viable options to increase the communication distance compared to single-cell wireless links. Finally, the power consumption and energy efficiency were modeled and discussed through several different case studies. An overview of the modeling methodology is shown in Fig. 5.52.

Based on this model, some general conclusions were drawn in the context of maximum data rate (capacity) as well as optimal energy efficiency. For single-cell wireless links, 16-QAM emerged as the optimal modulation scheme for speed, providing a good trade-off between spectral efficiency and SNR requirements. The maximum achievable data rate is mainly limited by signal power/noise and phase noise of the LO, and the optimal operating frequency decreases as the communication distance grows. Phased-array systems can significantly extend the range of wireless links but are limited in array size at higher frequencies due to the size of the ICs and numerous other scaling challenges. Finally, guided DWG links are a possible high-speed interconnect solution for short-to-medium lengths but are mainly confined by the dispersion of the channel and thus need higher order modulation schemes to push the data rate.

In terms of efficiency, 16-QAM emerges again as the most efficient modulation scheme in most situations. One exception is for high-speed dispersion-limited DWG links, where 16-QAM provides insufficient spectral efficiency to obtain the highest speeds. However, modulations more dense than 16-QAM consume significantly more power mainly caused by a lower PA efficiency and higher resolution data converters. Furthermore, in general the efficiency always degrades with increasing frequency. The optimal frequency is hence the one where just enough bandwidth is available to achieve the specified data rate. Increasing the distance also lowers the efficiency, as more signal power is required to compensate for the larger path loss. Lastly, in general, the power consumption is dominated by the RF front-end

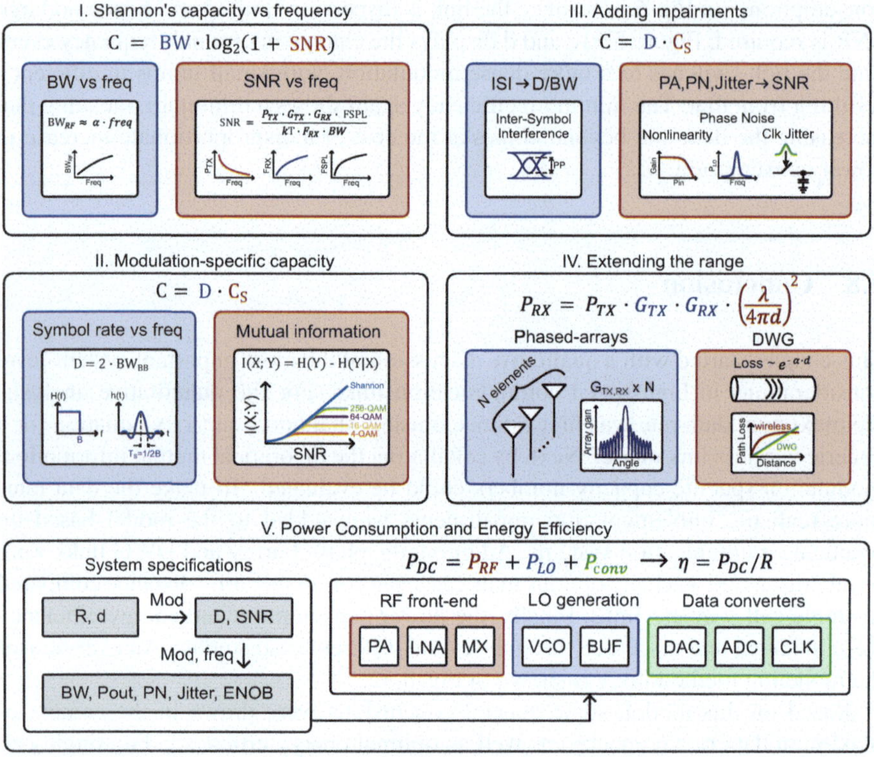

Fig. 5.52 Overview of the link modeling methodology

(mostly the PA) at long link distances with relatively small data rates, while the data converters are the most power hungry at shorter distances with high data rates.

While the current model grants considerable insight into the performance capabilities and limitations of high-speed links, it does not capture the complete picture of practical, highly complex communication systems. Equalization, for example, is very effective at (partially) compensating the ISI caused by the limited system bandwidth or excessive channel dispersion, or digital predistortion can be used to improve the average output power by pre-compensating the nonlinear behavior of the PA. Including all possible performance-enhancing techniques or algorithms in the model would be unfeasible. However, they can be included by altering existing parameters in the model, like increasing the system BW to include the effects of equalization, or decreasing the required power back-off to include the influence of digital predistortion.

Furthermore, the model can be extended further by considering other technologies. Indeed, as discussed in Sect. 1.4, SiGe or III/V technologies can provide significantly better high-frequency performance at the cost of less digital functionality. To increase the range of wireless links at high carrier frequencies, one can opt to implement the RF front-ends in a superior RF technology like InP without needing to realize phased arrays. However, to implement the data converters (and DSP), an additional CMOS die may be necessary, complicating the overall system and increasing its cost. A complete comparison of high-speed systems employing or combining different technologies is outside the scope of this work but may be necessary to obtain the most energy and/or cost optimal system.

As the desired data rate of communication systems continues to increase at a rapid rate, innovation will be necessary not only at circuit level, but also at higher system and architectural levels. While this chapter focused heavily on the system-level aspects of high-speed links, the next two chapters will delve deeper into architecture-level trade-offs and the implementation and measurements of practical circuit implementations.

Chapter 6
A High-Speed 390 GHz BPOOK Transmitter

This chapter details the design of a high-speed 390 GHz BPOOK transmitter in 28 nm CMOS. The objective of this project was to investigate transmitter architectures to exploit the large available bandwidth in the THz frequency range. The modulation, architecture, and circuit design are thus optimized for data rate as well as THz output power.

First an overview is given of possible THz transmitter architectures with their respective advantages and disadvantages. The central modulation of this system, binary-phase on–off keying or BPOOK, will be extensively studied in the next section. The subsequent sections will then cover the architecture, circuit design, and measurements results of the designed 390 GHz transmitter. Based on the transmitter's performance, the tripler-based architecture is reassessed and compared to its 130 GHz tripler-less transmitter equivalent. Finally, the chapter is concluded.

6.1 Overview of THz Transmitter Architectures

This section discusses different THz transmitter architectures suitable for high-speed communication. Figure 6.1 shows an overview of possible THz transmitters depending on the technology. Note that these block diagrams mainly focus on the THz signal generation and make abstraction of the data modulation, which can be performed along different parts in the chain.

6.1.1 Below-f_{max} Transmitters

Initially, suppose a technology is available with a high f_{max} above the operating THz frequency, for example, an advanced SiGe or III/V technology such as InP.

C. D'heer, P. Reynaert, *THz and Sub-THz CMOS Electronics for High-Speed Telecommunication*, Analog Circuits and Signal Processing,
https://doi.org/10.1007/978-3-031-64439-9_6

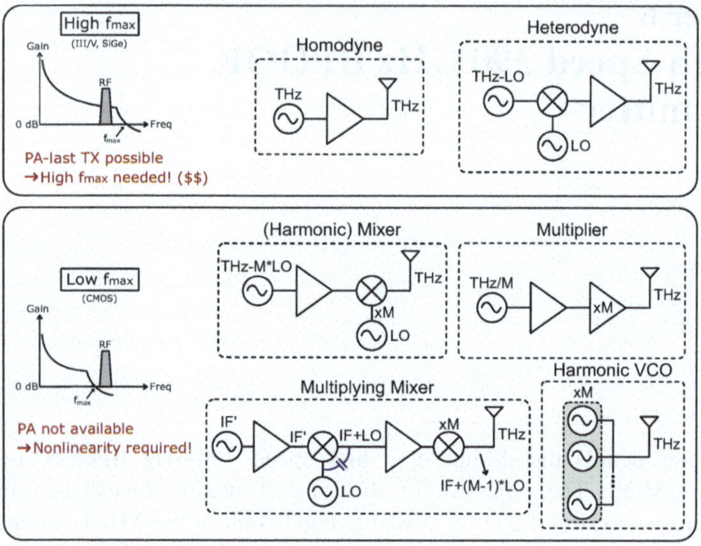

Fig. 6.1 Overview of different THz transmitter architectures

Fabricating chips in these technologies can be quite expensive and the potential yield is rather low. Furthermore, integration with advanced digital logic is limited or completely impossible, requiring an additional CMOS chip to perform these operations, which increases system complexity and cost. Nevertheless, such technologies enable the use of conventional PA-last architectures leading to relatively high output power and ordinary design trade-offs. Popular architectures include the homodyne (or direct-conversion) and the heterodyne architecture, as shown in the top part in Fig. 6.1.

6.1.2 *Above-f_{max} Transmitters*

However, with the goal of designing a THz system which is economically viable in large volumes, other technologies must be considered. CMOS technologies offer a high production yield and low unit cost while enabling large-scale integration with digital circuits. The main disadvantage is the low technology f_{max}, which is typically below the THz frequency range (around 280 GHz in 28 nm CMOS). This limits which architectures can be realized due to the unfeasibility of a power amplifier at THz frequencies. In order to generate power above f_{max}, unconventional architectures have to be devised often relying on nonlinear techniques.

6.1.2.1 Mixer-Last Transmitter

The bottom part in Fig. 6.1 shows possible architectures which enable above $f_{\max}$ signal generation. Firstly, a mixer-last architecture provides a straightforward design while maintaining linear operation. However, such a THz mixer must be passive and will be able to generate very little power. Furthermore, significant LO power needs to be generated to drive the mixer which increases the power consumption of the TX. For this reason, the mixer is often driven by a sub-harmonic of the LO signal, and the mixing product caused by a certain harmonic of the LO signal is then extracted at the output. Even though this drastically reduces the power consumption of the LO generation circuitry, the conversion gain and output power of the mixer become even lower compared to its fundamental operation.

6.1.2.2 Multiplier-Last Transmitter

Secondly, a multiplier-last architecture proves to be a viable option as it can provide a larger output power than a (harmonic) mixer without needing an additional LO signal, thus increasing the power efficiency of the TX. However, since the THz signal is extracted from the harmonics generated by a nonlinear device, the behavior is inherently nonlinear. This nonlinearity can possibly distort the modulated signal and must be taken into account when determining the modulation scheme of the system. For example, a tripler-last architecture is used to implement a 240 GHz QPSK transmitter in [50]. Note that a QPSK constellation is largely unaffected by the tripling operation.

6.1.2.3 Multiplying Mixer-Last Transmitter

Next, the two previous methods can be combined to realize a multiplying mixer-based transmitter. Here, the combination of an LO signal and a modulated IF signal is fed through a nonlinear block which generates various intermodulation products. In case the third harmonic is prioritized, the block is called a cubic mixer [138], and the intermodulation products can be expressed as follows:

$$(\mathrm{LO}+\mathrm{IF})^3 = \mathrm{LO}^3 + 3\,\mathrm{LO}^2\,\mathrm{IF} + 3\,\mathrm{LO}\,\mathrm{IF}^2 + \mathrm{IF}^3. \tag{6.1}$$

From these products, the desired signal is contained in the term "3 LO^2 IF" since it is linear with respect to the modulated IF signal, which means that the constellation shape ideally remains undistorted while being translated to a high frequency. The undesired terms can be filtered out by the matching networks and by inserting additional traps (shorts) at specified frequencies. Additionally, the LO and IF frequencies should be chosen to make this filtering practically possible. A multiplying mixer can possibly provide more output power than a conventional mixer while maintaining superior constellation linearity compared to a multiplier-last architecture. The biggest drawback is the significant system and design complexity associated with a multiplying mixer transmitter, particularly the large number of stages in the signal chain, LO generation, and intricate filtering operating. More design examples and details of a quadratic, cubic, and quintic mixer can be found in [101, 138–140].

6.1.2.4 Harmonic VCO-Based Transmitter

Lastly, a harmonic VCO can be used to directly generate a THz signal above f_{max}. Similar to previous nonlinear blocks, a higher harmonic of a fundamental VCO is extracted by providing specific filter networks. Since the output power of a single VCO is quite low due to its harmonic operation, multiple VCOs are often coherently combined to increase the effective THz output power. Despite being an efficient THz signal generation method, performing high-speed signal modulation is not trivial with this method as switching VCOs is quite slow and THz modulators are very lossy. This topology is thus very popular for imaging applications but not ideal for telecommunication purposes. An example of a 640 GHz 4×2 oscillator array is described in [141].

6.1.3 Chosen THz Transmitter Architecture

As a CMOS implementation is desired for reasons discussed before, a nonlinearity-based above-f_{max} topology is necessary. A multiplier-last design is chosen for its simplicity and power efficiency. Choosing the harmonic number of the multiplier entails a trade-off mainly between input frequency and output power. Higher harmonics contain less signal power but provide a larger multiplication factor. For example, a frequency quadrupler requires a lower frequency input compared to a doubler but has significantly lower conversion gain and output power. In practice, harmonic numbers beyond three contain insufficient power at THz frequencies in CMOS. Furthermore, an odd-number multiplier is preferred since it can be implemented with a fully differential circuit, which eases the design and modeling considerably.

In the end, a tripler-last transmitter is chosen, which provides a decent conversion gain with a manageable input frequency (130 GHz for a 390 GHz output). Additionally, the tripling operation is compatible with multiple modulations, as illustrated by working out the expression of a generic modulated signal multiplied by a tripler:

$$(A\,e^{j\omega t+j\phi})^3 = A^3 e^{j3\omega t+j3\phi}. \tag{6.2}$$

While the amplitude of the signal is cubed, the frequency and phase are multiplied by three. A constant envelope QPSK signal, for example, will thus remain undistorted by a tripling operation, as demonstrated in [50]. However, even though the relative location of the constellation points is unchanged, the actual phase transitions are three times as long and thus have to transition three times as fast in the same symbol period. This phenomenon manifests itself as bandwidth expansion in the frequency domain and is an important undesirable consequence of nonlinear operations on modulated signals.

6.2 Binary-Phase On–Off Keying

In this section, the unconventional modulation scheme called binary-phase on–off keying or BPOOK is discussed. Its unique bandwidth compressing property is analyzed along with ways to encode and decode the signal using practical circuits.

6.2.1 Modulation Considerations for THz Systems

Choosing the modulation scheme entails the inevitable trade-off between bandwidth, SNR requirement, and circuit complexity, as discussed in Chap. 2. However, a THz system in CMOS imposes specific limitations due to its above f_{max} operation, as discussed in the previous subsection. More specifically, even though bandwidth may be plentiful at THz frequencies, signal power is not and thus the receiver must operate with a limited signal SNR. Furthermore, since the THz transmitter typically includes a nonlinear output stage to generate the above-f_{max} signal, the linearity specification becomes critical.

The most simple modulation choices are binary modulations in amplitude (OOK) or phase (BPSK). Despite their straightforward implementation, the limited spectral efficiency of 1 bit/symbol inhibits optimal exploitation of the large THz bandwidth. Going to the other extreme, i.e., M-QAM modulations (16-QAM, 64-QAM, etc.), offers a much increased modulation density at the cost of very stringent linearity requirements and demanding THz circuit complexity. Even the simplest variety, 4-QAM or QPSK, requires quadrature LO generation and coherent detection, which significantly increases the power consumption and overall vastness of the system.

However, in between the simplicity of OOK/BPSK and the complexity of QPSK, there exists the modulation binary-phase on–off keying or BPOOK. By relying on the duobinary encoding technique, the spectral efficiency of QPSK can be achieved while keeping the circuit complexity close to that of a simple OOK scheme. As a result, an acceptable spectral efficiency can be attained while being compatible with a nonlinear transmitter and requiring minimal SNR for a practical BER. The effectiveness of BPOOK has been demonstrated before at mm-Wave frequencies [142, 143] and is further validated in this work at THz frequencies.

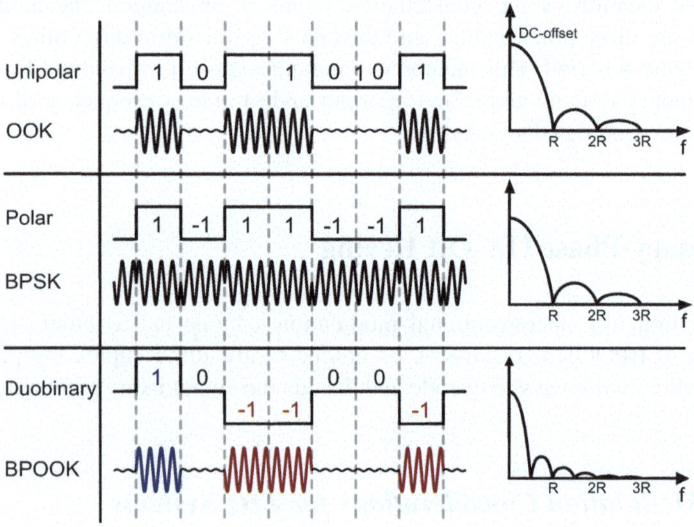

Fig. 6.2 A comparison between OOK, BPSK, and BPOOK modulation

6.2.2 BPOOK and Duobinary Encoding

Figure 6.2 shows the modulating signal, the upconverted modulated signal, and baseband spectrum of an OOK, BPSK, and BPOOK signal. While OOK and BPSK only have two levels, BPOOK has three levels: "1," "0," and "−1." By following a specific algorithm [144], binary input data is converted into three-level, duobinary data. A binary "0" remains unchanged, while a binary "1" gets converted into either a "1" or "−1." This encoding introduces correlation in the data which reduces the bandwidth by a factor of two compared to OOK and BPSK. This correlating of the data can also be interpreted as adding controlled ISI to the data which acts as a preemphasis that is compensated again in the detector. An extensive description of the encoding algorithm and derivation of the bandwidth compression property are given in Appendix A.3.

By modulating this duobinary data onto an RF carrier a BPOOK signal is obtained. No polar baseband signal is necessary as this can be achieved with two binary data inputs in a differential architecture, as shown in the designed system. Several observations about the BPOOK signal can be made. Most importantly, this BPOOK signal has half the bandwidth of an OOK/BPSK signal modulated at the same data rate. The spectral efficiency of the BPOOK modulation is thus equal to that of QPSK. To quantify the effect of this bandwidth compression, binary and duobinary signals are both filtered by a brickwall filter cutting of at half the data rate ($1/2T_b$) [145]. While a binary signal loses about 52% of the power and as a result suffers from a lot of ISI, the duobinary signal only loses 10% and its eye remains largely unaffected [145].

Also, this particular BPOOK signal has the same envelope as an OOK signal, alleviating nonlinearity concerns and improving the efficiency of the transmitter, as the power amplifier can be driven fully into saturation. Furthermore, compared to an OOK signal, no LO power is present in the output spectrum (no DC-offset in duobinary signal) leading to more efficient use of RF signal power, which is scarce at THz frequencies. Finally, as will be discussed, a simple envelope detector suffices to demodulate this BPOOK signal, simplifying the receiver considerably compared to its coherent counterparts [142]. It must be noted that the duobinary encoding is undone after rectification thereby increasing the bandwidth again to its original value. Overall, employing BPOOK can reach the same spectral efficiency as QPSK while avoiding a power consuming IQ modulator and coherent detector.

6.2.3 BPOOK Encoding and Decoding Techniques

There are various ways to introduce the desired duobinary correlation into the signal and decode the original information again. The work in [143] adopts the duobinary encoding method from [145] to perform BPOOK (de)modulation at 94 GHz. An encoder converts binary input data into a 3-level duobinary signal which is then upconverted onto a 94 GHz carrier yielding a 3-level duobinary-encoded ASK signal. On the receiver side, the 94 GHz BPOOK signal is incoherently detected by an envelope detector after which a 3-level comparator distinguishes the different levels. Since the duobinary signal is purely modulated onto the amplitude of the carrier, yielding duobinary amplitude shift keying modulation, the signal is sensitive to nonlinearities and contains significant LO power. Furthermore, the 3-level comparator is quite challenging to design with good sensitivity at high speeds. This type of architecture does not fully exploit the possible advantages of the BPOOK modulation.

A superior BPOOK (de)modulation approach is proposed in [142]. A BPOOK encoder first converts the binary input into two distinct binary streams, which combined to make up a differential duobinary signal. After upconversion a BPOOK signal is created with only one nonzero amplitude level and no LO feedthrough. After rectification in the receiver, a simple sign-check comparator can extract the original data. This BPOOK approach is preferred due to its robustness against nonlinearity, the absence of LO feedthrough, and simplistic detecting method. As visualized before in Fig. 6.2, this type of BPOOK will be employed in this design. Despite the simplicity of the modulator, detector, and decoder, the encoder can be tricky to design at high data rates due to stringent timing requirements in the logic blocks and the need for an oversampled clock signal [142]. The designed THz transmitter in this chapter only includes the modulator and mm-Wave/THz front-end, while performing the duobinary encoding off-chip.

6.3 Transmitter Circuit Design

6.3.1 Transmitter Architecture Overview

As discussed earlier, due to the absence of transistor gain at 390 GHz in 28 nm CMOS (f_{max} ≈ 280 GHz), conventional transmitter architectures with a power amplifier as final stage cannot be implemented. Therefore, a frequency multiplier is required to generate the THz signal. A fully differential frequency tripler is chosen as it naturally suppresses even order harmonics and does not corrupt the BPOOK data. The tripling operation does lead to bandwidth expansion, which limits the achievable data throughput.

Figure 6.3 shows a high-level block diagram of the proposed transmitter. A 130 GHz external local oscillator (LO) is brought on-chip and modulated by a switch-based modulator. The modulated 130 GHz signal is then amplified by a power amplifier and converted into a 390 GHz signal by a frequency tripler.

6.3.2 BPOOK Modulator

A switch-based modulator is proposed that supports OOK, BPSK, and BPOOK modulation with two binary data inputs. The external LO signal is first converted into a differential signal by means of a transformer-based balun and then fed into two single-pole double-throw (SPDT) switches which are cross-connected to the output, as shown in Fig. 6.3.

The working principle of the modulator is visualized in Fig. 6.4. For BPSK signaling, a 180° phase shift can be attained by switching both switches simultaneously to invert the differential output signal. For OOK, a low amplitude differential output can be created by switching only one of the switches. Finally, for BPOOK, both the low amplitude and two-phase high amplitude states are combined to create a three-level BPOOK signal. Owing to the differential, direct-modulation topology, there is no need for a DAC with additional negative supply to generate the desired three-level signal. Compared to a traditional mixer-based modulator, this switch-based phase modulator achieves a higher conversion gain as it is driven by a sharp-edged

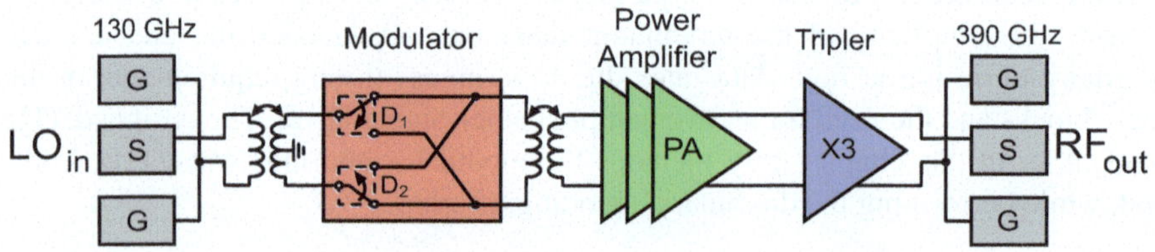

Fig. 6.3 Block diagram of the 390 GHz transmitter

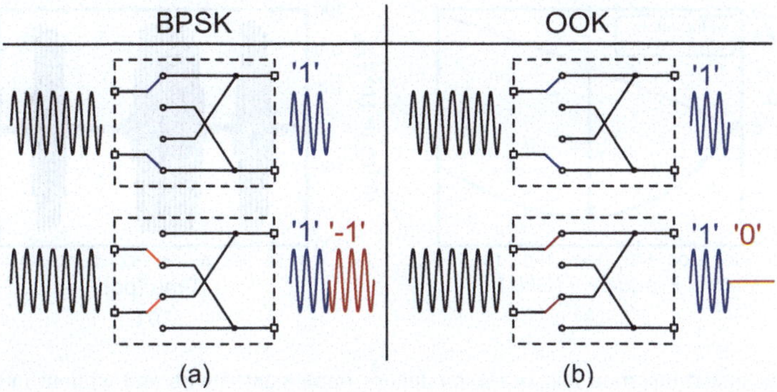

Fig. 6.4 Working principle of the modulator performing binary-phase modulation (**a**) and on–off amplitude modulation (**b**)

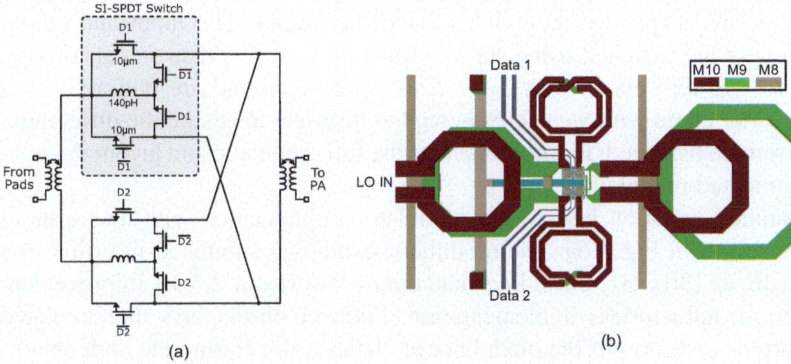

Fig. 6.5 Circuit schematic (**a**) and layout diagram (**b**) of the modulator

digital signal instead of a mm-Wave LO signal with limited swing and slower edges. Compared to conventional current-steering OOK modulators, this switch-based amplitude modulator can obtain much improved modulation depth without having to sacrifice modulation speed.

To reduce the capacitive coupling of the LO to the output, which increases the insertion loss of the modulator, inductors are added in parallel with the switches to resonate out the capacitance of the switch when turned off. To minimize the number of inductors per switch, a switched-inductor SPDT (SI-SPDT) switch is implemented [146], as shown in Fig. 6.5a. The SI-SPDT switch is composed of two transistor switches, a single inductor and two extra switches which reconfigure the inductor between the two output ports. The sizes of the switches are optimized for maximal modulation speed while keeping the insertion loss minimal, the latter of which is a trade-off between on-state resistance and off-state capacitance. A compact, double-turn inductor of 140 pH is used to resonate out the off-state capacitance of the switches at 130 GHz. The whole layout of the modulator is

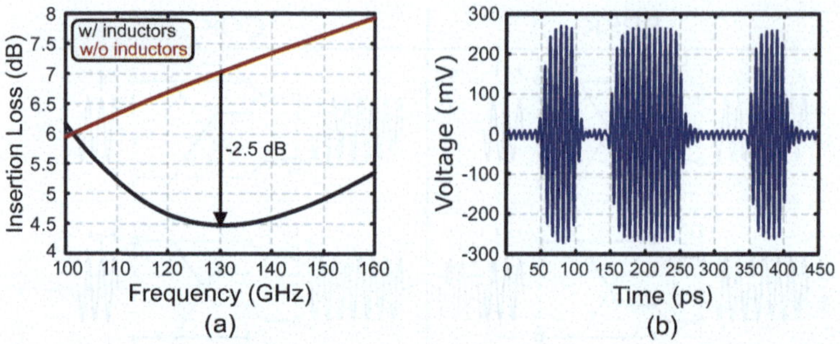

Fig. 6.6 Simulated insertion loss of the modulator implemented with and without inductors (**a**) and transient modulator output modulated at 20 Gb/s (**b**)

made as symmetrical as possible while minimizing resistive losses in the signal path and undesired capacitive coupling toward the output. The input and output of the modulator are matched using stacked transformers. The data signals driving the modulator are terminated with 50 Ω resistors on-chip and are buffered using a CMOS inverter chain with weak cross-coupled inverters to ensure the differential data inputs remain balanced. Figure 6.5 shows the full schematic and layout diagram of the implemented modulator.

The simulated insertion loss of the modulator implemented with and without inductors is shown in Fig. 6.6a. The modulator exhibits a simulated insertion loss of only 4.5 dB at 130 GHz, excluding matching networks, a 2.5 dB improvement compared to an inductorless implementation. Figure 6.6b displays the simulated transient output waveform of the modulator at 20 Gb/s, illustrating the high obtainable switching speed.

6.3.3 Power Amplifier

The power amplifier must have a sufficiently large bandwidth to support a high data rate and deliver high output power to the tripler without compromising reliability. As depicted in Fig. 6.7, the power amplifier consists of three stages each implemented as a pseudo-differential pair with neutralizing cross-coupled MOM capacitors. The layout of each neutralized transistor pair is carefully optimized for maximal power gain at 130 GHz, as described in Sect. 4.3.

To deliver maximum power to the tripler, the sizes of the transistors are determined to maximize the gain and output power at 130 GHz. More concretely, the sizes are increased along the chain to ensure the last stage can be fully saturated. This last stage is sized for maximal output power while providing an acceptable amount of power gain, resulting in a transistor width of 60 μm. For larger transistor sizes the gain and output power degrade significantly due to the increase of interconnect parasitics within each unit cell, and the interconnect inductance can

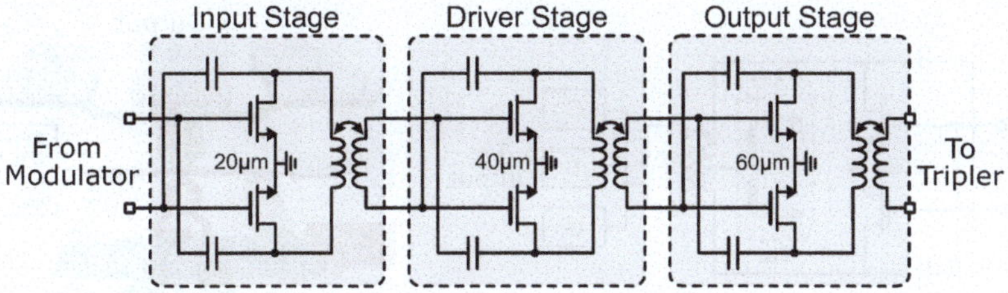

Fig. 6.7 Circuit schematic of the power amplifier

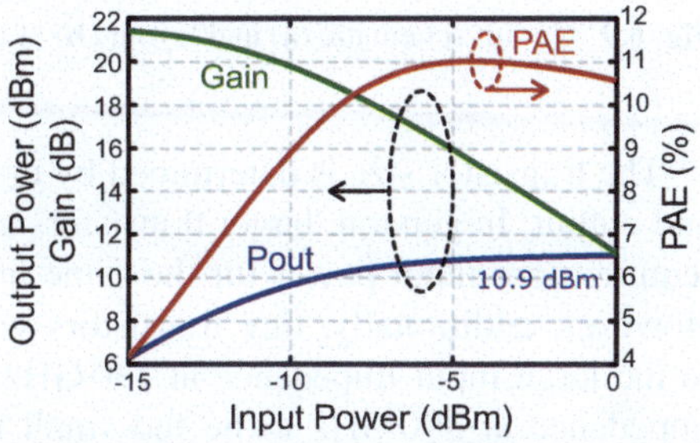

Fig. 6.8 Large-signal simulation results of the power amplifier at 130 GHz

cause instability issues. All stages are biased in class AB to obtain efficient, high gain performance.

Furthermore, stacked transformer-based matching networks are utilized to achieve wideband, large-signal power matching, based on loadpull simulations. The center taps of the transformers are used to connect the supply and bias voltages, where large resistors are inserted in series of the bias lines to ensure CM stability of the amplifier chain. Figure 6.8 shows the large-signal simulation results of the power amplifier at 130 GHz, predicting a saturated output power of 10.9 dBm including the output matching network toward the tripler.

6.3.4 Tripler

The frequency tripler converts power at the fundamental (130 GHz) to power at the third harmonic (390 GHz). The tripling operation preserves the modulation but expands the signal bandwidth for the same data rate. As seen in Fig. 6.9a, a neutralized pseudo-differential pair is employed as tripler which is biased such that the nonlinear behavior of the transistors is enhanced and generation of harmonics is maximized. For low input powers, the transistors should be biased in the subthreshold region to maximize nonlinearity. However, for large input swings biasing the transistors in deep class A is more beneficial to fully exploit the nonlinear current characteristics of the transistors over the full range (from subthreshold to triode operation). In this case, the tripler is biased in class A at 600 mV while being driven by a large input swing from the PA.

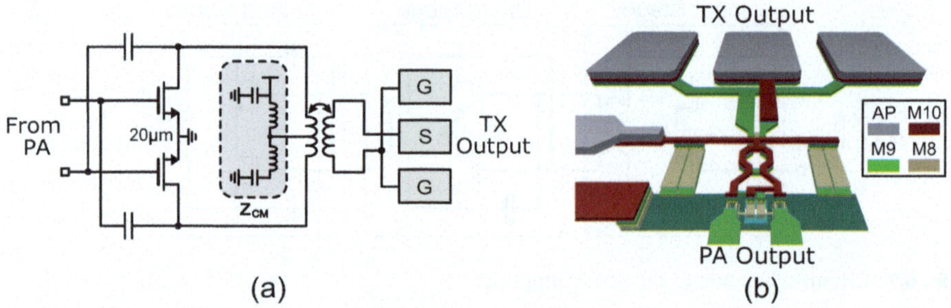

Fig. 6.9 Circuit schematic (**a**) and layout (**b**) of the frequency tripler

The transistor size is determined by opposing specifications at the tripler's input and output. In general, larger transistors are able to provide more output power but require more input power for the same input swing and hence nonlinear behavior. However, additionally, tiny transistors will increase the input matching loss due to the large input impedance at 130 GHz, while large transistors cause the output impedance at 390 GHz to be too small to efficiently match to the 50 Ω load at 390 GHz. Combining all of these factors, a size of 20 μm is chosen for the transistor pair. A similar transistor layout is utilized as for the power amplifier, optimizing the high-frequency power gain of the transistor pair.

Furthermore, the optimal tripler load impedances at 130 GHz and 390 GHz were determined for maximal third harmonic power generation and extraction by means of a loadpull simulation. These impedances are transformed into the 50 Ω load using a stacked transformer network which is co-designed with the output ground–signal–ground (GSG) pads. A shunt MOM capacitor is added between the output traces to resonate out the inductive load provided by the pads.

To further boost the power at the third harmonic, the common-mode impedance at the second harmonic must be well controlled. As analyzed in [99], the generation of power at the second harmonic has a destructive effect on generation of power at the third harmonic. To suppress the second harmonic, a small impedance is provided to the tripler at 260 GHz by adding two custom MOM capacitors creating a series resonance with the inductive common-mode line of the transformer (half a winding with part of the supply line). Figure 6.9 shows the schematic of the tripler with implemented common-mode network and the complete layout including output GSG pads. The full tripler network including the neutralized transistor pair, output transformer, output pads, and CM capacitors is simulated in a 3D EM solver (HFSS) to validate the performance at these high frequencies.

The simulated conversion gain of the tripler saturates at an input power of 10 dBm with an output power of −5 dBm, as shown in Fig. 6.10a. Figure 6.10b illustrates the impact of the MOM capacitors on the tripler's output power at different harmonics for an input power of 10 dBm. Simulations show a significant increase in output power at the third harmonic as a result of the suppression of the second harmonic leading to the simulated output power of −5 dBm at 390 GHz.

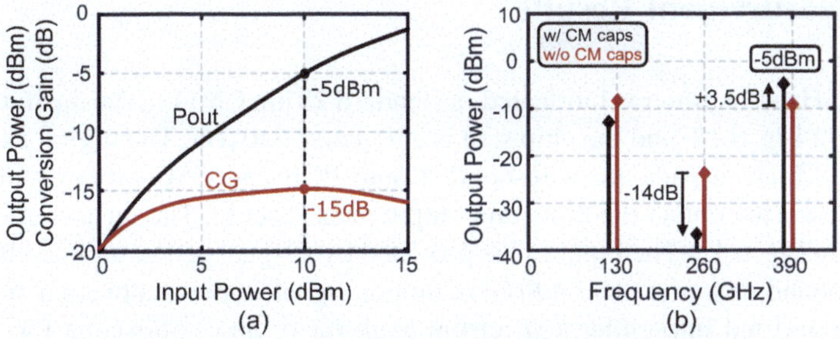

Fig. 6.10 Simulated tripler output power and conversion gain (**a**) and tripler output power at different harmonics with and without added CM capacitors at 10 dBm input power (**b**)

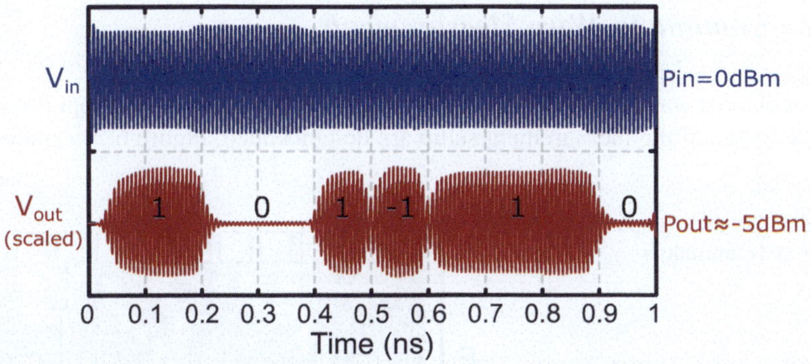

Fig. 6.11 Transient simulation of the transmitter LO input at 130 GHz and BPOOK output modulated at 10 Gb/s

6.3.5 *Full Transmitter Simulations*

Continuous wave simulations of the complete transmitter show a saturated output power around −5 dBm which is limited by the maximum conversion gain of the tripler. Even though the power amplifier can generate more than enough power, the maximum input swing of the tripler is limited to about twice the supply voltage due to reliability concerns which limits the achievable output power.

Full transmitter transient simulations were performed to assess the modulated performance, including all layout parasitics. Transient simulations are, however, very challenging due to the presence of many harmonics which can impede simulation convergence, especially with broadband modulated signals. Furthermore, EM simulations of large structures at THz frequencies can often lead to nonphysical S-parameter results, which further exacerbates transient simulation convergence. Nevertheless, Fig. 6.11 displays the 130 GHz LO input and BPOOK transmitter output modulated at 10 Gb/s. It is clearly visible that the output not only contains power at the third harmonic (390 GHz), but also at the fundamental (130 GHz) which will be filtered output by subsequent blocks in the THz system.

6.4 Measurement Results

The 390 GHz transmitter is fabricated in standard 28 nm CMOS. The die micrograph is shown in Fig. 6.12, and the chip size is 720 μm × 540 μm. The digital data input, supply and biasing pads are wire-bonded onto PCB and GSG probes with 50 μm pitch are used to contact the input and output pads on-chip. The measurement setup is shown in Fig. 6.13. The input LO is provided by a signal generator in combination with a D-band extender. The 390 GHz output signal is downconverted by a WR-2.2 mixer and fed into either a spectrum analyzer or an oscilloscope for analysis. Differential data signals are generated by an arbitrary waveform generator (AWG).

6.4.1 Continuous Wave Measurements

For continuous wave measurements, the inputs to the modulator are kept constant, and the losses of the measurement setup are de-embedded. Figure 6.14a shows the

Fig. 6.12 Die micrograph of the 390 GHz transmitter

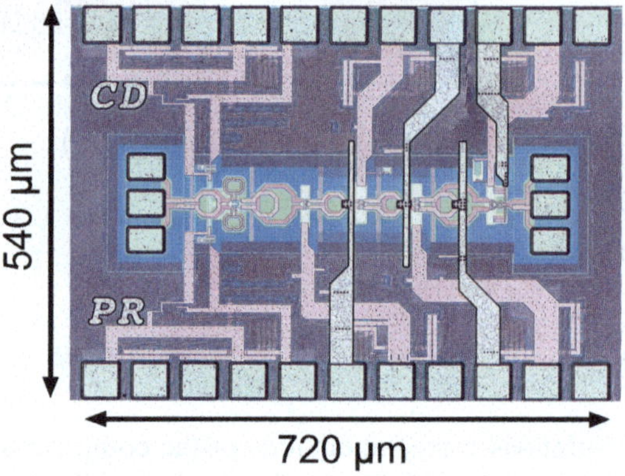

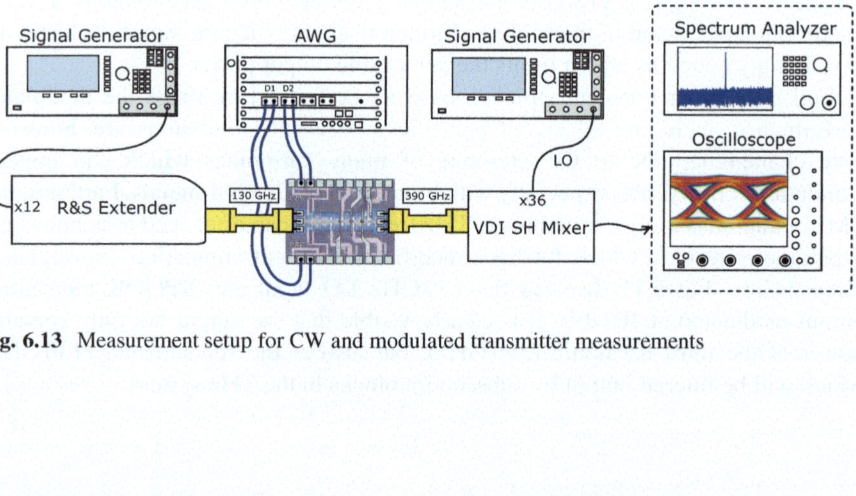

Fig. 6.13 Measurement setup for CW and modulated transmitter measurements

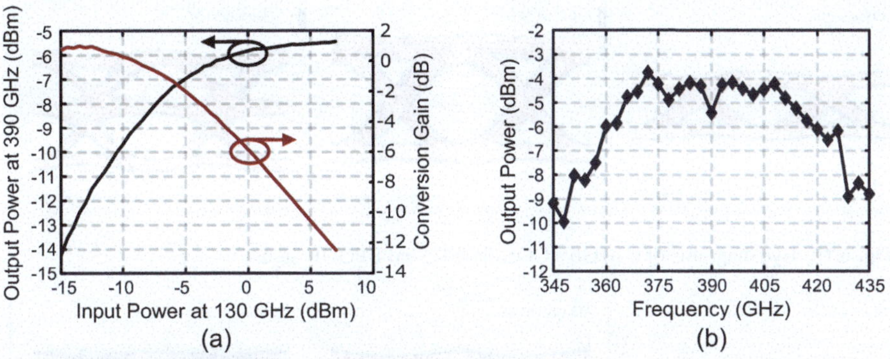

Fig. 6.14 Measured output power at 390 GHz and conversion gain versus input power at 130 GHz (**a**) and measured output power across frequency (**b**)

measured output power at 390 GHz and conversion gain of the transmitter versus input power at 130 GHz. A maximum output power of −5.4 dBm is achieved with 7 dBm of input power, which is limited by the frequency extender. Figure 6.14b shows the measured output power versus frequency for 7 dBm of input power. More than 60 GHz of bandwidth is available at 390 GHz, including the input balun which does not limit data throughput.

6.4.2 *Modulated Measurements*

For modulated measurements, the two differential data inputs depend upon the modulation scheme. For OOK, a pseudo-random binary sequence (PRBS) with length $2^7 - 1$ is generated and applied to one data input, while the other input is kept constant. For BPSK, both data inputs are set to the same PRBS. Finally, for BPOOK, the PRBS is encoded into a three-level duobinary sequence from which the two required binary sequences are derived. The output signal is downconverted into baseband from which eye diagrams are generated. Figure 6.15 shows the real-time eye diagrams of all three supported modulations at 10 Gb/s, while Fig. 6.16 shows the eye diagrams at their respective maximum achieved data rate. With OOK and BPSK signaling, a maximum data rate of 18 Gb/s is achieved, while using the more spectrally efficient BPOOK pushes the data rate up to 28 Gb/s. As expected, the peak-to-peak amplitude of an OOK signal is about half that of a BPSK signal, and the duobinary encoding scheme is clearly visible in the eye of the BPOOK signal. Furthermore, Fig. 6.17 clearly demonstrates the bandwidth compression property of the duobinary encoding by showing that the BPOOK null-to-null bandwidth is half that of a BPSK signal with the same data rate. The whole chip consumes 114 mW from a 1 V supply.

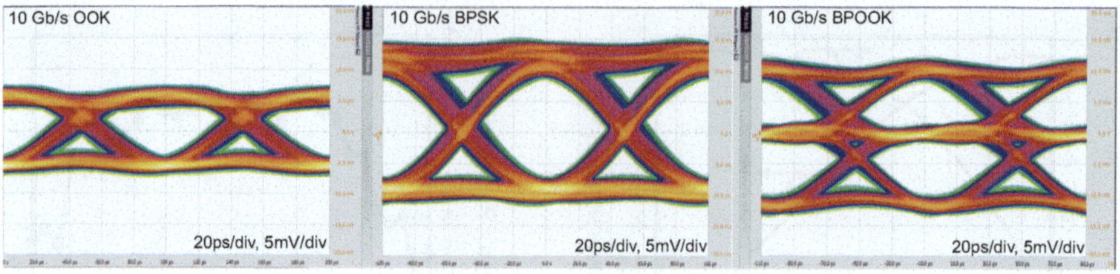

Fig. 6.15 Eye diagrams of a 10 Gb/s OOK, BPSK, and BPOOK signal

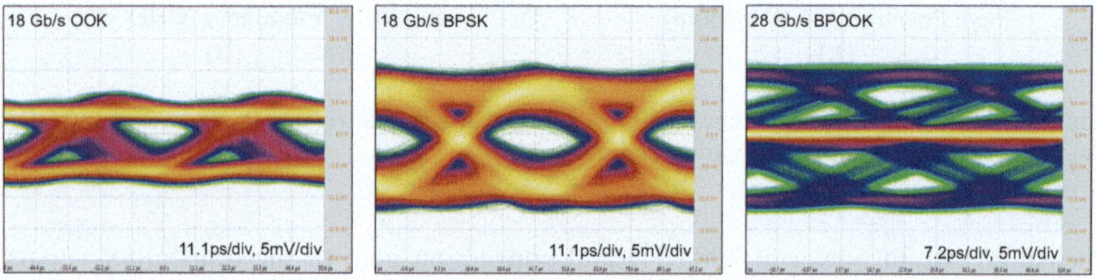

Fig. 6.16 Eye diagrams of a 18 Gb/s OOK, 18 Gb/s BPSK, and 28 Gb/s BPOOK signal

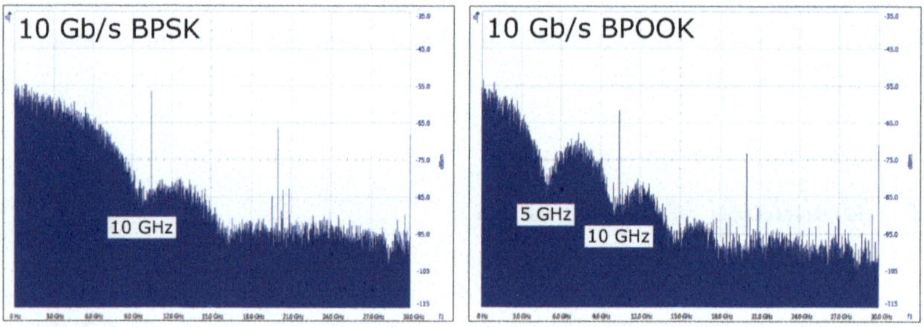

Fig. 6.17 Downconverted spectrum of a 10 Gb/s BPSK and BPOOK signal

6.4.3 *Performance Limitations*

After inspecting the eye diagrams in Fig. 6.16, it is clear the eyes are mainly degrading at higher speeds due to inter-symbol interference (ISI). Even though the SNR slightly degrades at higher data rates due to a larger integrated noise power, it is not the main limitation of the maximum achievable data rate in this setup. ISI is mostly a result of limited bandwidth in the system relative to the bandwidth of the signal. Part of the bandwidth limitation comes from the designed circuits, where the optimal response has been shifted down in frequency compared to simulations, and another part comes from the measurement equipment (downconverter, data cables, and probes).

This bandwidth limitation is exacerbated by the nonlinear operation of the tripler which causes bandwidth expansion of the modulated signal. Assuming the cubic characteristic leads to an expansion of three, a 30 Gb/s BPOOK with a null-to-null RF bandwidth of 30 GHz would roughly occupy a bandwidth of 90 GHz after the tripling operation, which is obviously a lot more than the available system bandwidth. This could be avoided by performing the modulation after the tripler directly at the THz frequency but would incur significant losses and thus limit the output power and efficiency of the THz transmitter.

The modulated measurement results were further restricted by the spectral purity of the D-band extender which provided the input LO signal. This extender contains an internal $\times 12$ multiplier chain and generates not only the desired 12th harmonic output signal, but also considerable spurs spaced at multiples of the fundamental. These spurs are also modulated by the transmitter and seriously contaminate the measured downconverted baseband signal, especially at higher speeds.

Lastly, since a coherent detector setup was utilized, phase coherence between the LO signals is incredibly important to reliably demodulate the generated signal. Even though both external signal generators are locked via a stable 10 MHz reference, there always exist some low-frequency phase drift between the sources. This phase error is multiplied by the multiplication factor of the LO generation chain. The downconversion mixer, for example, utilizes a $\times 36$ multiplication factor to generate the 390 GHz LO signal and thus severely suffers from LO phase drift relative to the LO signal used for the transmitter. This phase drift results in a limited acquisition time where the LO signals are coherent enough to achieve adequate demodulation.

6.4.4 *To Triple or Not To Triple?*

To triple, or not to triple—that is the question. As discussed, a tripler-last architecture is a great choice when trying to realize a THz transmitter, but why THz? One of the benefits of going to very high frequencies is the large available bandwidth. While this is true in the sense that bandwidth is available from a regulatory perspective, the available bandwidth is not effectively used. This is because the modulation is performed on the sub-THz IF, and then the tripler causes BW expansion when converting the signal to THz frequencies. In effect, the available BW of the sub-THz IF is utilized, not from the THz RF. Of course, this bandwidth expansion can be avoided by modulating the THz wave directly or by employing an exotic topology such as the multiplying mixer which preserves linearity despite its nonlinear behavior. In both cases the achievable THz output power will be much lower compared to the traditional multiplier-last approach. The main benefit of THz communication systems turns out to be the small size of the THz antenna, which is not implemented in this system but could lead to very compact integrated systems.

To assess the performance of the designed transmitter without the effects of the nonlinear bandwidth expansion, the same transmitter was fabricated without the frequency tripler. Like the discussed 390 GHz transmitter, this 130 GHz BPOOK transmitter includes a modulator and a three-stage power amplifier driving the GSG output pads, as shown in Fig. 6.18. A die micrograph of this chip, fabricated in

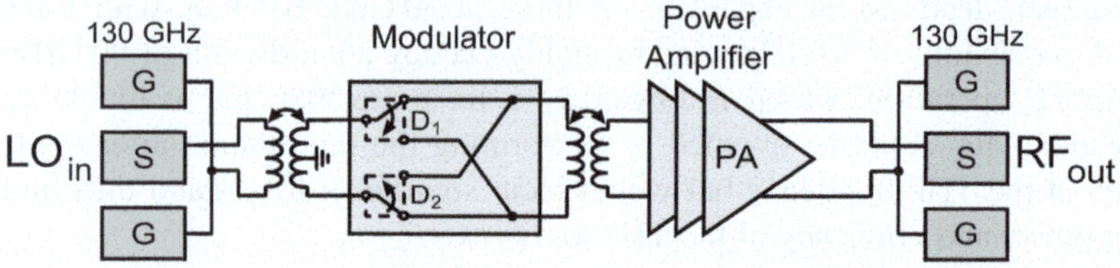

Fig. 6.18 Block diagram of the 130 GHz transmitter without the frequency tripler

Fig. 6.19 Die micrograph of the 130 GHz transmitter

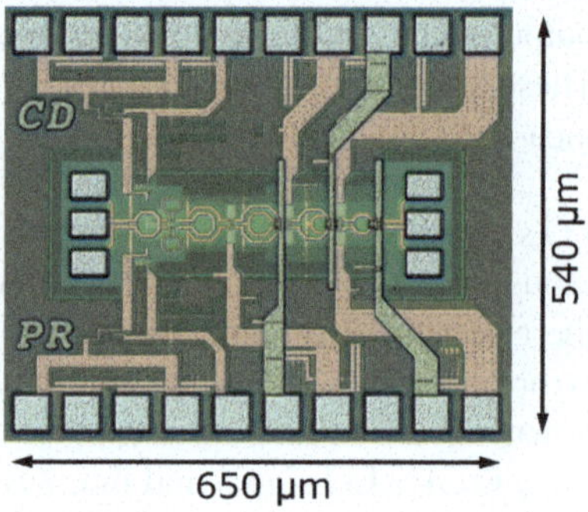

28 nm CMOS, is depicted in Fig. 6.19. Modulated transient simulations showed significantly higher data rates than were possible using the 390 GHz transmitter. Unfortunately, these results were never validated in measurements. In conclusion, going to THz frequencies generally does not provide any benefits from a bandwidth standpoint but could potentially lead to very small, integrated links.

6.5 Conclusion

This chapter presented a 390 GHz transmitter suitable for high-speed, short-distance communication. First, the most promising THz transmitter architectures were discussed and the BPOOK modulation scheme was analyzed. The circuit design of the proposed transmitter was then described along with measurement results, and the overall architecture was re-evaluated.

The 390 GHz transmitter was designed in 28 nm CMOS with a novel switch-based modulator capable of generating OOK, BPSK, and BPOOK signals. The transmitter generates a maximum output power of −5.4 dBm at 390 GHz and achieves a maximum data rate of 18 Gb/s with OOK/BPSK and 28 Gb/s with BPOOK modulation. The high achieved symbol rate proves the high-speed capability of the modulator and spectral efficiency of the BPOOK modulation. Table 6.1 compares the performance of the designed transmitter with other state-of-the-art THz and sub-THz silicon-based transmitters.

Table 6.1 Comparison of (sub-)THz silicon-based transmitters

	This work		CICC18 [147]	JSSC15 [50]	ISSCC17 [140]	JSSC18 [48]	JSSC12 [148]
Technology	**28 nm CMOS**		65 nm CMOS	65 nm CMOS	40 nm CMOS	130 nm SiGe	130 nm SiGe
Modulation	**OOK/BPSK**	**BPOOK**	QPSK	QPSK	32-QAM	BPSK	OOK
Frequency (GHz)	**390**		300	240	302	240	434
Output power (dBm)	**−5.4**		−6	0	−5.5	−0.8	−18.5
Power consumption (mW)	**114**		180[a]	220[a]	1400[a]	375[a]	120[a]
Data rate (Gb/s)	**18**	**28**	30	16	105[b]	25	10
Transmitter efficiency (pJ/bit)	**6.3**	**3.8**	6[a]	14[a]	13.3[a]	15[a]	12[a]

[a] Includes LO multiplier chain
[b] Achieved with off-chip modulation and equalization

Chapter 7
A 135 GHz Direct-Digital 16-QAM Wireless and DWG Link

This chapter discusses the complete design of a 135 GHz direct-digital 16-QAM wireless and dielectric waveguide (DWG) link in 28 nm CMOS. The aim of this project was to develop a complete sub-THz link over a wireless and dielectric waveguide channel achieving very high data rates while trying to minimize the energy consumption of the system.

First the challenges of conventional high-speed transceiver architectures are discussed, and the direct-digital system architecture is introduced. The complete 16-QAM system is described next, and system-level specifications are determined. The circuit design of the transmitter and receiver is then elaborately discussed along with chip-level simulations. Design considerations of the high-frequency board are illustrated, including the on-board antenna and coupler. Finally, measurement results of the standalone chips and complete link are discussed, and the performance of the whole system is concluded.

7.1 The Direct-Digital System Architecture

As discussed before in Chap. 2, there are numerous architectures suitable for high-speed modulation and demodulation in the sub-THz frequency range. However, as data rates start to reach 50 Gb/s and above, certain challenges start to emerge and cannot be ignored. In this section, the most conventional architectures are examined in the context of ultra-high-speed communication, and a new architecture is proposed to tackle the most severe problems: the direct-digital architecture.

C. D'heer, P. Reynaert, *THz and Sub-THz CMOS Electronics for High-Speed Telecommunication*, Analog Circuits and Signal Processing,
https://doi.org/10.1007/978-3-031-64439-9_7

7.1.1 Challenges of Conventional Architectures

7.1.1.1 Simple Architectures

Initially, frequencies above 100 GHz were explored using simple modulation schemes to limit the circuit complexity. Thanks to the high available bandwidth at these high carrier frequencies, relatively high data rates could still be achieved despite the low spectral efficiency of the modulation. Some popularly employed modulation schemes include ASK [47, 130, 149–152], FSK [52, 153, 154], BPSK [48], and QPSK [50, 155–157].

Particularly interesting are the architectures requiring just an incoherent detection method. Figure 7.1 shows a standard ASK transceiver, including a VCO with a switched PA as an amplitude modulator and an envelope detector for amplitude detection. An incoherent FSK transceiver is shown in Fig. 7.2. While a digital-driven VCO is used to perform frequency modulation, the original data can be recovered employing a time-delay-based quadrature detector [52]. In both cases, no accurate frequency reference is needed in the receiver for demodulation, which significantly simplifies the overall architecture. Most often, the modulation is performed in a binary fashion leading to straightforward 2-ASK and 2-FSK implementations. Multilevel modulation can also be realized to achieve slightly higher spectral efficiency (e.g., 4-ASK [130] and 4-FSK [52]), but the nonlinear detection methods make the demodulation rather challenging at high symbol rates.

As multilevel (de)modulation considerably complicates the required circuits, BPSK can be further extended to QPSK in order to double the modulation density with binary signaling. This results in rather straightforward circuit implementations but requires coherent detection to demodulate the data.

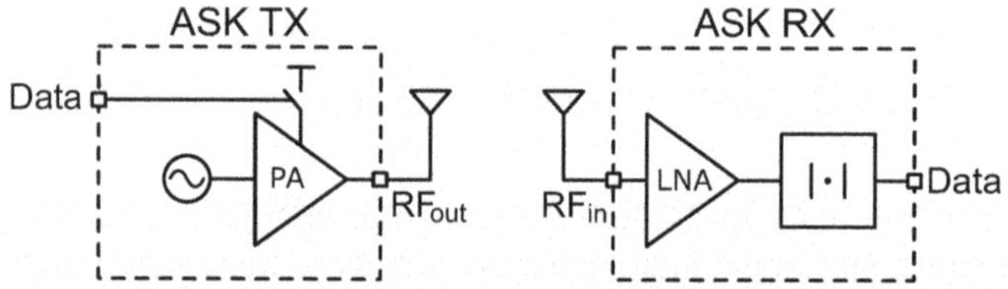

Fig. 7.1 Block diagram of a classic ASK transceiver architecture

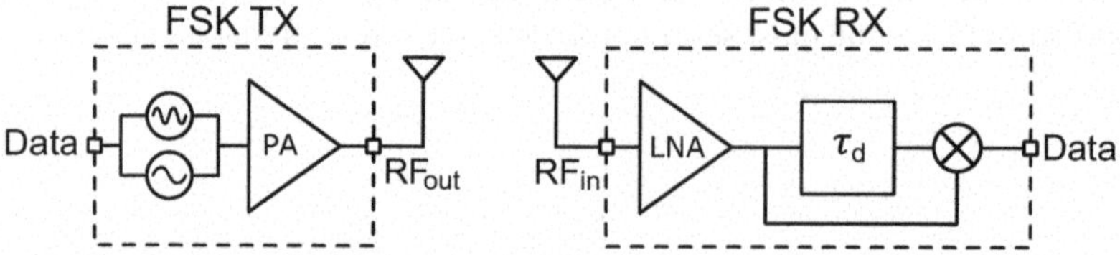

Fig. 7.2 Block diagram of a possible FSK transceiver architecture

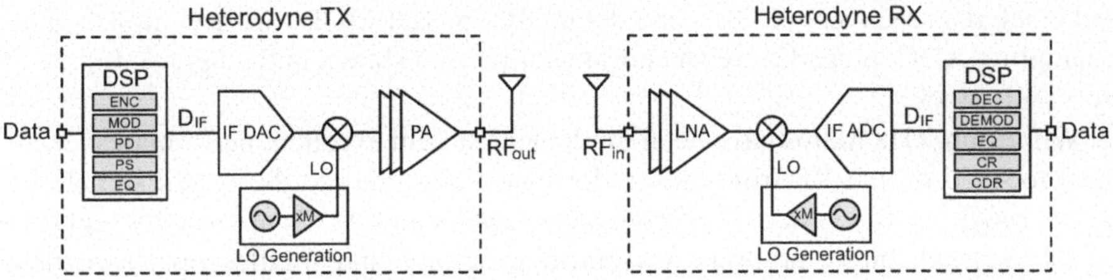

Fig. 7.3 Block diagram of the heterodyne transceiver architecture

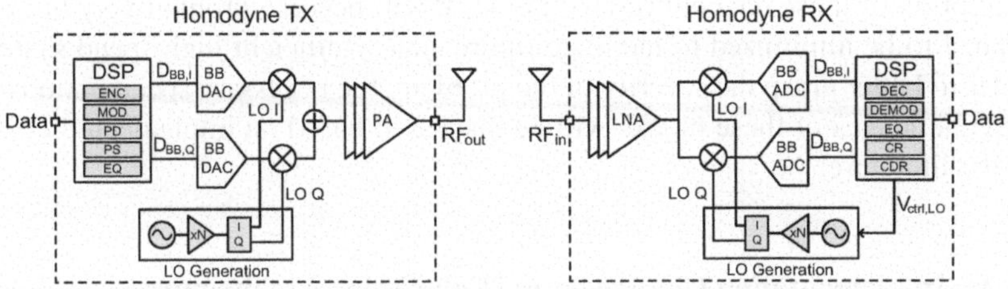

Fig. 7.4 Block diagram of the homodyne or direct-conversion transceiver architecture

Despite being very attractive circuit solutions, practical limits on circuit bandwidth combined with the low spectral efficiency of these modulations (1-2 bit/sym) restrict efficiently achievable data rates. As the desired data rate reaches 30 Gb/s and above, more dense modulation schemes have to be considered at the cost of more sophisticated circuit topologies.

7.1.1.2 High-Order Architectures

Recent publications have demonstrated sub-THz transceivers achieving data rates as high as 100 Gb/s while adopting denser modulation schemes ranging from 16-QAM to 256-QAM [101, 103–105, 109, 110, 158–164]. The two most used transceiver system architectures to achieve these high-order modulation schemes are the heterodyne transceiver (Fig. 7.3) and the homodyne (or direct-conversion) transceiver (Fig. 7.4).

In case of high-order complex modulations, the data interface consists of either I/Q multilevel baseband signals (homodyne) or a complex modulated IF signal (heterodyne). Optionally, a heterodyne architecture can be adopted with baseband data interfaces and double up/downconversion steps in the analog domain [104, 158, 159], optionally with a shared LO signal like in a sliding-IF architecture. On the transmit side, all possible signal processing (e.g., encoding (ENC), predistortion (PD), pulse shaping (PS), and equalization (EQ)) and high-order modulation (MOD) are performed by a DSP block followed by one or multiple DACs with a reconstruction filter (not shown in the figure). On the receive side, all necessary signal processing (e.g., decoding (DEC), equalization (EQ), carrier recovery (CR), and clock data recovery (CDR)) and demodulation (DEMOD) are accomplished by a sampling ADC, preceded by an anti-alias filter (not shown in the figure), followed by a DSP block.

While sub-THz radios have been implemented with excellent performance, most have focused on the RF front-end performance and ignored the mixed-signal and digital blocks of the system. The high-speed DACs and ADCs are usually replaced by an external, bulky arbitrary waveform generator and oscilloscope, and most signal processing is applied offline and not included in the power budget of the system. However, as data rates continue to increase, the complexity and power consumption of these data converters and DSP will become prohibitively high and thus need to be minimized to facilitate future energy-efficient high-speed systems. In order to better grasp the severity of the problem, the necessary specifications and design challenges of these blocks will be discussed based on implemented circuits from the literature.

7.1.1.3 Implementation Challenges in High-Order Architectures

The necessary specifications of the data converters are mainly determined by the architecture, the modulation density, and the complexity of the DSP algorithms. Firstly, in a heterodyne architecture, the complex modulated IF signal is significantly more challenging to accurately generate or digitize than the multilevel baseband signals in a homodyne system. This is because the IF signal contains more precise amplitude as well as additional phase information compared to just a fixed number of amplitude levels in (unprocessed) baseband signals, thus requiring a higher resolution. Furthermore, the data converters need to sample significantly faster in case of a heterodyne system, due to the larger occupied signal bandwidth of an IF-modulated signal. While a baseband signal can be adequately sampled at baud rate, an IF signal needs to be sampled at least at four times the baud rate to avoid aliasing effects.[1] Secondly, a more dense modulation scheme will require a higher converter resolution to reliably generate and distinguish between a greater number of distinct symbols. Thirdly, a higher resolution and oversampling ratio may be needed depending on the requirements dictated by the applied signal processing. For example, pulse shaping will require more levels and oversampling in the DAC, depending on the complexity of the desired pulse shape. Likewise, RX-side equalization algorithms will require a more accurate, often oversampled, representation of the signals to perform optimally. The complexity of these algorithms is typically determined by the system characteristics and desired performance. More severe circuit or channel impairments, often exacerbated with higher data rates, will require more advanced compensation mechanisms (PD, PS, EQ) and more complicated carrier and clock data recovery algorithms to get reliable system performance,

[1] The smallest alias-free IF is assumed equal to the baud rate in order to fully contain the null-to-null bandwidth of the signal.

placing a heavier burden on the data converters and digital back end. A more elaborate discussion of the DSP implementation challenges in broadband, real-time applications is given in [165].

To get a more tangible description of the situation, the specifications of the necessary data converters and DSP can be estimated for a future 100 Gb/s 16-QAM system adopting the two traditional architectures mentioned before. In a heterodyne system, the 25 GBd 16-QAM modulated signal on a 25 GHz IF needs to be sampled at least at 100 GS/s with a minimal 8-bit resolution. While the fastest reported DACs [166, 167] and ADCs [168, 169] achieve around this speed, their high-frequency effective number of bits (ENOB) is limited (4–5 b) and their analog bandwidth insufficient (20–30 GHz), while consuming over 1 W of power in a single converter. In a homodyne system, the 25 GBd PAM-4 signals are typically oversampled twice to support basic pulse shaping and make carrier/clock recovery more robust. A resolution of 8-bit (ENOB $\sim$ 6 b) is commonly desired for PAM-4 equalization and carrier/clock recovery algorithms [170]. State-of-the-art DACs [136, 137, 167, 171, 172] and ADCs [133, 134, 169, 172] are able to obtain the desired performance each with minimally around 250 mW of power consumption, resulting in a total data converter consumption of 1 W for a single-channel 100 Gb/s 16-QAM system. In both architectures, the design of the required data converters becomes extremely challenging [173, 174], and their power consumption immense. The power-dominating role of the data converters in high-speed links was also shown in Chap. 5.

Finally, estimating the power consumption of the DSP is challenging and highly dependent on the specific algorithms, implementation, and processing speed. As processor clock rates are usually limited to about 500–600 MHz, a high degree of parallelism is required to cope with the enormous data throughput which increases with modulation speed and oversampling rate. Digital baseband chips found in the literature, which implement (de)modulation, pulse shaping, and frequency-domain channel equalization for a 10 Gb/s 16-QAM link, already consume over 2 W of power [175, 176]. Extending such DSP-intensive systems to even higher data rates would greatly degrade the energy efficiency of the whole transceiver.

Based on these numbers, it is clear that realizing ultra-high data rates using traditional architectures would lead to a prohibitively high power consumption and impractically complex circuit implementations. Most published sub-THz radios, although impressive, ignore these challenges by performing the high-speed data conversion using a bulky, expensive, and power-hungry arbitrary waveform generator and a real-time oscilloscope and by performing signal processing offline in non-real time. Hence, simplifying the data converters and DSP is essential in order to enable practical ultra-high-speed, energy-efficient integrated transceivers. To achieve this, innovations on an architectural level are necessary.

7.1.2 The Ideal Direct-Digital System

In order to improve the energy efficiency of communication systems, recent trends in the literature have shown a move toward more integrated mixed-signal processing systems, minimizing the complexity of the data converters and DSP. Wireline PAM-4 transmitters have been proposed using a limited number of tunable DAC slices with embedded FFE filtering to reduce the power compared to higher resolution DAC-based transmitters [177]. Mixed-signal wireline PAM-4 receivers [178, 179] have been shown to prevail over their ADC–DSP counterparts [180, 181] in efficiency and area, although having more difficulties operating at high speed over high-loss channels. To improve the efficiency of ADC-based receivers, using adaptive ADC thresholds [182] or additional equalization in the analog front-end can reduce the required resolution of the ADC for the same BER performance. In limited-reach optical networks, "coherent-lite" systems [183–185] aim to take advantage of the limited amount of channel impairments by moving vital signal processing steps to the analog domain to simplify the otherwise power-dominating ADC and DSP blocks. Lastly, low-power mixed-signal baseband circuitry has been developed toward 60 GHz wireless for mobile applications [186, 187]. These ideas can be expanded toward more energy-efficient sub-THz transceivers.

Figure 7.5 shows the envisioned concept of a mixed-signal high-speed sub-THz transceiver. On the transmit side, all signal preprocessing and high-order modulation are carried out by a direct-digital modulator with embedded mixed-signal predistortion, pulse shaping, and equalization circuitry. On the receive side, signal demodulation and all signal postprocessing are executed by a direct-digital demodulator with equalization, carrier recovery, and clock data recovery circuits realized in the mixed-signal domain. Consequently, the necessary DSP is greatly simplified and no high-speed, high-resolution DACs or ADCs are required, minimizing the power overhead caused by these blocks. This reduction in power consumption not only directly benefits the environment, but it also reduces the energy expenses and costs associated with thermal cooling of the system, which is especially troublesome in high-density applications such as data center interconnects. Furthermore, heavy DSP can substantially increase the latency of the system, which can be detrimental in certain low-latency applications.

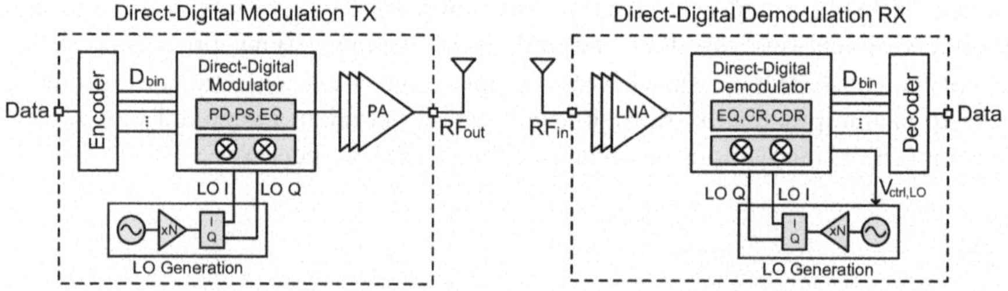

Fig. 7.5 Block diagram of the envisioned concept of a mixed-signal direct-digital transceiver architecture

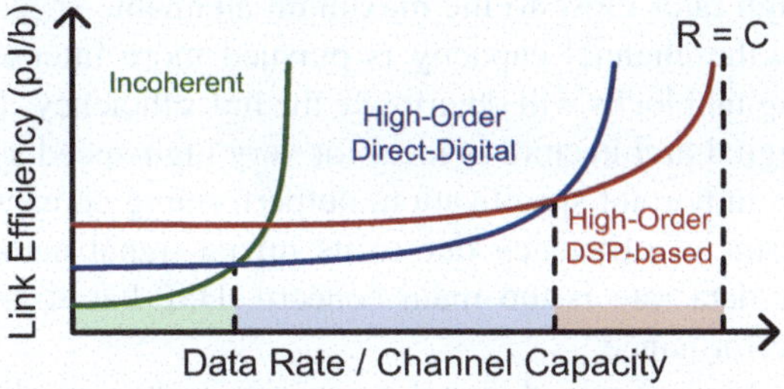

Fig. 7.6 Conceptual representation of the link efficiency versus relative data rate for three of the main transceiver architectures

Even though a fully integrated mixed-signal transceiver will outperform a DSP-based transceiver in terms of power and area, the flexibility of a DSP-heavy system could be preferred in case of more severe system impairments. For ultra-high data rates in combination with an impaired channel or limited circuit performance, mixed-signal processing might not suffice and intensive DSP may be needed to cope with these imperfections and obtain the desired system performance, albeit at the cost of an immense power consumption. The optimal system architecture thus depends on the application, as it usually does. The proposed mixed-signal transceiver is preferred for all high-speed applications which aim to minimize power consumption, area, and/or latency. Applications, which solely desire to maximize data throughput, with little concern for power or cost, will benefit more from a DSP-based approach, such as ultra-high-speed point-to-point links for backhaul applications. However, as mentioned before, applying mixed-signal processing techniques could even benefit DSP-based systems, making all research effort toward this mixed-signal direct-digital system profoundly valuable.

To further illustrate this reasoning, Fig. 7.6 shows the conceptual link efficiency (typically quantified in pJ/b) for three main architecture types versus normalized performance. The performance is quantified in terms of data rate relative to the maximum channel capacity. As the data rate approaches the maximum capacity, more severe system impairments must be avoided or compensated, which has a large influence on the complexity and efficiency of the system, as was previously discussed in Chap. 5 with Fig. 5.4. Incoherent architectures with rather simple modulations can achieve superior link efficiencies as long as the data rate remains relatively low with limited system impairments. As the desired speed becomes larger, the limited spectral efficiency results in an inefficient implementation due to the excessively large necessary bandwidth and need for equalization. The high-order direct-digital architecture uses a higher spectral efficiency to obtain faster rates, albeit with an efficiency penalty from the required coherent detector and direct-digital blocks. For very high data rates, one must resort to mixed-signal processing techniques to overcome inevitable channel and circuit impairments, yielding an increasingly poorer link efficiency. Finally, a DSP-based high-order architecture is able to achieve the highest data rates thanks to highly advanced digital algorithms. However, due to the required data converters and DSP processors, a big price in power consumption must be paid, which is only worthwhile for ultra-high rates close to the maximum attainable speed imposed by the channel. Again, as the channel capacity is pursued more intensely, the advanced converters and digital blocks will deteriorate the link efficiency. To conclude, the high-order direct-digital architecture is ideal for very high-speed systems where energy efficiency is an important specification, outperforming conventional high-order architectures in terms of efficiency due to its mixed-signal nature. For systems where flexibility or data rate is the main concern, DSP-based architectures still provide the best performance.

Various mixed-signal processing techniques have been published in the pursuit of the high-speed mixed-signal transceiver. An energy-efficient mixed-signal pulse shaping approach with spectral filtering was proposed in [188] which is compatible with higher order direct-digital modulation at multi-GBaud speeds. In [152], a current-mode digital predistortion was integrated in the PAM-4 modulator to correct the nonlinearity of the noncoherent detector. Different mixed-signal multilevel equalization circuits have been demonstrated ranging from continuous-time linear equalizers [178, 179, 189] to complex-tap feed-forward and decision feedback equalizers (FFEs/DFEs) [186, 190, 191] capable of compensating inter-symbol interference (ISI) caused by bandwidth limitations, dispersion in DWGs, or line-of-sight multipath effects, all without power-hungry DSP. Finally, mixed-signal clock and data recovery circuits are widespread in wireline receivers [192], and analog carrier recovery can be accomplished with a variation of the Costas-loop [51, 151]. However, another important aspect of a viable commercial product is the implementation of calibration loops which determine the optimal biasing conditions to ensure adequate performance in all PVT conditions. Since calibration usually heavily relies on DSP, achieving the same level of reliability within the proposed mixed-signal system with minimal DSP will require more research and circuit-level innovation.

Nevertheless, direct-digital high-order modulation and demodulation are arguably the biggest remaining challenges to realize a high-performance mixed-signal transceiver in the sub-THz band. While some high-order direct-digital modulation solutions exist, all require offline processing techniques to achieve acceptable EVMs at high modulation speeds. Direct-digital demodulation receivers are nonexistent for modulations more dense than 8-PSK. An introduction and overview of published direct-digital (de)modulation techniques will be given next as a segue into the proposed direct-digital 16-QAM architecture which is the focus of this chapter.

7.1.3 Direct-Digital Modulation and Demodulation

Direct-digital (de)modulation transmitters and receivers have been proposed in the sub-THz band to perform high-order modulation and demodulation, respectively, in the analog/RF domain with binary data interfaces. This would obviate the need for

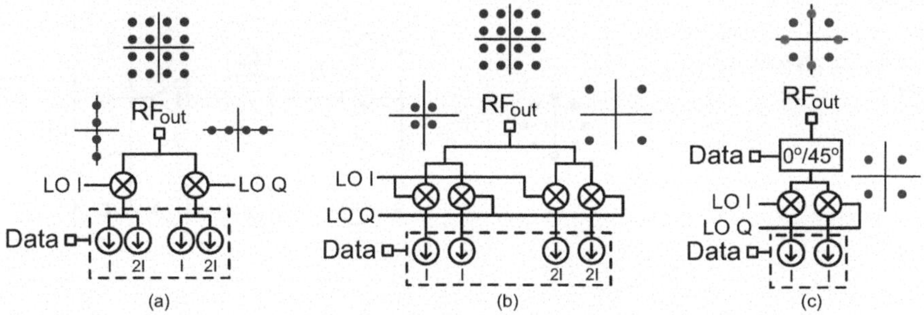

Fig. 7.7 Overview of some high-order direct-digital modulation methods: a 16-QAM RF-DAC (**a**), 16-QAM direct-digital TX [173] (**b**), an 8-PSK direct-digital TX [193] (**c**)

multi-bit data converters to significantly simplify the overall architecture and reduce global power consumption while still taking advantage of the high data rates granted by high-order modulations.

7.1.3.1 Direct-Digital Modulation Transmitters

On the transmit side, RF-DACs (Fig. 7.7a) have been a prominent way to generate 16-QAM constellations [108, 162, 194] and can readily be extended to transmit more complex constellations like 64-QAM [194]. Optionally, by stacking multiple amplifying transistors a power RF-DAC can be implemented [194], which directly provides a high output power modulated signal and obviates the need for an extra power amplifier block. RF-DACs typically consist of a current-steering DAC, where the magnitude of the current is controlled by the digital input code, and a mixer to upconvert that current to the desired RF frequency. As the RF impedances vary significantly with changing input code, there is significant data-dependent distortion, which degrades the EVM at fast switching speeds. Furthermore, sizing the different current sources exposes a fundamental trade-off between linearity and switching speed. To obtain good linearity, large current sources are necessary to minimize mismatch which limits the switching speed of the sources. All of these impairments limit the achievable density of the modulation and require digital predistortion or equalization to obtain an acceptable EVM [194].

Alternatively, a 16-QAM constellation can be constructed by adding two QPSK constellations with a factor two in magnitude [173], as shown in Fig. 7.7b. By adding more QPSK signals with suitable magnitudes, any higher-order M-QAM constellation can theoretically be generated. However, limited constellation tunability in the presence of process variations, along with load-modulation-induced distortion, will limit the achievable EVM without offline compensation. Furthermore, since the high-frequency LO needs to drive a higher number of modulators, the power consumption of the LO generation block will be significantly larger due to the increased load. Finally, a 8-PSK signal can be generated (Fig. 7.7c) by adding two

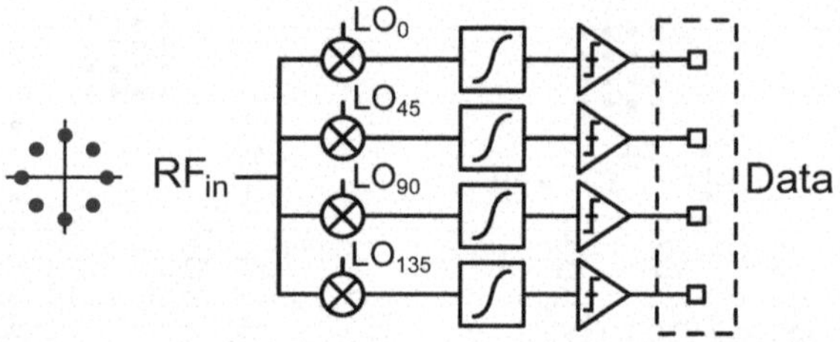

Fig. 7.8 A 8-PSK direct-digital demodulation principle as presented in [174]

QPSK constellations shifted 45° in phase using a switchable phase shifter [193]. In this case, mainly LO imbalance-induced distortion will limit the achievable performance in this architecture.

Despite showing good promise to generate high-order constellations with binary data inputs, all architectures require offline equalization or digital predistortion to avoid error vector magnitude (EVM) degradation at high speeds and are thus not suitable for the envisioned mixed-signal direct-digital system.

7.1.3.2 Direct-Digital Demodulation Receivers

On the receive side, a demodulation architecture has been proposed capable of directly converting a modulated 8-PSK signal into its non-return-to-zero (NRZ) output bits (Fig. 7.8) using an RF-correlation method with only sign-check comparators and little digital logic [174]. Direct-digital demodulation receivers for even more spectrally efficient modulations have not yet been reported at multi-GBaud speeds.

To work toward the goal of a fully integrated mixed-signal sub-THz transceiver, the rest of this chapter demonstrates a fully integrated direct-digital 16-QAM link in 28 nm CMOS operating at 135 GHz. A direct-digital modulation transmitter is proposed based on a Cartesian architecture with separate phase and amplitude modulation capable of generating a low-EVM 16-QAM constellation at high speeds [106]. In addition, a direct-digital demodulation receiver is introduced supporting efficient 16-QAM demodulation to NRZ outputs completely in the analog domain by combining a direct-conversion RF front-end with analog PAM-4 decoders [107]. A 135 GHz 16-QAM link is established over a wireless and over a DWG channel using off-chip Vivaldi antennas, without employing any offline predistortion, pulse shaping, or equalization.

7.2 System Description and Specifications

The goal of this project is to develop a high-speed (>50 Gb/s, ideally 100 Gb/s) sub-THz link with excellent energy efficiency (<10 pJ/b). As discussed in Chap. 5, there exist a great number of degrees of freedom, on a system-level and circuit-level, to achieve such a system. In this section, the most important system-level choices are discussed, the desired specifications are listed, and an overview of the designed system is described.

7.2.1 System-Level Considerations

7.2.1.1 Technology

As discussed in Sect. 1.4, an important choice is the technology in which the integrated circuits are to be fabricated. While III/V and SiGe technologies provide a substantial benefit in terms of high-frequency RF performance, CMOS technologies excel in mixed-signal and digital circuit performance. As the intended mixed-signal system will be integrated with digital-like circuitry, a more advanced CMOS node seems more suitable. Furthermore, standard CMOS processes can provide a high yield and low unit cost at high production volumes, showing greater economic promise. While smaller CMOS nodes have improved the performance at high frequencies for the longest time, this is no longer the case since the introduction of the FinFET transistor. The optimal technology is thus determined by a trade-off between RF performance and digital performance. All things considered, 28 nm bulk CMOS is a good choice as it offers great mixed-signal circuit performance with a decent f_{max} around 280 GHz.

7.2.1.2 Carrier Frequency

Choosing the carrier frequency has been extensively discussed in Chap. 5. Increasing the carrier frequency grants more available circuit bandwidth, while the attainable SNR in the system drops. As the frequency approaches the f_{max} of the technology, the obtainable output power significantly drops and the circuit noise drastically increases. Practical systems therefore operate well below the technology's f_{max} to avoid a low power efficiency and poor link budget. In 28 nm CMOS, the D-band (110–170 GHz) seems to strike a good balance between available bandwidth and active device performance. For this system, a carrier frequency of 135 GHz was chosen.

7.2.1.3 Modulation Scheme

Different modulation schemes have been investigated in Chap. 2. While simple modulations (ASK, BPSK, FSK, QPSK) can lead to compact systems achieving moderate speeds with good energy efficiencies, practical limits on circuit bandwidth restrict efficiently achievable symbol rates. Higher order modulations thus have to be considered as the desired data rate reaches 30 Gb/s and above to limit the necessary symbol rate. However, denser modulation schemes (16-QAM, 64-QAM, 256-QAM) have more stringent SNR requirements and give rise to more complicated circuit implementations, both of which are challenging to achieve at high frequencies. For the target data rate (100 Gb/s in a perfect world), 16-QAM leads to a realistic circuit bandwidth specification and SNR requirement at 135 GHz. The model described in Chap. 5 further confirmed 16-QAM as the optimal modulation scheme for maximum capacity.

7.2.1.4 Architecture

Working at 135 GHz enables the circuits to exhibit a large absolute bandwidth for a given relative bandwidth. However, working at sub-THz frequencies close to the f_{max} of the technology and working with very high BW signals pose significant challenges to obtain an energy-efficient implementation. There are two key elements to ensure high-speed performance at sub-THz frequencies with a good energy efficiency, both of which influence the choices that have to be made at an architectural level.

Firstly, efficient RF design close to f_{max} is paramount since transistor gain is a scarce and expensive resource. An efficient design is obtained partly by optimizing the design at circuit-level, which will be described in more detail in later sections, as well as making adequate choices on an architectural level to limit the complexity of the system. A direct-conversion architecture is preferred to limit the number of blocks in the system and avoid any IF filtering or gain stages that restrict the achievable bandwidth of the system.

Secondly, minimizing the complexity of the data converters and DSP blocks is crucial for efficient high-BW systems, as discussed in the previous section and Chap. 5. This is achieved by implementing a direct-digital (de)modulation architecture. By performing the (de)modulation directly in the digital domain, the modulation bandwidth is further maximized by avoiding any BW-limiting BB stages. Furthermore, simple nonlinear buffers can be employed to regenerate the steep edges of the binary data signals as opposed to power-hungry linear amplifiers needed to amplify sensitive multilevel signals in the case of a homodyne or heterodyne architecture.

7.2.2 *System Overview*

An overview of the proposed 16-QAM system is depicted in Fig. 7.9. A direct-digital 16-QAM transmitter and receiver chipset is developed and interfaced using an off-chip antipodal Vivaldi antenna via both an OTA wireless and DWG channel.

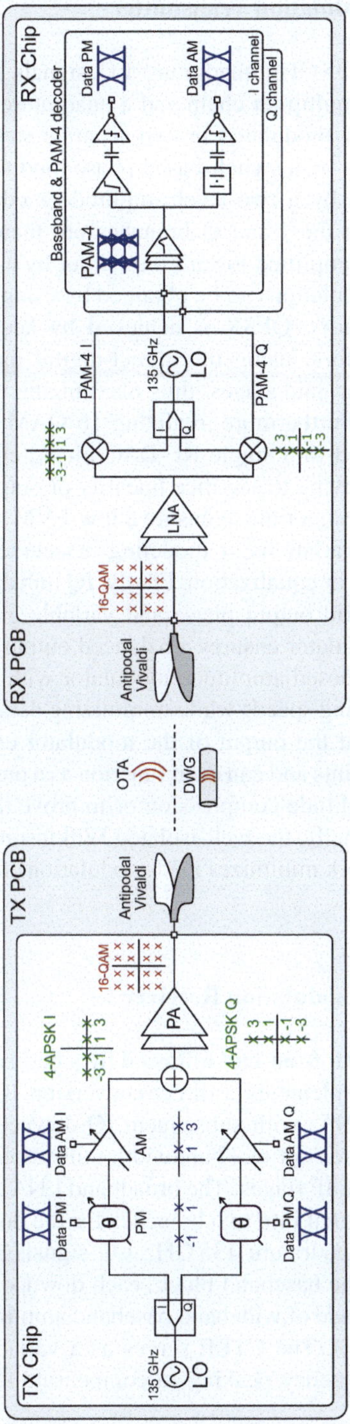

Fig. 7.9 Overview of the proposed 135 GHz direct-digital 16-QAM link

7.2.2.1 Direct-Digital Modulation Transmitter

The transmitter generates 135 GHz quadrature LO signals from a low-frequency reference using an on-chip multiplier chain and a quadrature hybrid. Both of these LO signals are subsequently modulated in two separate steps. First, the phase of each LO signal is modulated by a switch-based phase modulator (PM) after which the amplitude is modulated by a two-level amplitude modulator (AM). Both 4-APSK modulated signals in the I and Q branches are then combined to form a 16-QAM constellation and amplified toward the output by a power amplifier (PA). Additionally, star-QAM modulation can be obtained by using identical data streams for both amplitude modulators. QPSK is achieved by keeping both amplitudes constant. As mentioned before, using this direct-digital modulation architecture avoids any BB/IF filtering or gain stages, thus maximizing modulation bandwidth to achieve high data rates. Furthermore, realizing 16-QAM modulation with separate modulators, as opposed to a single RF-DAC block, enables optimization of modulation speed without having to sacrifice linearity or output power.

Special care is taken at design time to ensure a low-EVM 16-QAM constellation in the presence of process variations or modeling inaccuracies without having to resort to digital predistortion or equalization. Firstly, IQ imbalances can be corrected thanks to a tunable IQ hybrid output phase and variable gain in I and Q stages. The symmetrical phase modulator ensures a balanced output with high modulation speed. Furthermore, the proposed amplitude modulator with built-in kickback cancelation enables high switching speeds while minimizing data-dependent distortion. The adjustable gain levels at the output of the modulator ensure optimal spacing between the constellation points and can even function as a predistortion mechanism to compensate possible amplitude compression or improve the performance of the amplitude demodulation. Finally, the well-isolated Wilkinson power combiner with broadband amplifying network minimizes load-modulation effects and ripples in the frequency response.

7.2.2.2 Direct-Digital Demodulation Receiver

The receiver consists of an RF front-end followed by a baseband and demodulation block. The RF front-end implements a direct-conversion IQ receiver comprising of a low-noise amplifier (LNA) with subsequent IQ downconversion mixers. The homodyne architecture ensures optimal bandwidth utilization around the carrier by removing bandwidth-limiting IF stages. The broadband LNA amplifies the 16-QAM input signal and splits the signal into two branches for the downconversion mixers. The mixers are driven by quadrature 135 GHz LO signals generated similarly to those in the transmitter. In the baseband block, each downconverted PAM-4 signal is first conditioned by a cascade of wideband baseband amplifiers and a continuous-time linear equalizer (CTLE). The CTLE serves as a variable-gain amplifier and provides programmable frequency peaking to compensate the bandwidth-limiting

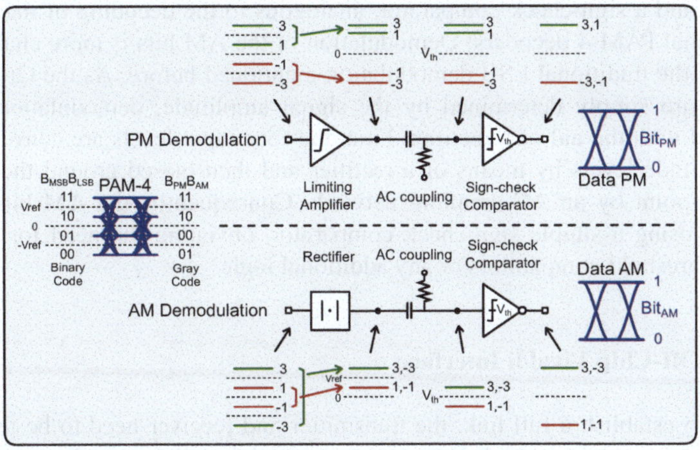

Fig. 7.10 Overview of the PAM-4 encoding and demodulation scheme

circuits in the receiver chain. The amplified baseband signals are finally sent to a PAM-4 decoder to efficiently demodulate the PAM-4 data from the I and Q channels of the 16-QAM signal.

PAM-4 decoders are commonly used in non-ADC-based PAM-4 wireline receivers to recover the most- and least-significant bits (LSB/MSB) from the received PAM-4 signal [178, 179, 195, 196]. Traditionally, a minimum of three data slicers is employed to extract the binary coded MSB and LSB. While only a simple sign-check comparator is needed to decode the MSB, the LSB needs two comparators with two distinct nonzero thresholds which need to be generated adaptively to ensure optimal decoding performance. The design of such comparators, combined with the necessary adaptive threshold generation, is challenging at high speeds and consumes significant power. Furthermore, additional logic is necessary to convert the thermometer-code outputs to binary code. In this system, a simplified rectifier-based demodulation method is proposed, inspired by [195].

Figure 7.10 illustrates the PAM-4 encoding and demodulation scheme. Instead of traditional binary coding, Gray coding [197] is used to ensure minimal bit errors in the presence of noise or other constellation impairments. This specific encoding scheme is hard-wired into the modulator and significantly simplifies the decoder. For both (I/Q) components in the transmitted 16-QAM signal, the amplitude modulator encodes the AM bit in the amplitude level, and the phase modulator maps the PM bit onto the phase of the modulated signal. Consequently, the AM bit denotes a high/low amplitude level, while the PM bit indicates a positive/negative phase. In the receiver, the demodulator separately extracts the Gray-coded amplitude modulated (AM) bits and phase modulated (PM) bits from both (I/Q) downconverted PAM-4 signals, as illustrated in Fig. 7.10. The PM bits are determined by the signal polarity and can thus be readily decoded by a limiting amplifier and a sign-check comparator, analogous to the decoding of the MSB in conventional PAM-4 decoders. Demodulation of the AM bits is more challenging, similar to the traditional LSB demodulation mentioned before. As the Gray-coded AM bits are simply determined by the signal amplitude, demodulation can be performed with the aid of a rectifier. First, the four input levels are converted into two amplitude levels by means of a rectifier and then biased around the optimal threshold point by an AC coupling network. Consequently, the AM bits can be extracted using a simple sign-check comparator, obviating the need for multiple nonzero threshold comparators or any additional logic.

7.2.2.3 Off-Chip Vivaldi Interface

In order to establish a full link, the transmitter and receiver need to be interfaced via a communication channel. In this project, a wireless link as well as a dielectric waveguide link was constructed to compare the performance in different scenarios. In the case of a wireless link, antennas are required to convert the electrical signal into an electromagnetic (EM) wave that can be transmitted through space. In case of a dielectric waveguide, the electrical signal needs to be coupled into the guiding channel in the correct waveguide mode. For both the free-space antenna and the waveguide coupler, an off-chip antipodal Vivaldi structure is designed to fulfill its respective function. The TX/RX chips are connected to a high-frequency PCB via flip-chip bumps where the Vivaldi antennas are implemented. More details on the design of the antennas and the interface will be given later in this chapter.

7.2.3 System Specifications

To estimate the performance of the proposed link, the expected link budget is determined based on simulated values in case of a 5 cm wireless link as shown in Fig. 7.11. For optimal 16-QAM operation, the PA is assumed to operate at 1-dB compression. The propagation loss of 39 dB includes an FSPL of 49 dB and TX/RX antenna gains of 5 dBi each. A similar propagation loss is obtained supposing a 7 m DWG channel with a channel loss of 4 dB/m and fiber coupler losses of 5 dB each side. A receiver noise figure of 10 dB is assumed, and its input is considered interferer-free, which is a fair assumption at these high frequencies. The targeted link budget provides sufficient SNR to support 16-QAM operation with a BER below 10^{-12}.

However, at very high symbol rates an extra SNR penalty is expected due to the ISI caused by the limited system bandwidth. This penalty can be calculated given the frequency response of the system, the applied symbol rate, and modulation scheme, as discussed extensively in Chap. 5. Additional ISI can be induced by the impairments of the channel. In the wireless channel, only limited multipath ISI is experienced due to the high operating frequency, while the frequency-

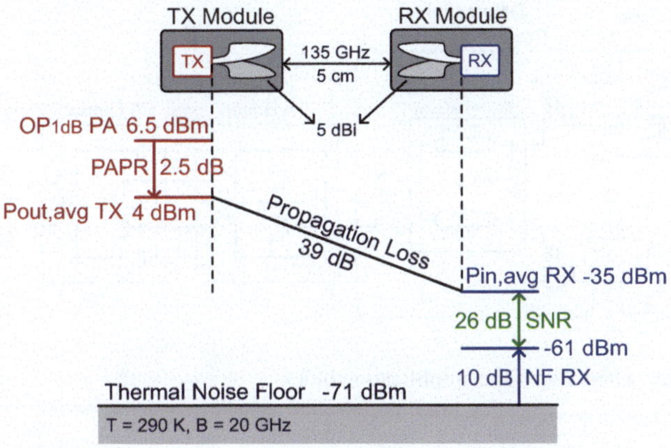

Fig. 7.11 Link budget diagram of the proposed 5 cm wireless 16-QAM link

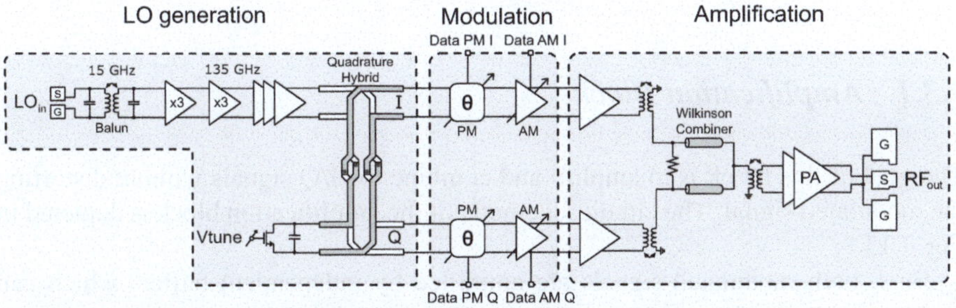

Fig. 7.12 Block diagram of the 135 GHz transmitter

dependent loss and dispersion of the DWG can significantly limit the performance [126], especially at longer ranges. Furthermore, the link budget does not include nonidealities such as IQ imbalances, LO spurs, phase noise, large-signal effects, and modeling inaccuracies. However, as discussed earlier, various mechanisms are implemented on circuit-level to limit performance degradation caused by circuit impairments and ensure robust, high-speed operation despite the lack of digital processing. The implemented circuit techniques and design considerations will be discussed elaborately in the following sections.

7.3 Transmitter Circuit Design

This section discusses all design details of the transmitter architecture and individual blocks. The transmitter block diagram is shown in Fig. 7.12. First, in the LO generation block the 135 GHz quadrature LO signals are generated using a $\times 9$ multiplier chain and quadrature hybrid. Both LO signals are subsequently each

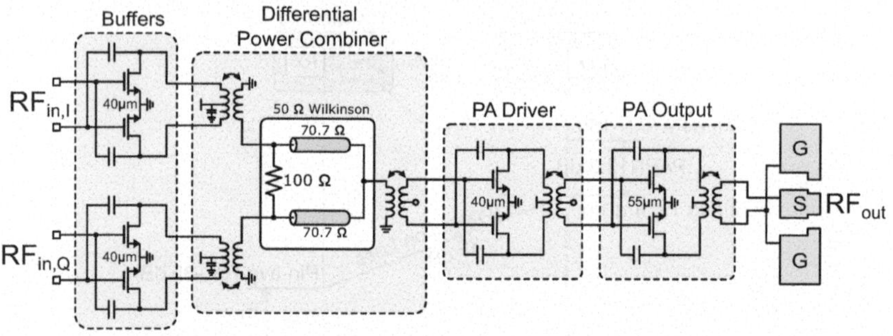

Fig. 7.13 Circuit schematic of the amplification block

modulated in phase and amplitude in the modulation block and finally buffered, combined, and amplified toward the transmitter output in the amplification block.

7.3.1 *Amplification Block*

The goal of this block is to amplify and combine the I/Q signals without distorting the modulated signal. The circuit schematic of the amplification block is depicted in Fig. 7.13.

First, both modulated signals are amplified by independent buffers which can be biased separately to compensate possible magnitude imbalance between the channels before the signals are combined. This magnitude imbalance could arise from IQ magnitude imbalance of the quadrature hybrid or gain mismatch between the modulators of both branches. These buffers also provide more isolation between the modulators' outputs, on top of the intrinsic combiner isolation. The outputs of both buffers are subsequently added using a balun-based differential Wilkinson power combiner with ideally low loss, good input isolation, adequate common-mode suppression, and broadband frequency response. Finally, the output signal is amplified by a two-stage power amplifier (PA) toward the GSG output pads. This PA should be able to provide enough gain and generate a large output power with minimal nonlinearities.

7.3.1.1 Balun-Based Differential Power Combiner

The power combiner must combine the modulated signals in the I and Q branches to form a modulated 16-QAM signal. The combiner must achieve this with minimal signal loss, minimal magnitude or phase error between the inputs, and over a large bandwidth. Furthermore, since two distinct signals are combined, both inputs must be isolated from one another to avoid load pulling effects. Without input isolation, a

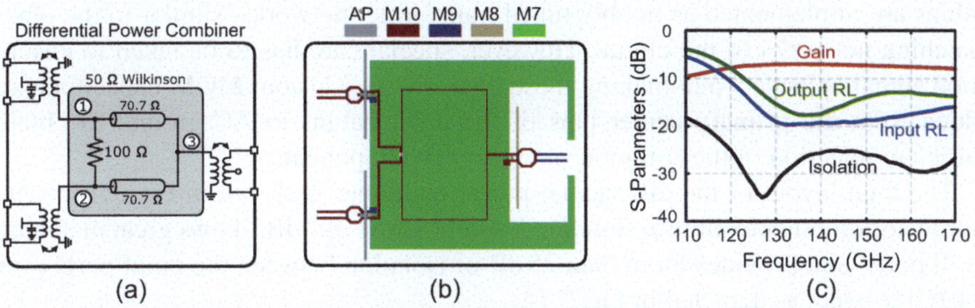

Fig. 7.14 Differential Wilkinson power combiner: circuit schematic (**a**), layout (**b**), and simulation results (**c**)

varying input signal will lead to a change in load impedance for the other input, resulting in signal-dependent signal distortion and an EVM degradation of the transmitter output.

It can be proven that a 3-port network cannot be simultaneously lossless, reciprocal, and matched at all ports [65]. In order to design a 3-port power combiner which is reciprocal and perfectly matched, some loss thus has to be introduced. Furthermore, such a lossy combiner can be made to have isolation between its input ports as well. A popular topology for such a combiner is the Wilkinson combiner (Fig. 7.14a), which appears lossless when all ports are matched, so only reflected power is dissipated in the resistor. The S-parameters of an ideal Wilkinson combiner can be written as

$$S_{11} = S_{22} = S_{33} = 0 \quad \text{(matched)} \tag{7.1}$$

$$S_{12} = S_{21} = 0 \quad \text{(isolation)} \tag{7.2}$$

$$S_{32} = S_{31} = S_{23} = S_{13} = -j/\sqrt{2}. \tag{7.3}$$

Driving the combiner with two identical signals results in an output 3 dB higher than its inputs, so no power is dissipated in the resistor. Driving the combiner with quadrature signals (as is the case when adding the I and Q components of the QAM-modulated signal) leads to combiner loss of 3 dB, which is unavoidable for a matched combiner with isolated inputs.

Since the inputs and output are differential signals, a differential power combiner is required. However, a differential Wilkinson combiner has inevitable signal crossovers in the layout, which leads to discontinuities in the characteristic impedance of the transmission lines. These discontinuities translate into an increased magnitude and phase imbalance and poor port matching. To avoid a degraded combiner performance, a balun-based differential Wilkinson power combiner is designed [173]. This combiner consists of a 50 Ω single-ended Wilkinson combiner (without signal crossovers) with baluns at the input and output ports to perform the conversion between single-ended and differential operation. The baluns are implemented as doubly tuned transformer networks, similar to the other matching networks of the circuit. However, special care has to be taken to ensure good signal balance when making these conversions. Custom MOM capacitors are added from the primary center taps of the input baluns to AC ground to obtain sufficient rejection of the common-mode signal components.

The final layout of the differential power combiner is shown in Fig. 7.14b. The complete combiner exhibits a simulated insertion loss of 7 dB, shows great matching at all ports, and provides more than 25 dB of isolation between the input ports over a 40 GHz band, as depicted in Fig. 7.14c.

7.3.1.2 GSG Pads

The GSG (Ground–Signal–Ground) pads at the output of the transmitter chip function as an interface to bring the RF output outside the chip. For this chip, we are interested in two interface options. Firstly, in order to connect the transmitter output with an external, on-board antenna to form a full link, the GSG pads need to be compatible with a flip-chip interface. However, characterizing the standalone performance of the transmitter is not trivial in this configuration as the antenna and interconnects cannot be easily de-embedded. So secondly, we would like to be able to measure the transmitter with GSG probes. This way, the transmitter can be accurately measured because the external losses from the measurement setup (probes, waveguides, mixer, cables) can be de-embedded up to the actual transmitter output.

To make the GSG pads compatible with both a flip-chip process and GSG probes, a custom design is required. First of all, flip-chip bumps with a diameter of 60 μm and a pitch of 160 μm (constrained by the PCB technology) have to fit onto the pads. Second of all, GSG probes with a 50 μm pitch have to be landed onto the pads with enough margin to account for misalignment and adequate skating distance. These spatial specifications have to be fulfilled while bearing the electrical performance in mind. Specifically, the loss of the pads needs to be minimized as well as the variation of the input impedance of the pads. Both can be optimized by carefully designing the dimensions of the pads and interconnect to the output of the last stage. The most detrimental factor is the inductance of these interconnect lines, which can be minimized by making the lines as short as possible and increasing the width of the lines. Widening the lines increases the capacitance between the signal and ground nodes, but this can easily be absorbed in the output matching network of the transmitter. Series inductance, on the other hand, degrades the loss and bandwidth of the network. The final layout of the GSG pads and interconnect lines is shown in Fig. 7.15. Compared to GSG pads optimized for probes only, the designed pads exhibit a slightly higher loss of only 0.4 dB while also accommodating a flip-chip interface.

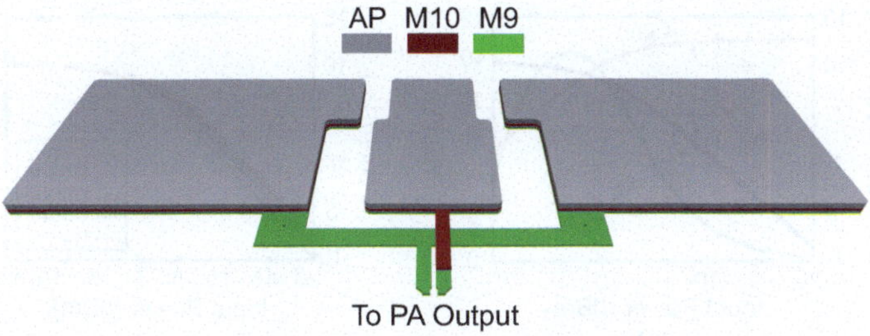

Fig. 7.15 Layout of the custom designed GSG pads with interconnect lines

7.3.1.3 Power Amplifier

The PA is the most critical block determining the output power and linearity of the transmitter output. An important source of nonlinearity is the variation of the gain and phase response over input power. Particularly, for high input powers, the gain of the PA typically drops (AM–AM distortion or gain compression) and the output phase starts to vary (AM–PM distortion), leading to EVM degradation as explained in Sect. 2.3.3. These distortions generally worsen as the PA output power increases, and the employed modulation schemes gets more dense. A straightforward approach to improve the linearity would be to scale back the PA output power and thus operate in a power back-off (PBO) where the gain and phase response are linear across the whole input range. However, this leads to a decrease in signal output power (degrading the link budget) and a decline in power-added efficiency (PAE). A better approach is to make clever design choices to reach a good optimum between PA performance (Pout, gain, PAE) and PA distortion (AM–AM, AM–PM), which are all a function of transistor size, biasing, and amplifier topology.

Each amplifier stage is implemented as a pseudo-differential pair with capacitive neutralization (see Fig. 7.13), as it offers superior gain performance at high frequencies with favorable stability conditions (see Sect. 4.3). Neutralization was realized with custom EM simulated MOM capacitors with high-quality factors ($Q \approx 120/90$ for the driver/output stage at 135 GHz) in order to maximally boost the power gain. Due to the low transistor gain at the working frequency ($G_{\text{max}} \approx 7.5$ dB at 135 GHz), the operation point is chosen to maximize the power gain per stage, thus minimizing the number of required amplifier stages in the whole chain (which benefits the TX bandwidth as well as efficiency). In order to also minimize the power consumption, the PA is operated in class AB with a bias voltage around 400 mV.

First, the output stage is sized to provide a large output-referred 1-dB compression point (OP1dB), which is the approximate peak output power when operating for 16-QAM signaling. At the same time, this stage must be able to provide acceptable power gain with an ideal impedance to be matched to the GSG pads with minimal loss. Since AM–PM distortion is limited when operating close to class A above 100 GHz, it can be safely ignored while sizing the amplifier. When increasing the transistor size, the OP1dB initially enhances as the signal swing decreases for the

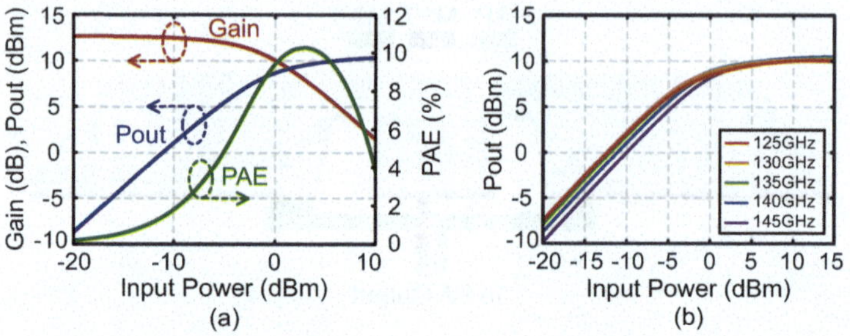

Fig. 7.16 Large-signal simulation results of the two-stage power amplifier at 135 GHz (**a**) and output power curves over frequency (**b**)

same input power, but leads to a decrease in gain for very large sizes due to the considerable layout parasitics. A larger transistor size also comes with a decrease in output impedance due to an increase in transistor and layout parasitic capacitance. So the choice of transistor size for gain and OP1dB must be co-optimized with its output impedance to minimize the loss of the output matching network, which is directly subtracted from the gain and output power of the amplifier.

The driver stage is then sized for maximal power gain with adequate output power to drive the output stage. It is particularly important that the driver stage starts to compress later than the output stage in order to avoid linearity degradation. Concretely one wants to ensure that the OP1dB of the driver stage (including matching losses) is a couple of dBs bigger than the (input-referred) P1dB of the output stage. Additionally, it is desirable to have the output impedance of the driver stage be close to the input impedance of the output stage to guarantee a broadband, low-loss matching network in between the stages.

The buffer amplifiers provide a similar function to the driver stages, requiring maximal power gain without limiting the linearity of the whole chain. Since the power combiner between the buffers and driver amplifier adds quite a bit of loss, the buffer stages do not need a OP1dB much larger than the P1dB of the driver stage. For this reason, the buffer and driver stages are sized identically.

As depicted in Fig. 7.16a, the two-stage PA shows a simulated Psat of 9.6 dBm with a maximum PAE of 10.3%, and a small signal gain of 12.7 dB. The OP1dB is 6.5 dBm at 135 GHz and remains higher than 5 dBm over a 40 GHz bandwidth, as illustrated in Fig. 7.16b. At 1-dB compression the AM-PM distortion is less than 0.5°.

7.3.1.4 Frequency Response

The second big source of nonlinearity is magnitude and group delay variation over frequency, which causes pulse distortion and by that inter-symbol interference (ISI), as explained in Sect. 2.3. These variations are minimized by designing for a broadband and ripple-free amplifier frequency response.[2] Furthermore, the aforementioned AM–AM and AM–PM distortion are by default also changing over frequency. Generally, optimizing the bandwidth of the amplifier will also limit the frequency variation of these distortions, but a more in-depth analysis of these effects might be interesting. Finally, the symmetry of the frequency response is important to avoid coupling between orthogonal I and Q channels of the modulated signal. This symmetry is again mostly minimized by optimizing the amplifier bandwidth although a more in-depth analysis is required. Particularly an asymmetry in the phase response could be a big limitation in high bandwidth transceiver systems [109]. In the end, all aforementioned frequency-dependent impairments are tackled by maximizing the bandwidth of all relevant networks.

A broadband frequency response is only important for the circuits that need to process a high bandwidth modulated signal. LO buffer amplifiers, which will be discussed later, do not need a wideband response because they only need to amplify a narrowband LO signal. Every block from the output of the amplitude modulator to the output of the GSG pads,[3] so the complete amplification block, will handle a modulated signal and thus need a large bandwidth.

All matching networks are implemented as tightly coupled 1:1 stacked transformers in combination with series inductor. As discussed in Sect. 4.4, the series inductors aid in providing a larger impedance transformation with low insertion loss compared to only transformers with a large diameter ratio. Parallel capacitors are added where necessary to establish the desired resonances (but are not drawn on the schematics for clarity). It should also be noted that the matching networks are designed for small-signal operation as the gain performance is priority and since, as discussed in Sect. 4.4, the characteristic should not degrade much when operating in large signal, due to a high similarity between small- and large-signal impedances above 100 GHz. Besides, for 16-QAM operation the PA will not be operated in saturation but in a power back-off, making the small-signal approximation even more appropriate.

In order to ensure a broadband frequency response over the complete amplification block, all different matching networks are co-optimized for maximum bandwidth and minimal insertion loss. By stagger tuning the matching networks (i.e., choosing different center frequencies) and by compensating the frequency-dependent gain roll-off of the active devices, a large bandwidth is obtained over the whole network, at the cost of more insertion loss in each stage. While co-optimizing all stages for the optimal global response, extra attention needs to be given to the ports of the Wilkinson combiner which all need to be well matched to ensure good operation with the desired input isolation. More information on the design methodology and modeling approach was given in Sect. 4.4. A layout diagram of the full amplification block is depicted in Fig. 7.17.

---

[2] While increasing the bandwidth of the gain (magnitude) response generally improves the flatness of the group delay response, particular design techniques have to be employed to really optimize the group delay flatness. In this design, the magnitude response was prioritized.

[3] Including the amplitude modulator itself, since it modulates an already phase modulated signal.

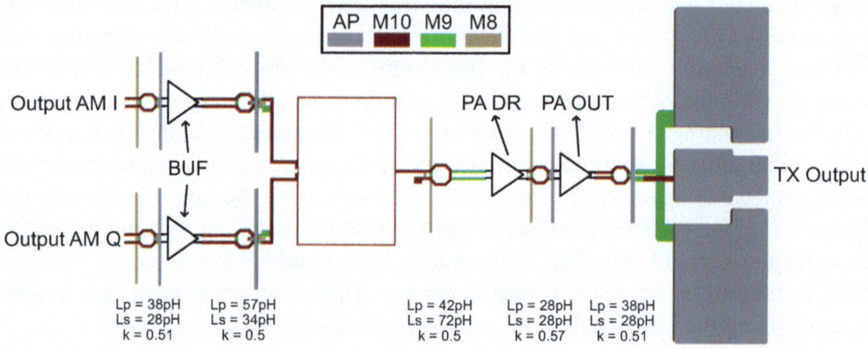

Fig. 7.17 Layout diagram of the full amplification block with transformer parameters (ground in M7 omitted)

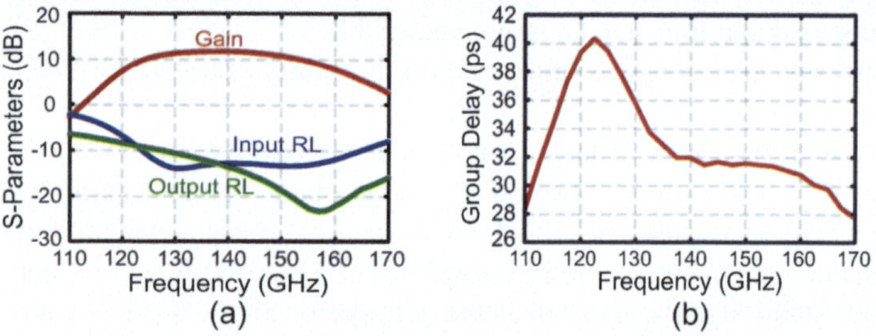

Fig. 7.18 Simulated S-parameters (**a**) and group delay (**b**) of the amplification block

Figure 7.18a shows the small-signal simulation results of the amplification block and the input isolation of the Wilkinson combiner. A simulated 3-dB bandwidth of 35 GHz is achieved from the output of the modulators to the transmitter output without any in-band ripple while providing 12 dB of gain from both inputs. Compared to centering each network at 135 GHz, stagger tuning provides a 8 GHz bandwidth improvement for 2.5 dB less gain. The group delay of the amplification block is depicted in Fig. 7.18b. Simulations show that the designed network is capable of processing 16-QAM signals up to 25 GBd.

7.3.2 Modulation Block

The modulators should perform digital phase and amplitude modulation directly on the quadrature LO signals using binary data inputs. The circuits must support high switching speeds while providing highly linear modulation with minimal signal distortion.

7.3.2.1 Some Thoughts on Digital Modulation

Performing signal modulation directly in the digital domain as opposed to the analog domain (using a DAC combined with a linear mixer) comes with a number of potential benefits. Direct-digital modulation of the carrier generally results in faster modulation because any intermediate filtering stages between the DAC and mixer are omitted. Furthermore, the binary data signals can be buffered using simple and efficient nonlinear buffers (e.g., CMOS inverters) as opposed to power consuming, linear amplifiers needed to amplify multilevel data signals.

Due to the different nature of modulation, the notion of linearity needs to be revisited. In analog modulation (i.e., a mixer), linearity characterizes how the different levels of the data input translate to the different amplitude levels of the RF modulated signal. In the case of digital modulation, the modulator is inherently nonlinear so linearity in this case simply refers to the relationship between the different constellation points of the modulated signal. In both cases linearity can be quantitatively described by the EVM, but the way a low EVM constellation is obtained can be fundamentally different.

Another aspect that needs some discussion is spectral shaping. Pulse shaping (see Sect. 2.3.4) is often performed by the DAC in order to limit the signal bandwidth. Pulse shaping can reduce the ISI introduced by bandwidth-limiting circuits and minimize interference with signals in neighboring frequency bands. Ideal digital modulation (i.e., assuming infinitely steep symbol transitions) yields a modulated signal equivalent to an upconverted modulated signal with a rectangular pulse shape. This signal would thus have a Sinc frequency spectrum with large sidelobes. While the large emission spectrum will pose a problem in numerous applications, as discussed in Sect. 1.3, it is irrelevant for DWG links and less of a problem in most use cases for sub-THz communication. This is because a wireless link at such high frequencies experiences large propagation losses and will make use of very directive antennas, which inherently limits possible interference with other links. An exception is multichannel operation within the same link, which will be inevitable if a bandwidth larger than 12.5 GHz is desired (Sect. 1.3), where special precautions need to be made to minimize the leakage into neighboring channels, such as dedicated RF filtering stages [104] or direct-digital pulse shaping techniques [188]. It is important to note, however, that pulse shaping generally requires a higher resolution and/or oversampled operation in the DAC, which burns significantly more power. Furthermore, the mixer must exhibit very linear behavior not to impair the generated pulse shape by the DAC.

In this design, digital modulation will be performed without oversampling by digital inputs with edges as steep as possible. While slower edges may decrease the signal bandwidth, it comes at a cost of slower symbol transitions and thus more ISI which degrades the EVM of the modulated output. Pulse shaping without introducing ISI is only possible when a Nyquist pulse shape is obtained, as discussed in Sect. 2.3.4, which is only possible with a higher resolution and sample rate. The proposed digital modulation method is fast, simple, and results in only a small potential ISI penalty because of the inevitable filtering of succeeding stages (if the circuit BW is designed to be large enough).

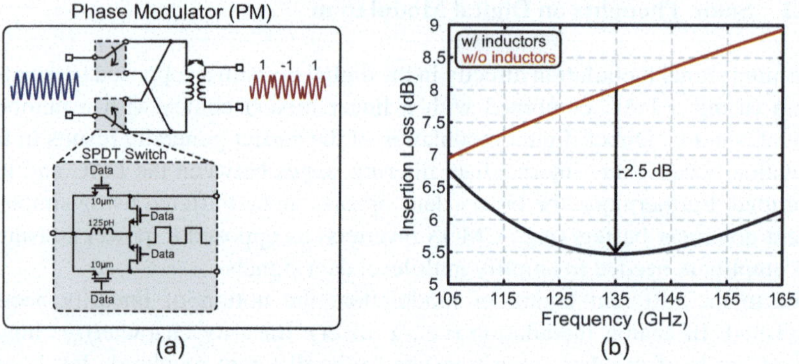

Fig. 7.19 Circuit schematic of the phase modulator (**a**) and simulated insertion loss of the modulator implemented with and without inductors (**b**)

7.3.2.2 Phase Modulator: Overview

Figure 7.19a shows the circuit-level implementation of the switch-based phase modulator. The modulator is very similar to the modulator introduced in [49] and is described in detail in Chap. 6. The main difference is in the data driving the modulator. Instead of driving the modulator with two distinct data signals to facilitate BPSK and OOK modulation, only one data signal is needed in this system to enable binary-phase modulation.

The phase modulator is realized by means of two cross-connected single-pole double-throw (SPDT) switches which can invert the differential input signal depending on the data input. Each switch is implemented as a switched-inductor SPDT (SI-SPDT) switch where an inductor is utilized to resonate out the off-state capacitance of the switches at 135 GHz and consequently reduce the insertion loss of the modulator. Furthermore, an SI-SPDT switch employs two extra switches to reconfigure the inductor between the two output ports to reduce the number of inductors per switch and thus minimize the modulator footprint.

The switch consists of ultra-low VT (ULVT) transistors to ensure minimal on-state switch resistance. A transistor size of 10 μm is chosen as a compromise between on-state resistance and off-state capacitance. A compact double-turn inductor of 125 pH is designed to resonate out the off-state capacitance of the switch. The compact and symmetrical layout of the phase modulator is shown later on in Fig. 7.23.

After parasitic extraction of the complete modulator layout and EM simulation of the inductors, simulations show a modulator insertion loss of 5.5 dB at 135 GHz. As shown in Fig. 7.19b, the loss of the modulator is reduced by about 2.5 dB compared to an inductorless implementation. Thanks to the switch-based modulator topology, only small switches need to be controlled and slow settling times are avoided, thus supporting a very high switching speed. Moreover, its symmetrical layout ensures

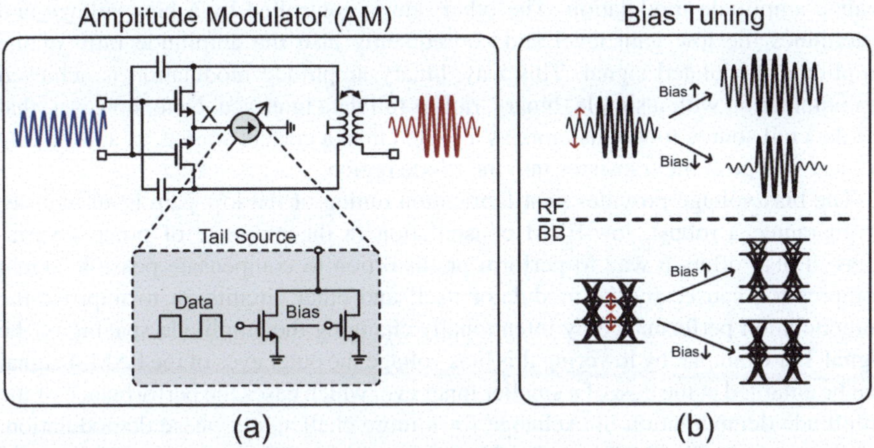

Fig. 7.20 Circuit schematic of the proposed amplitude modulator (**a**) and visualization of the effect of bias tuning on the RF/BB waveform (**b**)

minimal phase error or magnitude imbalance and provides equal impedances in both phase states leading to minimal data-dependent distortion.

7.3.2.3 Amplitude Modulator: Topology

The amplitude modulator is arguably the most challenging block to design in the direct-digital modulation transmitter. This modulator must be able to switch between two amplitude states at 20-25 GS/s driven by a binary, rail-to-rail data signal. The two gain states must differ with a factor of 3 or 9.5 dB to ensure optimal spacing between the 16-QAM constellation points. Ideally, the high gain state of the modulator is as high as possible to maximize the output power, and the low gain state can be tuned post-fabrication to compensate for any process variations or distortions in the rest of the transmitter. Finally, the input and output impedances of the modulator should stay as constant as possible while switching to minimize data-dependent distortions.

The initial circuit proposal for the amplitude modulator was based on a current-steering RF-DAC, as shown in Fig. 7.20a. A capacitively neutralized differential pair, driven by the LO signal, is combined with a switchable tail current source, controlled by the data signal. Since the modulator needs to be driven by a rail-to-rail, binary data signal, it is not possible to obtain two nonzero gain levels by controlling a single current source, as is often used in ASK modulators [150]. Here, a binary signal is often generated off-chip with carefully controlled amplitude and DC point to achieve the desired linear behavior. Instead, linear two-level amplitude modulation is achieved by combining two different tail sources. One is driven by digital data to switch between the high and low gain state in order to realize amplitude modulation. The other one is controlled by a bias voltage and determines the low gain level and consequently also the amplitude ratio of the amplitude modulated signal. This way, binary amplitude modulation is achieved in combination with a simple, binary, rail-to-rail data interface. Note, however, that the data tail source functions more as a switch than a current source, as a large gate voltage will push the transistor into the triode region.

The bias voltage provides post-fabrication tuning of the low gain level in order to guarantee a robust, low-EVM constellation in the presence of process variations. It also offers a way to perform predistortion to compensate possible signal compression caused by the modulator itself and other circuits, or to improve the demodulation performance by intentionally changing the amplitude spacing of the signal. For example, by lowering this bias voltage the outer eyes of the PAM-4 signal can be enlarged at the cost of a smaller inner eye, which eases the performance of the amplitude demodulation in exchange for a more challenging phase demodulation, as illustrated in Fig. 7.20b on the RF and baseband (BB) waveform. This could benefit the global system performance if the amplitude demodulation would be the bottleneck. Also note that the amplitude ratio is controlled by a DC voltage and does not switch at the baud rate as is the case in traditional RF-DACs. The switching nature of the amplitude modulator results in a high modulated output power which reduces the number of power amplifying stages required to reach the desired transmitted output power. This not only enhances power efficiency, but also improves modulation bandwidth by minimizing the number of bandwidth-limiting inter-stage matching networks. In addition, since the bias tail source ensures the modulator is never fully turned off (keeping the differential pair in saturation), long settling times of the tail node voltage are avoided, which improves the obtainable switching speed.

However, as Fig. 7.21a illustrates, due to the steep edges of the digital switching signal, serious kickback charge is injected into the tail (X) and gate (A,B) nodes of the modulator. Indeed, as the data signal changes abruptly, a fast change in voltage on the tail node is induced, which capacitively couples toward the gates nodes. Due to the high CM impedance at the gate node (desired for CM stability concerns), this would lead to significant unwanted settling of the common-mode input voltage whenever the modulator switches. If this long settling time is not respected, significant ISI is observed in the output (because the non-settled amplitude depends on the previous data symbols) and the dynamic range is degraded. Either way, this settling behavior severely impacts the speed and modulation depth of the modulation and needs to be addressed in order to achieve reliable, high-speed modulation.

The dynamic behavior of the modulator can be improved, inspired by the double-balanced Gilbert mixer, by adding a replica modulator which is driven by an inverted version of the data (Fig. 7.21b). As a result, data coupling to the gate nodes is canceled, drastically improving the switching speed. Note that the output of the replica is connected to an inductive dummy load. Connecting both outputs together would yield a constant gain amplifier, while cross-connecting the outputs would result in a phase modulator. Only by connecting the replica's output to a dummy load can we achieve the desired amplitude modulation without the settling behavior,

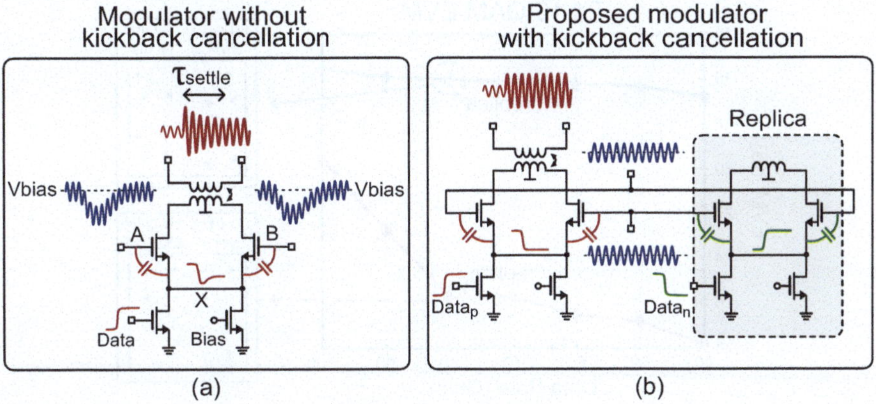

Fig. 7.21 Schematic and principle of the amplitude modulator without (**a**) and with (**b**) kickback cancelation (neutralization capacitors are omitted for clarity)

which leads to a fraction of the modulator's output power being lost. Combined with an increased capacitive load at the input, this reduction in gain is the price that needs to be paid for an improved switching behavior. Another advantage of adding the replica modulator is a more constant input impedance and supply current while switching (since exactly one of two data tail sources is turned on at all times), both resulting in significantly less data-dependent distortion. The output impedance does vary slightly in between gain states but only leads to a small deviation in amplitude, which can be compensated via the bias voltage of the constant tail source. A symmetrical layout combined with adequate sizing of the inductive load of the replica modulator, the phase response of the modulator varies negligibly in both gain states, leading to a highly linear constellation. The influence of the kickback effect and added replica is illustrated in Fig. 7.22, which shows modulated signal simulations of the complete transmitter, with and without replica amplitude modulator. As demonstrated by the 16-QAM constellations at 80 Gb/s, the kickback introduces undesired ISI and data-dependent distortion to the signal. Adding the replica improves the EVM significantly and facilitates high-speed digital 16-QAM modulation.

7.3.2.4 Amplitude Modulator: Design

Sizing the differential pair brings up a trade-off between gain, output power, and speed. Large transistors are able to deliver more output power and higher gain, at least until the layout parasitics start to dominate. Furthermore, their large transconductance (g_m) decreases the impedance seen at the tail node ($\approx 1/g_m$), which benefits the modulation speed. However, larger transistors will also add more capacitance on the tail node, increasing the time needed to (dis)charge that node

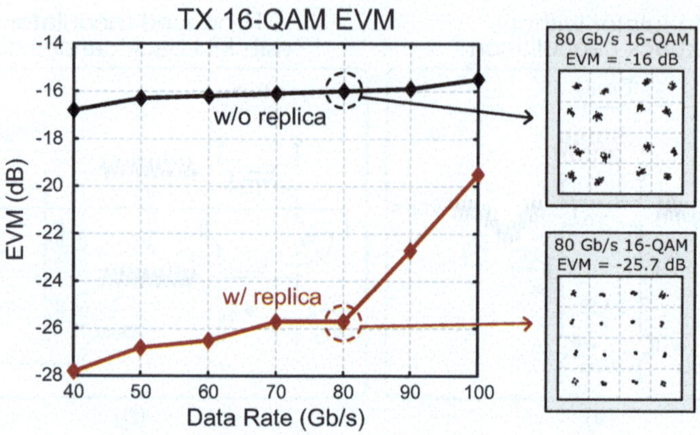

Fig. 7.22 Modulated 16-QAM simulations of the modulator block with amplification block, with and without replica amplitude modulator

and impacting the obtainable switching speed. The final transistor width of 30 μm provides a compromise between these factors. The DC bias was chosen rather high to maximize the power gain and transconductance of the transistor pair.

The tail sources can be sized independently because the low gain level is fully determined by the bias source (because the data source is turned off), and the high gain level is mostly determined by the data source (because it dominates the total current thanks to its large overdrive voltage). The data source is sized as small as possible to minimize the parasitic capacitance that needs to be charged, but large enough to ensure enough gain (or current) in the differential pair. An ULVT transistor flavor is employed to further maximize switching performance (by maximizing the overdrive voltage for a fixed gate drive). The size and bias of the bias source are co-optimized to provide a large tuning range around the desired gain level (9.5 dB below the maximum gain level) with minimal parasitic capacitance. The data and bias sources are sized to have widths of 30 μm and 10 μm, respectively.

The size of the dummy inductor only has a small effect on the modulator performance but is optimized for minimal variation in input impedance between both states. Since this inductor has asymmetrical feed lines, the center tap is slightly offset from the geometric middle to be in the electrical virtual ground. Finally, the layout of the complete modulator is drawn carefully to be as symmetrical as possible and minimize undesired coupling. The full layout is shown later on in Fig. 7.23.

7.3.2.5 Combining the Modulators

To complete the design of the whole modulator block, both modulators have to be combined. It turns out that putting the phase modulator in front of the amplitude modulator, and not vice versa, gives the best overall performance. To understand

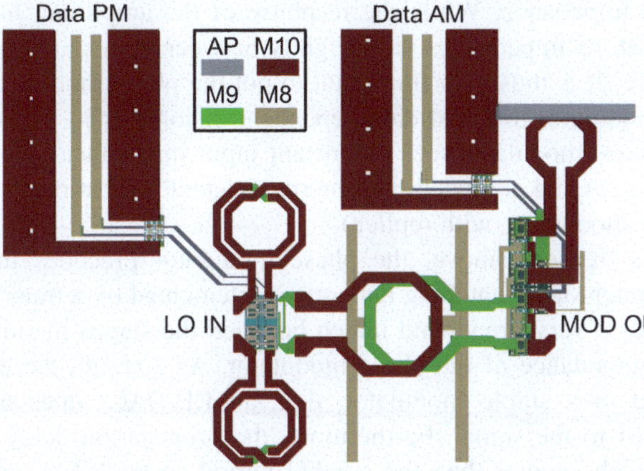

Fig. 7.23 Layout of the modulator block

this, three aspects of the modulator combination need to be analyzed: output power, modulation speed, and modulation linearity.

Firstly, it is important to maximize the modulated output power of the modulator block in order to minimize the number of subsequent stages and thus maximize the overall bandwidth and efficiency. For this reason, the modulator block must be driven by as much LO power as possible. Due its much lower input impedance, the phase modulator can perform much better with high input powers compared to the amplitude modulator. Combined with the fact that the phase modulator provides signal attenuation, putting the phase modulator first will result in the best overall linearity of the block, in terms of achievable output power as well as switching performance with high input swings.

Secondly, the modulation speed of the modulator chain has to be maximized. Since the first modulation in the chain undergoes an extra filtering operation, due to the matching network between the modulators, the fastest modulator should preferably be put first to balance the ISI contributions from both modulations. Due to the smaller transistor sizes and faster switching operation (passive switches i.o. current steering), the phase modulator is able to switch slightly faster than the amplitude modulator and should thus be put first. This effect is, however, quite insignificant up to very high switching speeds.

Lastly, the order of the modulators has an influence on the severity of certain distortions. The distortions of each isolated block are negligibly small. The phase modulator's output is extremely balanced, the phase response of the amplitude modulator is very well controlled, and its amplitude can be tuned. However, combining the blocks brings about new problems. After all, when simulating the modulators separately, a constant source and load impedance is assumed. Unfortunately, due to the limited isolation between input and output, the phase and magnitude response of the phase modulator is very sensitive to variations in source and load impedance. While the response of the amplitude modulator can be well controlled, its impedances vary slightly between gain states, especially the output impedance. It is thus very important to put the phase modulator in front of the amplitude modulator to avoid data-dependent distortions in the output signal. This way, the phase modulator sees a constant input impedance (provided by the last LO buffer stage) and is loaded by a pretty constant load impedance (provided by the amplitude modulator, with replica).

For all of the reasons above, the phase modulator precedes the amplitude modulator. The inter-stage matching network, implemented by a transformer-based network, provides a very broadband match between the stages mainly because of the low output impedance of the phase modulator. As a result, the extra filtering stage (compared to a single modulator, e.g., an RF-DAC) does not introduce any additional ISI to the signal. Furthermore, the propagation delay between the modulators is much smaller than the symbol period so no ISI is introduced due to a timing mismatch between both modulation operations. Finally, no significant coupling was observed, after EM simulations, between the transformer and the dummy inductor, nor between the modulator blocks in the I and Q channels. Transient simulations of the combination of these modulators show achievable modulation speeds up to 25 GBaud with excellent linearity, as shown in Fig. 7.22. The full layout of the complete modulator block (of a single I/Q branch) is shown in Fig. 7.23.

7.3.2.6 Data Interface

Both modulators work with rail-to-rail, binary data inputs and are insensitive to variations in the amplitude. This makes the data interface very straightforward and robust. Figure 7.24 shows a circuit diagram of the most important components of the data interface. When denoting the size of the inverters, Mx means the width is scaled up by a factor of M compared to a minimally sized inverter.

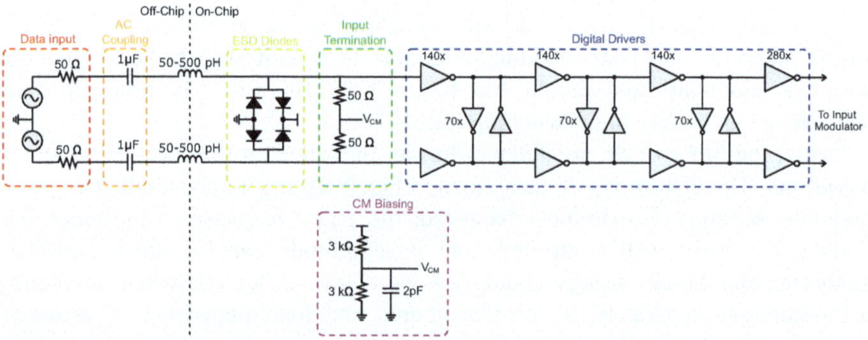

Fig. 7.24 Circuit diagram of the data interface

The AC-coupled data inputs are terminated to 50 Ω on-chip (100 Ω differential) and biased to half the supply rail with large resistors. All data signals are subsequently buffered using a pseudo-differential inverter chain consisting of simple CMOS inverters. These inverters are able to regenerate the steep data edges and compensate the ISI caused by the long interconnects and bondwire (or flip-chip)—ESD diode interface at the input. The inverter stages are equally distributed over the interconnect distance to minimize the ISI per stage, with the last stage being placed as close to the modulators as possible. Additionally, cross-coupled inverter pairs are inserted to ensure minimal phase error between the differential data signals, effectively providing common-mode rejection. The combination of the drivers with cross-coupled inverters is also referred to as a low-jitter latch and was inspired by [198]. All data paths are phase matched to experience identical delays, including interconnect parasitics, to avoid EVM degradation caused by internal delay offsets. Finally, the inverters are scaled up to drive the load of the modulators with full voltage swing. The data buffers are able to drive both modulators correctly (rail-to-rail without errors) up to 30 GBd so should not limit the achievable performance of the transmitter.

7.3.3 LO Generation Block

This block generates the 135 GHz quadrature LO signals from a 15 GHz external LO reference. The circuit schematic of the LO generation block is depicted in Fig. 7.25. A large LO output power is desired to maximize the output power of the modulators and limit the number of bandwidth-limiting amplifying blocks after modulation. Furthermore, a low IQ magnitude and phase imbalance is critical to guarantee a low EVM at the transmitter output.

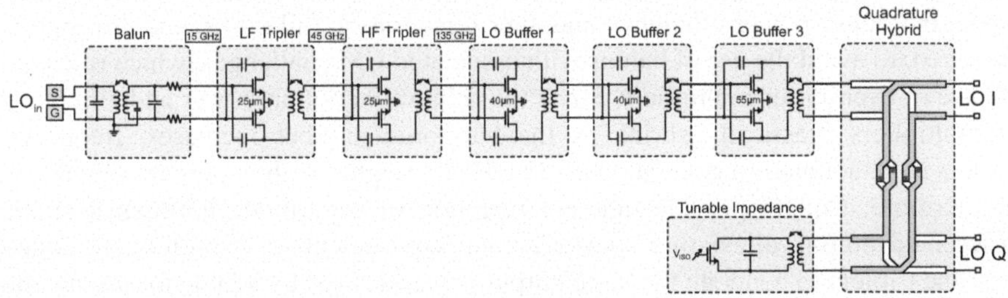

Fig. 7.25 Circuit schematic of the LO generation block

7.3.3.1 Input Balun

In order to employ a fully differential implementation for each circuit, the single-ended LO input at 15 GHz first needs to be converted into a balanced signal by a balun. Improper suppression of the common-mode (CM) will significantly decrease the conversion gain of the first stage in the LO chain and possibly limit the maximum LO output power. A transformer-based topology is slightly adapted to function as a good balun. A large capacitor is added from the center tap of the secondary winding to ground to provide a low impedant path for the CM signal to flow and thus enforce an AC ground at that node. The LO interface as well as return current path should be accurately modeled and EM simulated to ensure accurate CM rejection behavior.

Due to the desired flexibility in LO interface (bondwire and flip-chip), the series inductance connecting the PCB to the chip varies over large range (50 pH–1 nH). Designing the balun for good CM rejection and minimal insertion loss over such a large inductance range is quite challenging due to the resonating nature of the network. Eventually the network was designed to have a CM rejection better than 25 dB for both a bondwire and flip-chip interface.

7.3.3.2 Multiplier Chain

From a balanced 15 GHz LO signal, a high output power 135 GHz LO signal is generated by means of two frequency triplers and three LO buffers. A lower frequency LO reference could have been used, easing the LO interface and off-chip power requirements. However, this would require a larger multiplication factor for the multiplier chain, increasing the complexity of the LO chain and spur rejection concerns. 15 GHz is chosen such that the LO signal can be interfaced using standard bondwires, and the signal can be multiplied with a practical on-chip multiplier chain.

Nonlinearity-based frequency multipliers were used for their wideband operation compared to injection-locked frequency multipliers which exhibit limited locking range, but better power efficiency and spur rejection. A fully differential design is preferred, to avoid the use of baluns with associated CM challenges, which rules out the use of even-order multipliers. Since the achievable output power and efficiency of multipliers drastically shrink as the harmonic number increases, frequency triplers were adopted to generate the 135 GHz LO signal with maximum efficiency. Furthermore, triplers provide inherent rejection of even-mode harmonics at the differential output, alleviating spur rejection requirements. LO buffers are added after the triplers to generate the large output power desired to be able to saturate the modulators. Additionally, the buffers are critical to suppress undesired harmonics produced by the triplers. Given the fundamental input signal at $f_0 = 15$ GHz, the first tripler not only generates power at the third harmonic, $3f_0 = 45$ GHz, but inevitably also provides significant output power at the fundamental. Due to the presence of two frequency components, the second tripler then generates intermodulation products, spaced at multiples of f_0 from the desired component at $9f_0 = 135$ GHz. The generation of these spurious products is visualized in

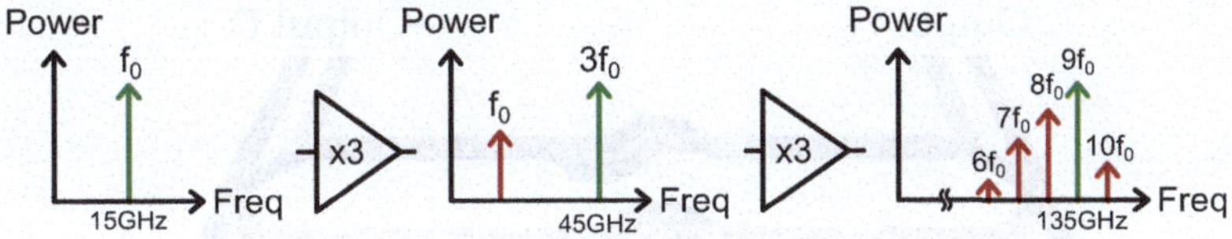

Fig. 7.26 Conceptual visualization of the generation of spurs in the multiplier chain

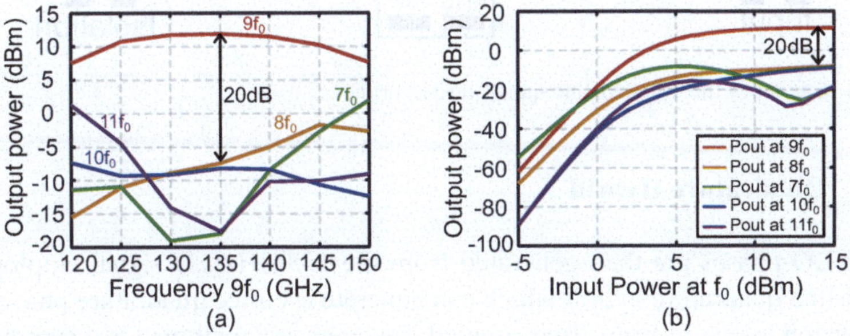

Fig. 7.27 Simulated output power of the desired tone ($9f_0$) and most important spurs ($8f_0, 7f_0, 10f_0, 11f_0$) at the output of the TX LO chain (**a**) versus 9th harmonic output frequency at saturation and (**b**) versus input power at 15 GHz

Fig. 7.26. From these products, the signal at $8f_0$ is the most problematic, being the largest of all spurs and located most closely to the main signal and thus most challenging to filter out. Most of the filtering is obtained by the matching networks of the LO buffers. However, designing these matching networks comes with a trade-off between spur rejection (for which a steep response is desired) and LO tunability (for which a broadband response is desired).

All tripler and buffer stages are implemented by means of a pseudo-differential pair with capacitive neutralization to boost the power efficiency. The triplers are optimized in size and bias to maximize third harmonic power generation, while the LO buffers comprise a conventional high gain power amplifier chain. A series resistance of 50 Ω is added at the input of the first tripler to decrease the Q-factor of the input impedance for easier and more robust input matching. Matching is implemented using high-k transformer-based matching networks to maximize LO chain efficiency. The matching networks after the triplers are designed to provide the optimal load impedance for maximum generation of power at the third harmonic, while the LO buffers are conjugately matched for maximum gain. Simulations of the LO chain over input power and frequency in Fig. 7.27 show a saturated output power over 11 dBm with 20 dB of spur rejection at high input powers. At low input powers the spur rejection deteriorates as the triplers require a large input swing to exploit the nonlinear behavior.

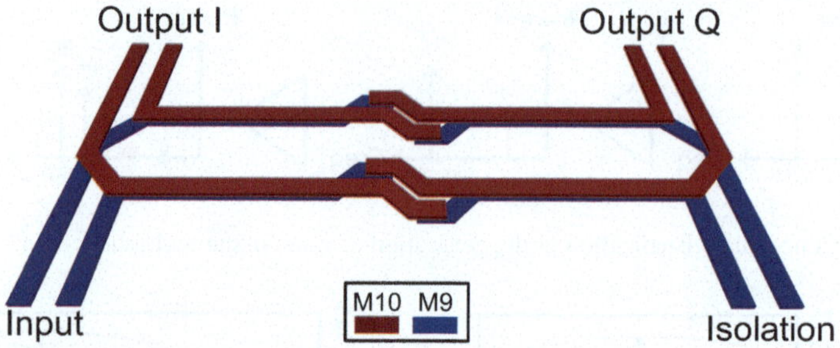

Fig. 7.28 Layout of the coupled-line quadrature hybrid

7.3.3.3 Quadrature Hybrid

The I/Q LO phases are then generated from the 135 GHz LO signal employing a coupled-line quadrature hybrid which can generate accurate quadrature phases with low insertion loss at D-band. Two coupled-line pairs are combined in a symmetrical layout to create a differential quadrature hybrid without the need for baluns [199]. Compared to other popular hybrid topologies such as the branch-line hybrid or polyphase filter, the coupled-line network leads to a more compact layout and exhibits lower insertion loss, while providing some tuning in magnitude and phase between the outputs.

The quadrature hybrid is designed to have minimal magnitude imbalance between the outputs and a phase difference as close to 90 ° as possible. A tunable impedance is designed at the isolation port of the hybrid, consisting of a fixed MOM capacitor and a tunable MOSFET resistor. Combined with a transformer matching network, this impedance is sized to provide a similar impedance as the one driving the input port, with maximal tunability to compensate any IQ phase error emerging due to process variations. A layout of just the quadrature hybrid is shown in Fig. 7.28.

The final simulated IQ magnitude and phase imbalances are negligible at the design frequency of 135 GHz and are below 0.7 dB and 5° across a 20 GHz bandwidth, as shown in Fig. 7.29a. A tuning range of $+/-2.5$ ° (Fig. 7.29b) is achieved with minimal additional magnitude error (<0.5 dB). Residual magnitude imbalance can be canceled via the tunable buffers in the I and Q branches after the modulators.

7.3.4 *Full Transmitter Considerations*

7.3.4.1 Floorplan Considerations

Some additional considerations are required when combining the different blocks into a complete transmitter layout. ESD diodes are placed, where necessary, close to the bondpads to minimize path resistance for the ESD current. Decoupling cells

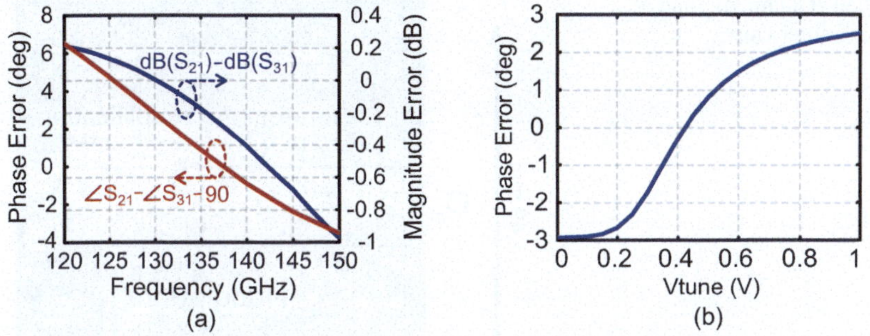

Fig. 7.29 Simulated phase and magnitude error of the quadrature hybrid outputs (**a**) and output phase error at 135 GHz versus tuning voltage of the impedance at the isolation port of the hybrid (**b**)

are added close to the transistor cores and are damped with small resistors for CM stability. Especially the digital supply has been decoupled by a very large amount of high-Q capacitance (>100 pF).

Due to the relatively large pitch necessary for the flip-chip interface (160 μm), the number of bondpads which fit onto the chip floorplan is limited. As a result, many blocks need to share supply, ground, or bias nets, which come with some necessary precautions. First of all, one must keep the nets of sensitive circuits as separated as possible. For example, the supply and ground lines of the digital drivers are isolated from all of the lines connected to high-frequency RF circuits. The supply/ground lines of the modulators are also kept separate. Secondly, connected nets can obviously not be tuned separately, so important nets which require individual tuning must be kept isolated. For example, the supplies of the different LO buffers can be connected, but the bias voltage to tune the amplitude modulator should be kept separate from other bias nets, even if they are similar in voltage. Lastly, one must be wary of possible CM stability problems when connecting nets between different circuit blocks. As discussed in Sect. 4.2, CM stability can be enhanced by inserting large resistors in the bias lines and by providing sufficient inductance in the supply/ground lines. The latter is accomplished by connecting the lines at the edges of the chip and thus inserting enough inductance before any possible CM loops are closed. The die micrograph of the transmitter, fabricated in 28 nm CMOS, is shown in Fig. 7.30.

7.3.4.2 Full TX Verification

First of all, the stability of the combined transmitter circuits must be verified. As discussed (Sect. 4.2), rigorous analyses are required to ensure every possible oscillation is observed. The stability of the transmitter was thus verified by means of admittance analysis and pole–zero identification, as described in [200].

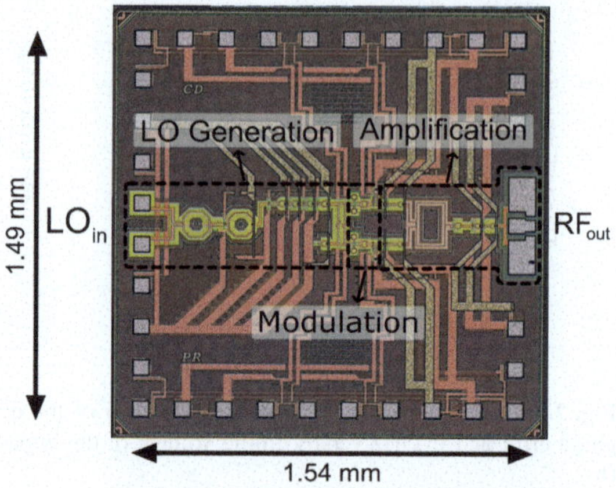

Fig. 7.30 Die micrograph of the transmitter chip

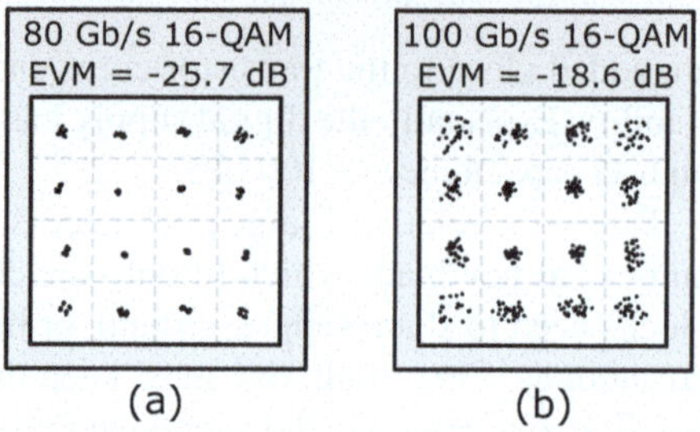

Fig. 7.31 Full TX simulated 16-QAM constellations at 80 Gb/s (**a**) and 100 Gb/s (**b**)

Next, modulated measurements are performed including the complete data driver network, modulators, and amplification block. These simulations show great 16-QAM EVMs up to 100 Gb/s transmission speeds to an ideal 50 Ω load. Figure 7.31 shows the simulated constellations at 80 Gb/s and 100 Gb/s data rates, clearly demonstrating the high-speed capability of the modulators and the broadband characteristic of the amplification network.

7.4 Receiver Circuit Design

Design details of the receiver architecture and circuit blocks are discussed in this section. The block diagram of the receiver is shown in Fig. 7.32. The 135 GHz 16-QAM input signal is first amplified, split, and downconverted into PAM-4 baseband signals in the RF front-end block. The 135 GHz quadrature LO signals driving the mixer are generated in the LO generation block from a 15 GHz reference. The baseband PAM-4 signals from each branch are amplified, equalized, and finally demodulated into binary outputs in the baseband and demodulation block.

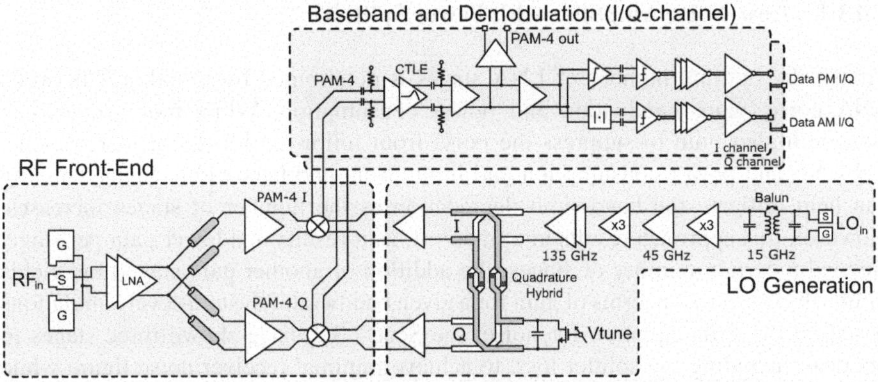

Fig. 7.32 Block diagram of the 135 GHz receiver

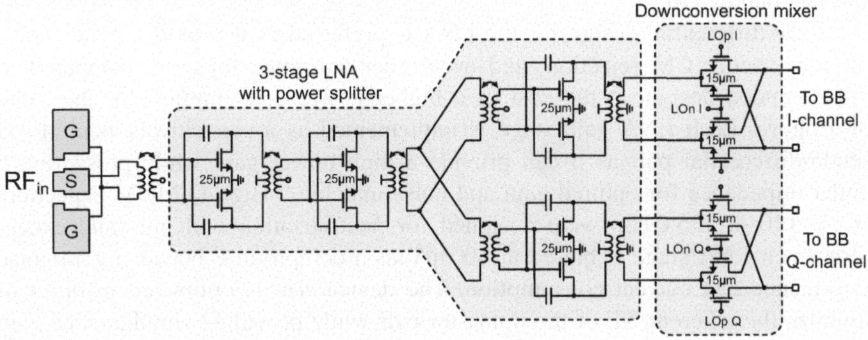

Fig. 7.33 Circuit schematic of the RF front-end

7.4.1 RF Front-End Block

The circuit schematic of the RF front-end is shown in Fig. 7.33 and includes a three-stage LNA with two-way power splitting and two downconversion mixers driven by quadrature LO signals. To obtain good receiver sensitivity, the LNA must exhibit a low intrinsic noise figure (NF) and provide sufficient gain to suppress the noise generated by subsequent stages in the receiver chain. Furthermore, a broadband response ensures minimal ISI at high symbol rates, and high linearity leads to a large dynamic range in the receiver. The GSG input pads are identical to the transmitter GSG output pads and are suitable for standalone characterization with GSG probes and full link tests with flip-chip-interfaced off-chip antennas, as discussed in the previous section.

7.4.1.1 Low-Noise Amplifier with Power Splitter

Firstly, the optimal number of LNA stages is determined by a trade-off between LNA noise, gain, bandwidth, and power consumption. While more stages can provide higher gain to suppress the noise from following blocks, it increases the LNA NF and power consumption and limits front-end bandwidth. Stagger tuning can help mitigate the bandwidth degradation as the number of stages increases. However, this approach results in a trade-off, as it results in a lower gain per stage. Beyond a certain number of stages, the addition of another gain stage thus yields diminishing returns in terms of gain for a given bandwidth. System-level simulations based on the Friis equation for noise, see Sect. 2.3.2, have shown three stages to be ideal, including the splitter loss, to achieve minimal receiver noise figure while achieving wide bandwidth operation. More stages would barely decrease the RX noise figure, while significantly increasing power consumption and limiting the RF front-end bandwidth.

A fully differential design for the LNA is preferred as it provides better noise immunity, better CM rejection, and even-order harmonic rejection compared to a single-ended design, at the cost of a higher power consumption for the same noise figure. Each LNA gain stage is implemented as a capacitively neutralized pseudo-differential pair as it can provide a high power gain while presenting a similar impedance for optimal gain and noise matching. High-Q MOM capacitors ($Q \approx 120$ at 135 GHz) were designed for neutralization with minimal excess noise. Each LNA stage is biased at 400 mV as a compromise between gain/noise performance and current consumption. The device size is optimized in order to minimize the inherent NF of the transistor pair while providing simultaneous gain and noise matching. Indeed, as shown in Fig. 7.34, the optimal source impedances for maximum gain and minimal noise lie close together on the Smith chart. Additionally, by making the input and output impedances of the stage comparable, identically sized stages can be cascaded with minimal inter-stage matching loss. A transistor width of 25 μm was chosen resulting in an ideal power gain (G_{max}) of 7.8 dB and a minimum noise figure (NF_{min}) of 4.5 dB per stage, including layout parasitics after optimization. Furthermore, the conjugate gain input match leads to an NF degradation of only 0.25 dB compared to the minimum attainable NF. The chosen transistor size also offers more than adequate linearity performance for the given system.

The signal is split into separate I and Q paths after the second stage by a parallel power splitter. A parallel splitter as opposed to a series splitter is chosen as this leads to a more favorable impedance transformation and thus more broadband network characteristic. A non-isolated splitter is designed, consisting of two transmission line branches with transformers at the input and outputs. The splitter exhibits an insertion loss of 5.2 dB, including input/output transformer loss and inherent 3 dB loss from the power split. All other matching networks are realized by means of stagger-tuned transformer-based networks and co-optimized for a flat, broadband frequency response, similar to the amplification network of the transmitter. While most networks employ 1:1 stacked transformers with optional series inductance,

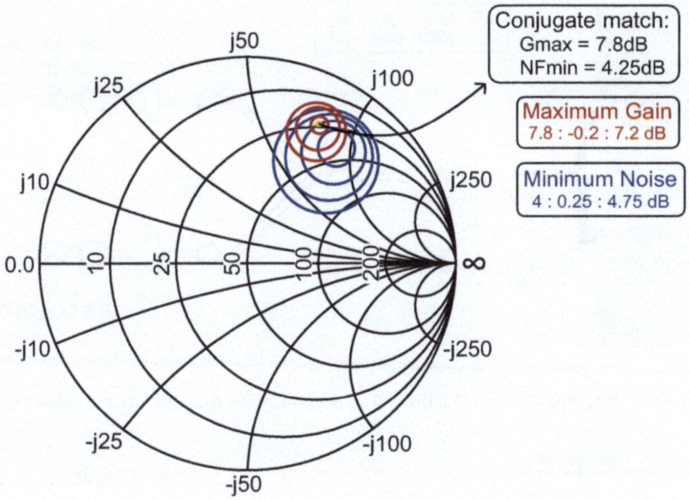

Fig. 7.34 Visualization of the maximum gain circles and minimum gain circles on the Smith Chart, indicating that a conjugate input match leads to only little degradation of the noise figure

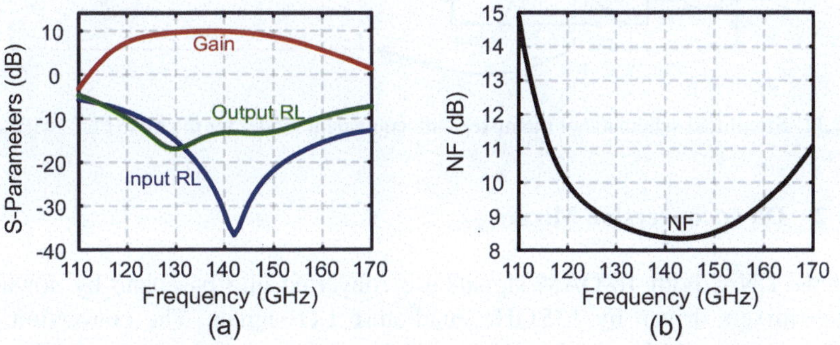

Fig. 7.35 Simulated S-parameters of the LNA (**a**) and simulated NF of the LNA (**b**)

the input network utilizes a 1:2 winding ratio to efficiently cope with the larger impedance ratio.

As shown in Fig. 7.35a, the LNA achieves 10 dB of simulated gain, including the splitting loss, with a 3-dB BW of 37 GHz from the 50 Ω GSG input to the input of the mixers. A minimum NF of 8.4 dB is obtained across the band, with 6.5 dB coming from the first stage (loss input matching + NF first stage), as depicted in Fig. 7.35b. Furthermore, the input return loss is lower than 10 dB across more than 20 GHz, and the LNA has an input-referred P1dB of −7 dBm. A layout overview of the RF front-end, from receiver input to the mixers, is shown in Fig. 7.36.

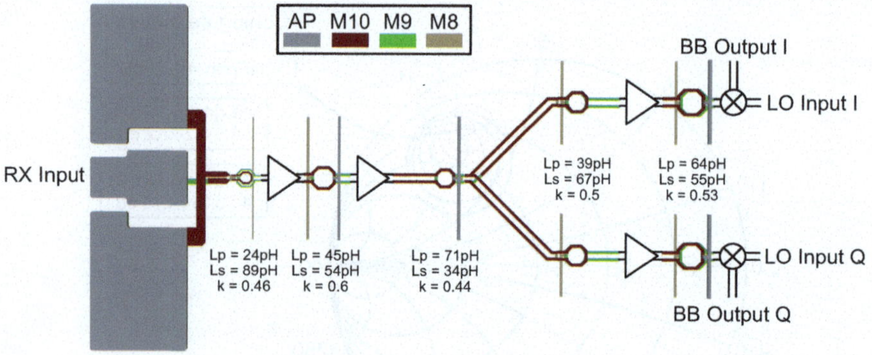

Fig. 7.36 Layout diagram of the full RF front-end block with transformer parameters

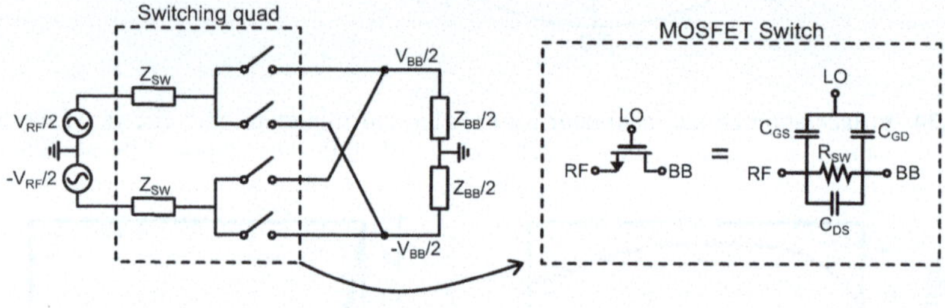

Fig. 7.37 Simplified schematic of the mixer with equivalent model for the MOSFET switches

7.4.1.2 Downconversion Mixers

After the LNA, both 16-QAM signals are converted into baseband by downconversion mixers driven by 135 GHz quadrature LO signals. The conversion gain of the mixer should be maximized to limit degradation of the receiver NF, and a broadband downconversion response is essential. Hence, a voltage-mode passive mixer is selected, which can provide a large bandwidth with minimal conversion loss and low power consumption. Furthermore, employing a double-balanced topology is crucial to ensure optimal signal balance.

Based on the simplified schematic in Fig. 7.37, the overall conversion gain of the voltage-mode, passive mixer is the difference in conversion when the switches are turned on versus when they are turned off, which can be approximated as follows:

$$CG(\omega) = \frac{V_{BB}(\omega)}{V_{RF}(\omega_c + \omega)} = CG_{\mathrm{on}}(\omega) - CG_{\mathrm{off}}(\omega) \tag{7.4}$$

$$= \frac{2}{\pi}\left(\frac{Z_{BB}(\omega)}{2Z_{SW,\mathrm{on}}(\omega) + Z_{BB}(\omega)} - \frac{Z_{BB}(\omega)}{2Z_{SW,\mathrm{off}}(\omega) + Z_{BB}(\omega)}\right), \tag{7.5}$$

where ω is the angular baseband frequency, ω_c the angular carrier frequency, Z_{BB} the impedance provided by the first baseband stage, and $Z_{SW,\text{on/off}}$ the impedance of the mixer switches when turned on/off. In this simplistic model, the last LNA stage is assumed to be conjugately matched to the mixer and is therefore simply modeled as a voltage source with a fixed, implicit source impedance. The factor $\pi/2$, called the commutation factor, is the conversion gain of an ideal, voltage-mode mixer when driven by a full swing rectangular LO signal. The switch impedance in the on- and off-state can be expressed based on Fig. 7.37:

$$Z_{SW,\text{on}} = R_{SW} = \frac{1}{\frac{W}{L}\mu C_{ox}(V_{GS} - V_T)} \tag{7.6}$$

$$Z_{SW,\text{off}} = \frac{1}{j\omega(C_{GS} \parallel C_{GD} + C_{DS})}, \tag{7.7}$$

where it is assumed the channel resistance dominates the impedance in the on-state and can be neglected in the off-state. From these equations, some clear design trade-offs can be deduced. First of all, the LO swing at the gates of the switches must be maximized to attain a large commutation factor. A small LO swing will cause the switches to turn on/off slower and will lead to signal loss. A large LO swing can be obtained by maximizing the LO power going into the mixer and by maximizing the gate input impedance, which is mostly determined by the parasitic capacitance at the gate nodes thus favoring a small switch size (W and L). To improve the power efficiency of the receiver, it is beneficial to minimize the required LO power driving the mixer for the desired conversion gain, as generating power at these frequencies is generally very power hungry. Secondly, the on-state switch resistance must be made as small as possible by choosing a large W/L ratio and large overdrive voltage, while trying to minimize the off-state impedance by limiting the parasitic capacitances of the switch. This trade-off is often quantified using the $R_{\text{on}}C_{\text{off}}$ product of the switch, where a smaller value is associated with a smaller $R_{\text{on}}/Z_{\text{off}}$ ratio and thus more optimal switch performance. Lastly, a high baseband input impedance is preferred for a large conversion gain but comes with an inherent trade-off for bandwidth at the output node of the mixer. It must be noted that the noise of the mixer is automatically minimized by optimizing the conversion and resistance of the switches.

The switches are designed to have minimal transistor length with ultra-low threshold voltage in order to attain maximal overdrive with minimal $R_{\text{on}}/Z_{\text{off}}$. Choosing the transistor width entails a compromise between on-resistance (which determines CG_{on}) and off-state capacitance (determining CG_{off}, the mixer input impedance, and the output bandwidth), eventually leading to an optimal width of 15 µm. The bias voltage is then chosen to minimize the switch $R_{\text{on}}/Z_{\text{off}}$ for a fixed LO power, arriving at 300 mV (around the threshold voltage of the switch). The final mixer design achieves a simulated voltage conversion gain of −1 dB with only 0 dBm of LO power driving the mixer.

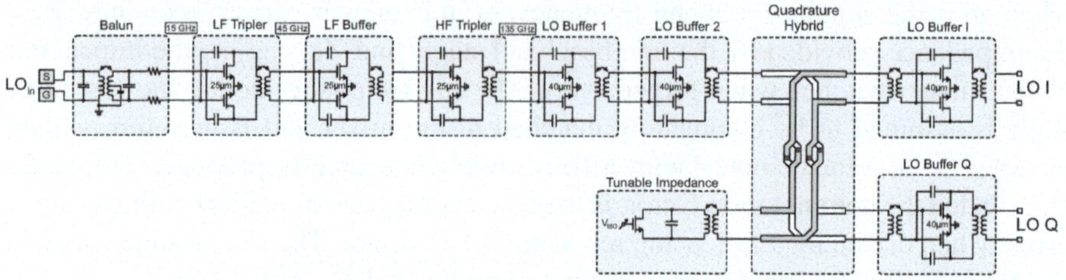

Fig. 7.38 Circuit schematic of the LO generation block

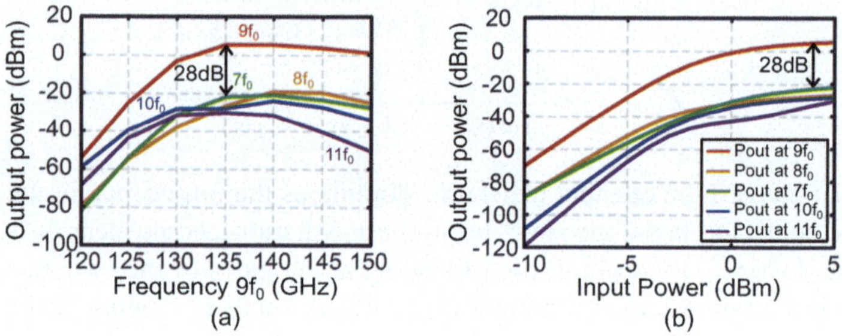

Fig. 7.39 Simulated output power of the desired tone ($9f_0$) and most important spurs ($8f_0$, $7f_0$, $10f_0$, $11f_0$) at the output of the RX LO chain (**a**) versus 9th harmonic output frequency at saturation and (**b**) versus input power at 15 GHz

7.4.2 LO Generation Block

Similar to the transmitter, the high-frequency quadrature LO signals which drive the mixers are generated from a 15 GHz external LO reference by means of two triplers, LO buffer amplifiers and a coupled-line quadrature hybrid. All tripler and LO buffer stages again adopt a pseudo-differential pair implementation with capacitive neutralization. Since the design trade-offs are similar to those discussed in the section of the transmitter (Sect. 7.3.3), only the most important differences are highlighted here (Fig. 7.38).

While IQ imbalances are still to be limited, the output power specification is less demanding compared to the TX LO chain and is co-determined with the mixer design, ultimately leading to a 0 dBm desired output power. Spur rejection is, however, of much greater concern due to a smaller number of filtering stages before reaching the mixer input. To this end, an additional 45 GHz buffering stage is inserted between both frequency triplers to provide extra suppression of the 15 GHz fundamental component coming out of the first tripler. As a result the level of the undesired intermodulation spurs out of the second tripling stage is reduced. The LO chain is able to supply more than the required LO power of 0 dBm to the mixers with 28 dB of spur rejection, as illustrated in Fig. 7.39.

The quadrature LO signals are generated using the same coupled-line quadrature hybrid described before with similar performance. Any possible IQ phase and magnitude imbalance can again be calibrated out with a variable impedance at the isolation port of the hybrid and tunable buffers in the I and Q branches following the hybrid.

7.4.3 Baseband and Demodulation Block

The baseband chain must provide high signal amplification over a wide frequency band. Furthermore, gain tuning and light equalization are desirable to condition the signal for optimal demodulation performance. The demodulation circuitry must finally be able to reliably extract the phase and amplitude bits from the amplified PAM-4 signal. The amplified PAM-4 signal is also brought off-chip to allow for characterization of the receiver without demodulation circuitry. Figure 7.40 shows an overview of the baseband and demodulation block with circuit-level implementations of various highlighted blocks.

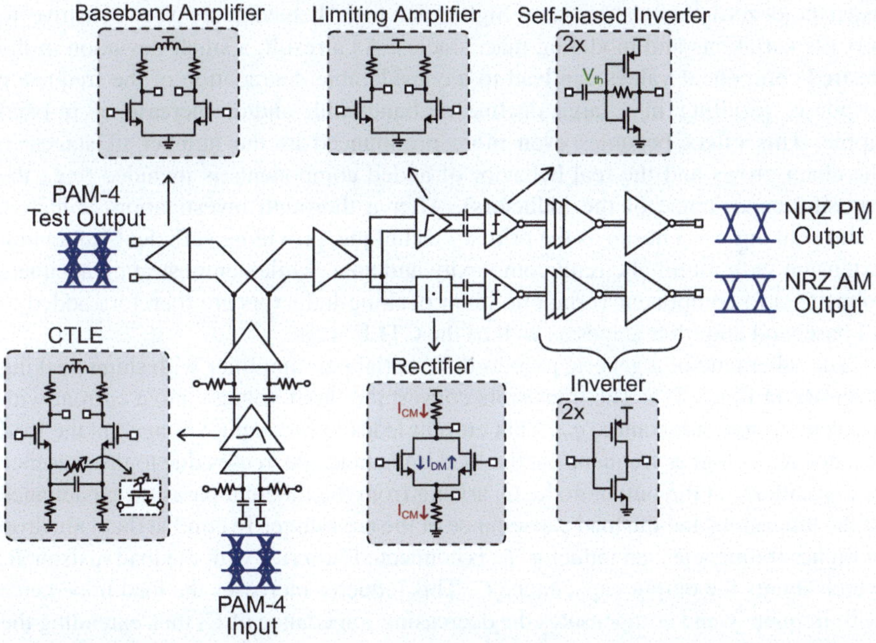

Fig. 7.40 Overview of the baseband and demodulation block with highlighted circuit-level implementations

7.4.3.1 High Gain–Bandwidth Amplification

The starting point for each baseband amplification stage is a resistively loaded pseudo-differential pair, which is the preferred topology due to its simplicity and hence high-frequency performance. An additional tail transistor could improve the common-mode rejection and offer additional gain control, but the loss in voltage headroom significantly degrades the attainable gain–bandwidth performance and hence a pseudo-differential topology is employed for this design. Given this topology, the maximum attainable gain–bandwidth (GBW) product per stage is bounded to about $g_m/2\pi C_L$. The design thus comes with an inherent trade-off between gain and bandwidth. Since a very large bandwidth is needed to support over 20 GS/s of symbol rate, only very limited gain per stage can be obtained. However, it is possible to break through this GBW limitation using bandwidth extension techniques and make sufficient amplification over 10+ GHz of bandwidth possible.

Numerous different techniques are described in the literature to extend the bandwidth of baseband amplifiers [201, 202] without having to sacrifice the gain, including shunt peaking, bridged-shunt peaking, series peaking, bridged-shunt–series peaking, and T-coil peaking. These techniques introduce additional poles and zeros in the frequency response to expand the bandwidth at the cost of additional circuit complexity and circuit area. Furthermore, complex techniques relying on exact pole–zero cancelations and high-order filter behavior are very sensitive to process variations and modeling inaccuracies. As a result, a small deviation in the desired component values can lead to a considerable degradation of the frequency response, resulting in a large decline of bandwidth and/or increase of in-band ripple. This effect becomes even more pronounced as the number of stages in the chain grows and the real behavior of added components is included (e.g., the parasitic capacitance of the inductors). After a thorough investigation, inductive shunt peaking was chosen as the primary technique, providing sufficient bandwidth extension with minimal circuit complexity and area, while demonstrating adequate robustness to component variation. Shunt peaking inductors are therefore added to all baseband amplifier stages as well as the CTLE stage.

The schematic of a generic pseudo-differential pair amplifier with shunt peaking is shown in Fig. 7.41a. The transistors convert the input voltage into a current with a certain transconductance, g_m. That current leads to an output voltage via the load resistor R. At higher frequencies, the load impedance decreases due to the presence of capacitance at the output node, C, arising from the inherent parasitic capacitance of the transistors and the load capacitance of the next stage. To combat that gain drop at higher frequencies, an inductor, L, is connected in series with the load resistor R, which shunts the output capacitance C. This inductor increases the load impedance with frequency and compensates the decreasing impedance of C, thus extending the bandwidth. This can be expressed more formally as follows:

$$A_v = g_m \cdot \left[(R + sL) \parallel \frac{1}{sC} \right] = g_m \cdot \frac{R + sL}{1 + s\mathrm{RC} + s^2\mathrm{LC}}. \tag{7.8}$$

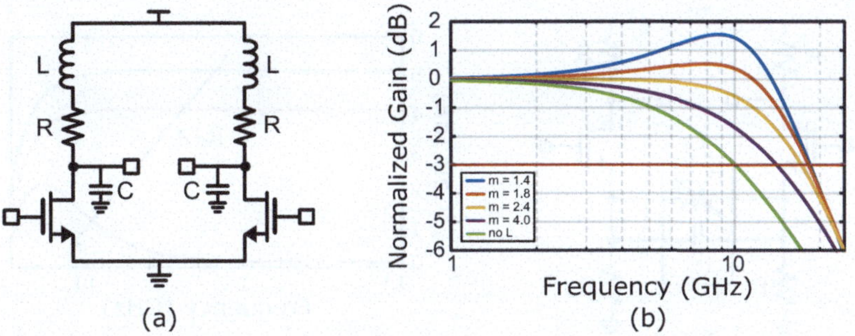

Fig. 7.41 Schematic of a pseudo-differential pair amplifier with shunt peaking (**a**) and normalized gain response for different inductor values (**b**)

The added inductor indeed introduces a zero in the amplifier response, which increases the 3-dB bandwidth. It is useful to normalize the gain response to the value at DC (i.e., R) and substituting the reference 3-dB angular bandwidth without compensation, $\omega_0 = 1/\mathrm{RC}$, and the variable $m = R^2C/L$:

$$A_{v,n} = \frac{1 + s/m\omega_0}{1 + s/\omega_0 + s^2/m\omega_0}. \tag{7.9}$$

The normalized gain response for different values of m, corresponding to different inductor values, is shown in Fig. 7.41b. It can be shown that a maximally flat bandwidth extension of 72% can be achieved for $m = 1 + \sqrt{2}$, while a larger maximum expansion of 84% can be achieved for $m = \sqrt{2}$ at the cost of 1.5 dB of peaking [201]. The exact value of the inductor can then be determined based on the desired response and values of R and C.

7.4.3.2 Continuous-Time Linear Equalizer

A CTLE was chosen as the preferred equalizer due to its simplicity, low power, and good tuning capability, despite its drawback of high-frequency noise amplification. The CTLE functions as a variable-gain amplifier and provides programmable frequency peaking to compensate for bandwidth-limiting circuits in the receiver chain. Tunable source degeneration is added to a pseudo-differential pair in order to attain the desired frequency peaking with minimal sacrifice of the DC gain. Based on the CTLE circuit in Fig. 7.42a, the CTLE gain can be expressed as

$$A_v = g_m \cdot \frac{1}{1 + g_m Z_S/2} \cdot R_L = g_m \cdot \frac{1}{1 + g_m R_S/2} \cdot \frac{1 + sR_SC_S}{1 + s\left[\frac{R_S}{1+g_m R_S/2}\right]C_S} \cdot R_L, \tag{7.10}$$

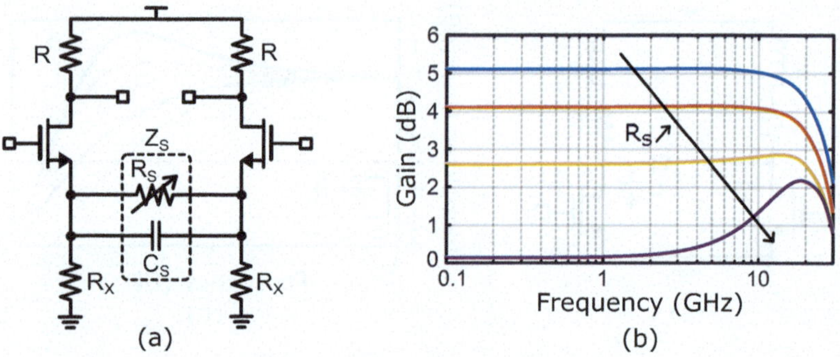

Fig. 7.42 Schematic of a continuous-time linear equalizer (**a**) and gain response for different source degeneration values (**b**)

which ignores the parasitic capacitance present at the output node for clarity but can easily be included in the load R_L. It can be seen that the source degeneration impedance ($Z_S = R_S \parallel 1/sC_S$) decreases the low-frequency gain while introducing a zero ($\omega_z = 1/R_SC_S$) and a higher frequency pole. Of course, the peaking effect will only be visible and useful if the frequency zero is located within the bandwidth as determined by other poles in the baseband chain.

Increasing the resistor value R_S will first only decrease the gain as the zero ω_z falls outside of the band. At larger values of R_S, the location of the zero and the amount of frequency peaking can be controlled. This CTLE behavior is visualized in Fig. 7.42b. Additionally, C_S could be made tunable (as a varactor) to have more degrees of freedom. However, the limited tuning range, significant parasitic capacitance, and series resistance of the varactor would degrade the performance of the CTLE. Therefore, only R_S is implemented as a tunable MOSFET resistor, while a high-Q MOM capacitor functions as C_S. The resistors R_X are added to determine the DC operation point of the transistors, instead of tail current sources, but have no direct influence on the differential characteristic of the CTLE. Furthermore, a shunt peaking inductor is added at the output to ensure that the CTLE peaking is not undone by the pole at the output.

7.4.3.3 Amplifier Chain Design

Using the discussed amplification and CTLE building blocks, the schematic of complete baseband amplifier chain, including all component values, is shown in Fig. 7.43. The CTLE stage is added after the first amplification stage, and the signal split toward the PAM-4 output and demodulation input occurs after the third baseband stage. It is important to think about how to connect the different stages. Directly connecting different stages, e.g., DC coupling, removes any biasing flexibility as the DC input voltage of each stage is determined by the DC output

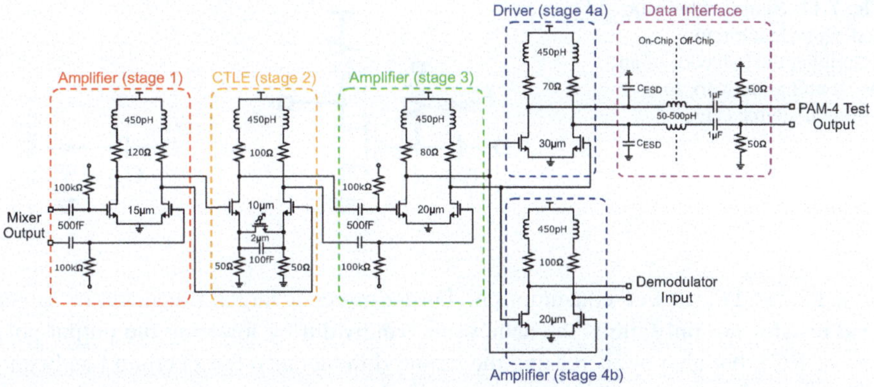

Fig. 7.43 Schematic of the baseband amplifier chain

voltage of the previous stage. AC coupling enables each stage to be biased and optimized separately, which allows for more flexibility and robustness. However, each AC coupling network introduces a low-frequency high-pass cut-off which potentially filters out an important part of the signal and can cause baseline wander (see Sect. 2.3.4). As a compromise between biasing flexibility and low-frequency signal attenuation, two AC coupling networks are inserted in the chain. Using two bias voltages, the amplifier chain can be adequately tuned for optimal operation, even in the presence of process variations.

When designing the AC coupling components, two elements have to be considered: the high-pass cut-off frequency and the signal attenuation. Based on the schematic in Fig. 7.44, the high-pass cut-off can be calculated based on C_{AC} and R_{BIAS}, and the signal attenuation due to a capacitive division with the input capacitance of the next stage can be determined as follows:

$$\omega_{HP} = \frac{1}{R_{BIAS} C_{AC}} \tag{7.11}$$

$$v_{gs} = v_{in} \cdot \frac{C_{AC}}{C_{AC} + C_{gs}}. \tag{7.12}$$

The size of C_{AC} needs to be significantly larger than the capacitive load to avoid significant signal attenuation but is limited by the parasitic capacitance to ground and series resistance which both lead to signal loss as well. Furthermore, R_{BIAS} must have an enormous size to attain the required low-frequency cut-off (typically $<$ 1 MHz), which is not straightforward to implement without again adding unwanted parasitic capacitance.

The transistor size, bias, and load resistor are co-designed for each stage to obtain the desired gain, bandwidth, and DC operating point (if needed). The gain and bandwidth can be roughly exchanged by adapting the load resistor R, while increasing the transistor size can produce a larger GBW product at the expense

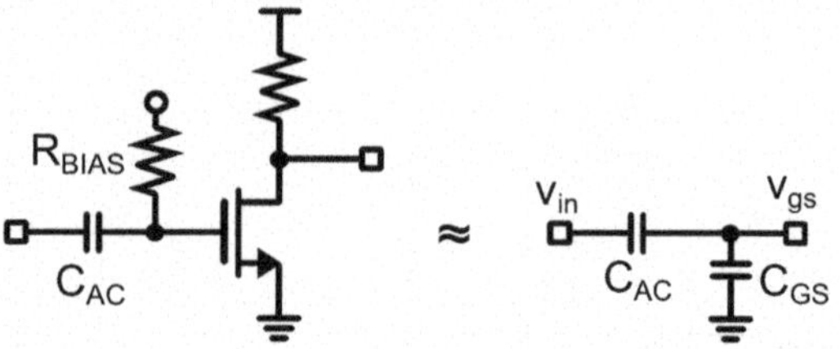

Fig. 7.44 Simple schematic and equivalent circuit describing the behavior of an AC coupling network to a common-source stage

of a higher DC power consumption. Furthermore, a larger transistor or larger load resistor not only limits the achievable bandwidth by lowering the output pole ($f_p \propto RC$), but also by increasing the required inductance for a certain bandwidth extension (since $m = R^2C/L$). After all, for very large necessary inductance values, the parasitic coupling capacitance between the coils will limit the SRF of the inductor and therefore the attainable frequency peaking. To ensure sufficient bandwidth in the stage driving the splitting node, the two loading stages must be limited in size and some gain is sacrificed in the driving stage. Finally, the last amplifying stage toward the PAM-4 test output must drive a 50 Ω load, large ESD diodes, and an inductive bondwire while coping with the most stringent linearity requirements in the chain and important return loss conditions. As a result, this final driver stage consumes a lot of DC current (large transistor size) with very little gain (small load resistor) to satisfy these demands.

7.4.3.4 Shunt Peaking Inductor Design

The optimal shunt peaking inductor size is designed to provide a maximal flat bandwidth. Overpeaking in the frequency response leads to overshoot in the time domain, which seriously degrades the signal integrity and must thus be avoided. Only the CTLE is designed to provide optional frequency peaking which can be controlled by an external voltage. An inductance value of 450 pH is chosen for the peaking inductors, identical for each stage. This leads to slightly different peaking behaviors depending on the exact load resistance and capacitance of the stage but ensures overpeaking is avoided. Only the buffer stage preceding the demodulator experiences a small amount of frequency peaking, which slightly improves the steepness of the data edges enhancing the demodulator performance, as will be discussed later. Note that the two peaking inductors can be integrated into a single, larger inductor with a center tap for the supply connection.

When implementing the peaking inductance, the main specifications to keep in mind are the inductance, area, and SRF. The desired inductance must be achieved with minimal area (to minimize the interconnect between amplifier stages) and large enough SRF (not to influence the amplifier frequency response by adding additional poles and reducing the effective inductance at higher frequencies). While a triple-turn spiral inductor can realize a large inductance in a compact area, its substantial parasitic inter-winding capacitance leads to an insufficient SRF, rendering this

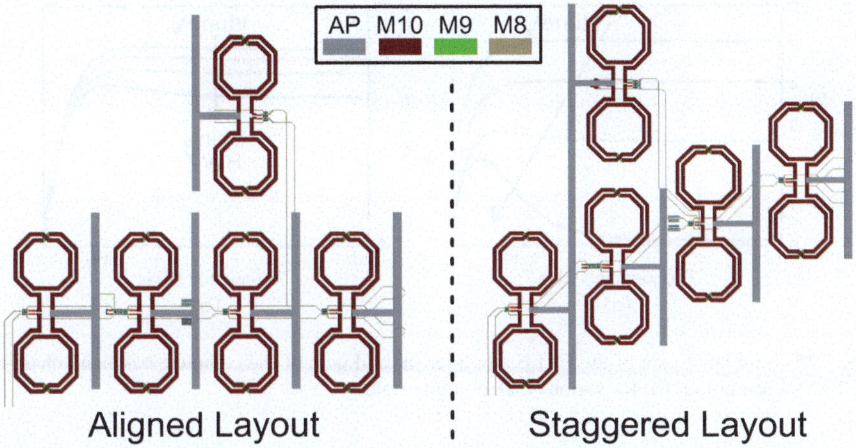

Fig. 7.45 Layout of all shunt peaking inductors: aligned approach (**a**) and staggered approach (**b**)

topology useless. Therefore, the peaking inductor is split into two smaller double-turn inductors which can easily achieve the desired inductance with satisfactorily high SRF. Hence, two double-turn spiral inductors of 450 pH are designed for each amplifier stage with an outer diameter of 90 μm and a simulated self-resonance frequency far beyond the baseband bandwidth.

The optimal placement, and more concretely, the spacing between the inductors, is determined by a trade-off between the mutual coupling and interconnect length between the stages. More coupling decreases the effective inductance of the peaking inductors (due to Eddy currents, see Sect. 3.3), and longer routing lines increase the load for each stage, both significantly decreasing the overall bandwidth of the baseband amplifier chain. Putting the inductors closer together increases the coupling but shortens the routing, and thus an optimal spacing can be found, which optimizes the overall bandwidth. A compact, staggered layout configuration for all inductors is adopted to minimize the bandwidth degradation caused by long inter-stage routing, as can be seen in Fig. 7.45. Compared to a straightforward, aligned layout, staggering increases the baseband bandwidth by over 10% toward the PAM-4 test outputs. Furthermore, all inductors are placed within a ground plane for greater shielding between the stages, thus allowing a greater inductance for the same interconnect length. The added ground plane does slightly decrease the Q-factor of the inductors (to a value of 10), which is not a problem since any series resistance can be absorbed in the resistive load of the amplifier.

Figure 7.46a illustrates the tuning of the CTLE amplifier to achieve a 5 dB gain with a flat frequency response, as well as the capability to introduce up to 2 dB of peaking at 19 GHz. The complete amplifier chain provides 23 dB of simulated gain with 22 GHz of baseband 3dB-BW to the input of the decoders. To the PAM-4 test outputs the chain achieves a maximum gain of 15.5 dB with 16 GHz BW, while the

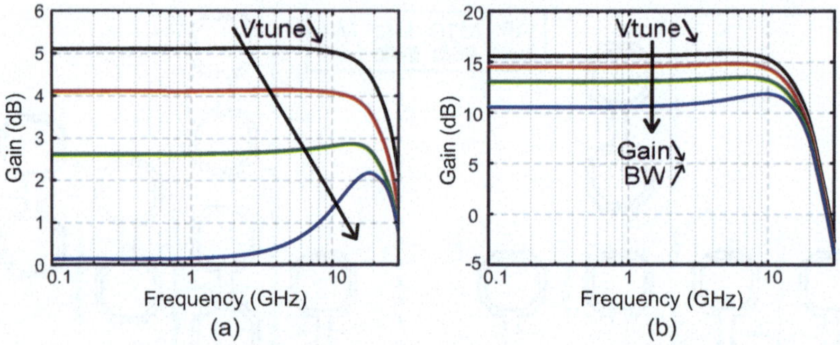

Fig. 7.46 Simulated gain of the CTLE amplifier (**a**) and gain of the complete baseband chain to the PAM-4 test output (**b**) for various CTLE tuning states

CTLE can provide a trade-off between gain and BW, as shown in Fig. 7.46b. The return loss at the PAM-4 test outputs is better than 10 dB and 6 dB over more than 8 GHz and 11 GHz, respectively. These results include the layout parasitics, on-chip ESD protection, bondwire inductance, and on-chip 50 Ω termination.

7.4.3.5 PAM-4 Demodulator

The demodulator or decoder needs to perform the aforementioned PAM-4 demodulation at very high symbol rates while limiting the circuit complexity for robustness and power consumption. To this end, a fully unclocked implementation is realized, thus working in continuous time. Removing all clocking circuits saves power and area while avoiding timing issues at high clocking speeds [196]. Moreover, all unclocked sign-check comparators can be implemented efficiently using a standard CMOS inverter, demonstrating a significant power advantage compared to CML comparators while supporting multi-GBaud operation. The schematic of the demodulation circuits is shown in Fig. 7.47.

To perform the demodulation of the amplitude bits, the PAM-4 input is first processed by a rectifier which maps the four levels onto two amplitude levels. After biasing the signal around the optimal threshold voltage, the amplitude bits can be recovered by a sign-check comparator. The rectifier is designed to provide a large rectified output swing while minimizing unavoidable spikes in the output during input zero crossings. As these voltage spikes only occur during certain symbol transitions, the value used for bit decision is not directly influenced. However, large spikes can impede fast level transitions possibly introducing ISI. To suppress the generation of these undesired spikes, the input signal of the rectifier must present steep edges to minimize the time around the differential zero crossing. This is obtained by maximizing the GBW of the baseband gain stages (for high swing and steep edges) and introducing a small amount of frequency overpeaking in the stage driving the demodulator (stage 4b in Fig. 7.43).

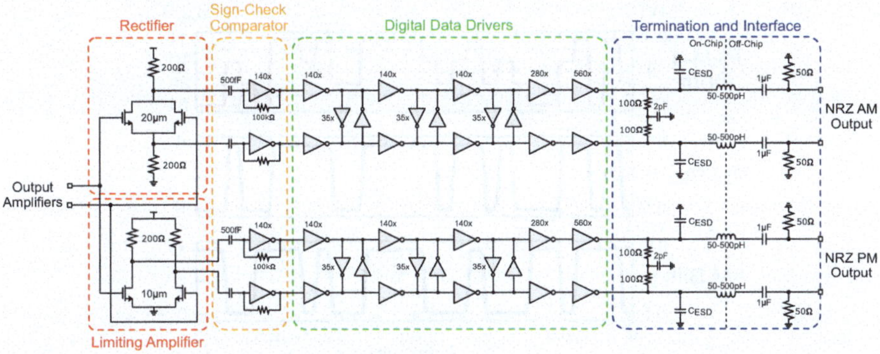

Fig. 7.47 Schematic of the PAM-4 demodulator

To extract the phase bits, the PAM-4 signals are fed into a limiting amplifier after which another sign-check comparator outputs a binary signal determined by the polarity of the PAM-4 input. The limiting amplifier is designed as a high gain, nonlinear baseband amplifier, without inductive peaking. This amplifier also provides delay matching between the AM and PM demodulation paths.

To make the bit decision, CMOS inverters are adopted to function as the sign-check comparator. After AC coupling, the input is biased around the optimal threshold by means of an inverter with resistive feedback. These inverters are sized to provide the maximal sensitivity given the expected input level. Finally, all demodulated NRZ signals are passed through a pseudo-differential inverter chain to provide clean, binary signals. The inverter chain is scaled up and matched to 50 Ω to drive the instrument load. Termination resistors of 100 Ω instead of 50 Ω are used to improve the return loss at higher frequencies, which degrades due to the load capacitance, while still providing satisfactory matching performance all the way down to DC. Consequently, the return loss remains better than 6 dB over more than 10 GHz. Weak cross-coupled inverters are once again added along the chain to provide CM rejection and thus maintain signal balance.

The combination of significant ESD diode capacitance at the output, large bond-wire inductance, and the resistive 50 Ω load leads to a sub-optimal binary output signal, with slower edges and a voltage swing smaller than the full rail. However, this does not induce any demodulation problems during the measurements, as NRZ signals are very robust against such impairments. PAM-4 signals, on the other hand, are much more sensitive and need power-hungry drivers and bandwidth extension techniques to combat these effects, as described before when designing the amplification chain.

Finally, some discussion is needed on the effects the comparator and digital drivers have on the signal. The comparator and nonlinear drivers will always generate a full swing, rail-to-rail signal at the output, even if the input is severely distorted and a wrong bit decision (bit error) is made in the process. This means that the demodulator performance cannot be evaluated by simply observing the

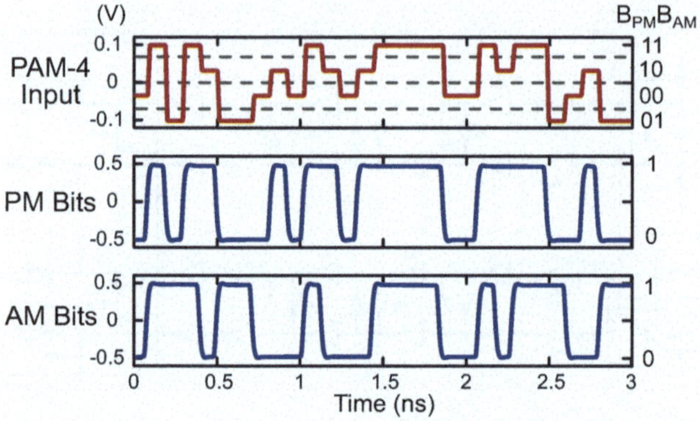

Fig. 7.48 Transient simulation of the baseband and demodulation block demonstrating PAM-4 demodulation at 10 GBaud

vertical eye opening at the output, which is often used as a metric for receiver performance. Instead, rigorous BER measurements must be performed to assess the performance. However, even though the output swing will always be rail-to-rail, yielding maximum vertical eye opening, impairments can still influence the quality of the eye. Any impairment, such as ISI, noise, IQ coupling, etc., which causes uncertainty in the zero crossing of the signal, will be translated by the comparator into uncertainty on the edges of the binary output. This is called jitter and can severely reduce the horizontal eye opening, which puts a heavier burden on the sampling clock.

The performance of the complete baseband and demodulation block is displayed in Fig. 7.48 demonstrating correct demodulation of the phase and amplitude bits from a PAM-4 input signal at a symbol rate of 10 GBaud. The layout of the baseband amplifier chain and demodulation circuits is shown in Fig. 7.49.

7.4.4 *Full Receiver Considerations*

7.4.4.1 Floorplan Considerations

When combining the different blocks into a full receiver, similar aspects as for the TX must be considered. Again, ESD diodes and damped decoupling cells are placed where needed. Special care is taken to provide maximal decoupling for the demodulation block, as its operation is quite sensitive to fluctuations of supply/ground voltage. The rectifier in particular is very vulnerable because its signal current does not flow in a fully differential loop, but through the supply/ground nodes. Consequently, a low impedant path must be provided outside the circuit between supply and ground to ensure stable supply/ground nodes for good operation. The low impedance should guarantee a stable voltage despite the generated current

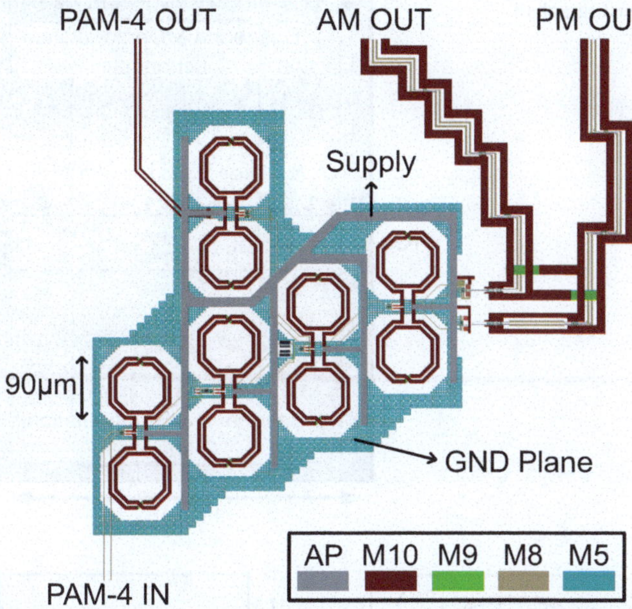

Fig. 7.49 Layout of the full baseband and demodulation block with staggered inductor placement

spikes caused by fast transitions in the signal. This is accomplished by using a lot of decoupling cells very close to the transistor core, minimizing routing parasitics which increase the impedance of the decoupling loop.

Different nets are combined with similar tactics as for the transmitter. In case of the baseband blocks, good isolation is required between the analog and digital supply rails to avoid the propagation of voltage spikes from the digital to the analog circuits, which can degrade the integrity of the signal. The die micrograph of the receiver, fabricated in 28 nm CMOS, is shown in Fig. 7.50.

7.4.4.2 Full RX Verification

Combining the RF front-end with the baseband chain, the complete receiver achieves a maximum conversion gain around 21 dB to the PAM-4 test outputs with a 3-dB RF BW of 26 GHz. The minimal double-sideband noise figure of the receiver is 9.3 dB with an integrated value of 9.6 dB over a 15 GHzband. Both simulations are shown in Fig. 7.51. Modulated full receiver simulations show error-free demodulation for 16-QAM inputs with symbol rates as high as 20 GBd, illustrating the high-speed capability of the receiver and PAM-4 demodulator. Finally, the stability (in DM and CM) of the receiver was verified using admittance and pole–zero analyses.

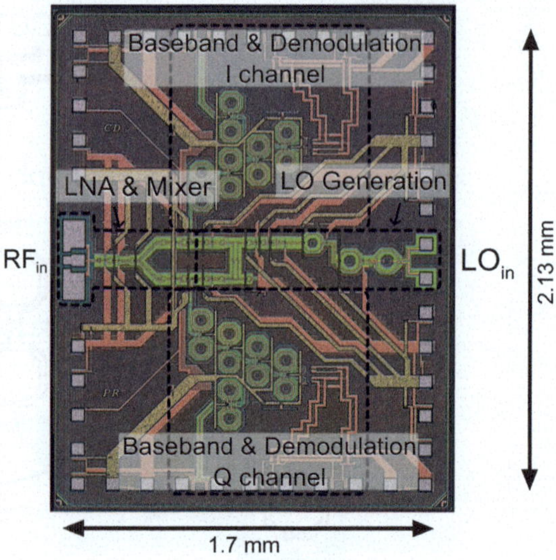

Fig. 7.50 Die micrograph of the receiver chip

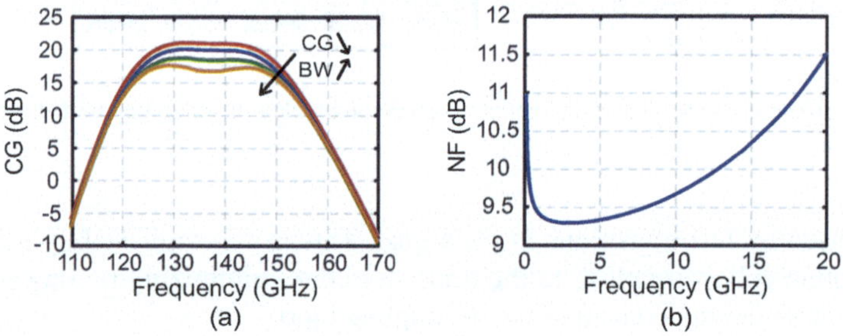

Fig. 7.51 Simulated receiver conversion gain to the PAM-4 outputs (**a**) and double-sideband receiver noise figure (**b**)

7.5 Prototype Board Design

In order to measure and characterize the chips, prototype boards (realized as printed circuit boards or PCBs) are necessary as an interface between the chips and measurement equipment. For standalone characterization, the high-frequency signals are interfaced using a GSG probe, but the high-speed data signals and DC signals have to be connected to external sources via a PCB. To perform link measurements, high-speed data and DC signals again have to be brought toward their respective sources, but the high-frequency signals are now interfaced via a free-space antenna or DWG coupler. The antenna and coupler are implemented on a PCB which highly complicates the design and required PCB technology. The most important design considerations for both boards will be discussed in this section.

7.5.1 Boards for Standalone Characterization

The PCBs for standalone TX/RX measurements should be able to accommodate probing of the chip while routing high-speed signals and many DC signals to the equipment. While DC signals can simply be transferred using simple copper lines, high-speed signals should be transmitted over dedicated transmission lines designed to carry high-frequency signals over a long distance with minimal distortion and reflections. Since the EM wave of the signal partly propagates through the substrate of the PCB, a low-loss dielectric is chosen as PCB substrate. Furthermore, only two layers are chosen as more layers are not needed and would drive up the fabrication costs needlessly. A two-layer PCB with Rogers 4350B substrate is chosen, which has a relative permittivity, ϵ_r, of 3.48 and a loss tangent or tan(δ) of 0.0039.

All transmission lines are routed differentially to ensure good immunity against noise and other interference signals. Close to the chip, differential microstrip lines are designed to provide a characteristic impedance of 100 Ω differently to avoid reflections at the interface. These lines then transition to 100 Ω grounded coplanar waveguide (GCPW) lines as they provide better shielding and experience less dispersion than microstrip lines. Furthermore, CPW can be easily be designed to have a characteristic impedance of 50 Ω, despite the thick substrate. These differential lines finally split into two single-ended GCPW lines to interface with the data connectors. The lines are routed as symmetrically as possible to ensure optimal signal balance. Asymmetrical or sharp bends are to be avoided to minimize DM-to-CM signal conversion. The transmission lines for the TX data inputs are delay matched to ensure the different data signals are at the modulator at the same time, to avoid EVM degradation. Due to a large number of data outputs, delay matching all data lines of the RX was not possible. However, for characterization purposes these outputs can be re-aligned in the digital domain during measurements. The transmission lines have been simulated in Momentum's 2.5D EM solver and verified using the 3D EM solver from Ansys—HFSS, demonstrating a 3-dB bandwidth over 25 GHz. The DC lines, including supply, ground, and bias nets, are places as far from the sensitive transmission lines as possible, and each DC pad is provided with a local ground pad for additional decoupling on the boards.

Figure 7.52 shows the layouts of the PCBs designed for TX and RX standalone measurements. The LO and high-speed data signals are interfaced off-board using 2.92 mm coaxial end-fire connectors, and the DC signals are connected using standard ribbon cables, as illustrated in Fig. 7.53a. As mentioned before, the high-frequency D-band signals (TX output and RX input) are probed by a GSG probe, as shown more clearly in Fig. 7.53b. The chip itself is glued onto the PCB and interface to the PCB via bondwires, as illustrated in Fig. 7.54a. Special attention is needed when probing not to touch the bondwires when approaching the chip, as can be observed from the close-up picture in Fig. 7.54b.

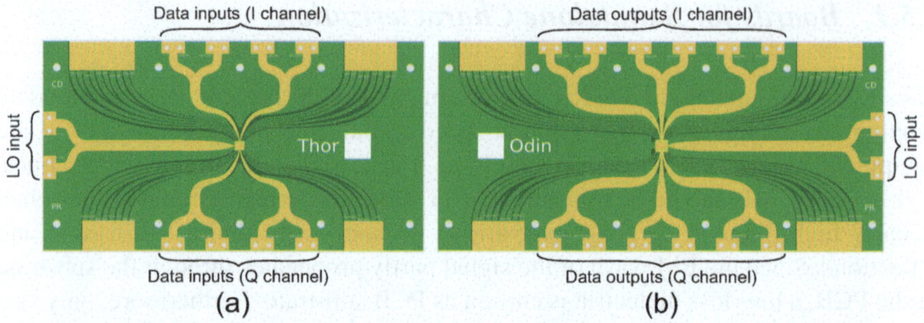

Fig. 7.52 PCB layouts for TX (**a**) and RX (**b**) standalone measurements

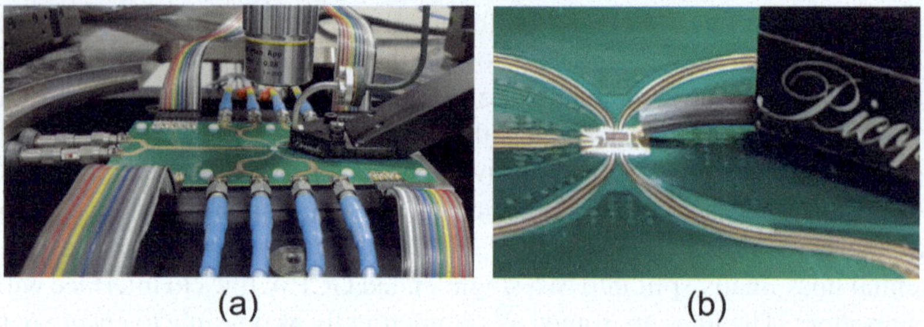

Fig. 7.53 Board setup during TX measurements with high-speed data cables and DC ribbon cables (**a**) and close-up picture of the GSG probe being landed on the TX chip (**b**)

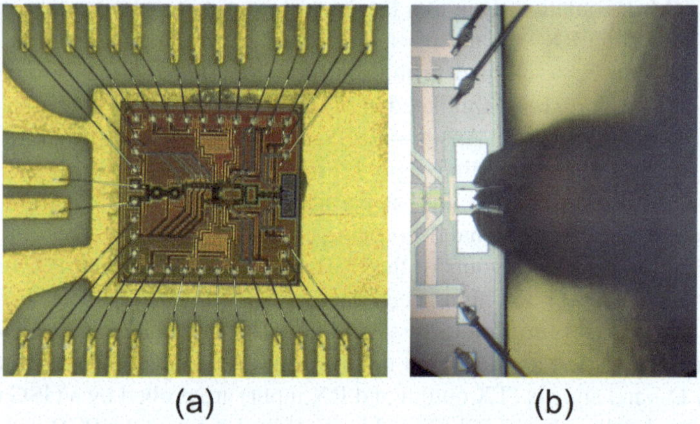

Fig. 7.54 Microscope picture of the chip interfaced to the PCB using bondwires (**a**) and a close-up microscope picture of the probe landed on the TX output pads (**b**)

7.5.2 Boards for High-Speed Link Measurements

For high-speed link measurements, GSG probes are not an effective interface since custom, expensive waveguide components are required to link both chips and the setup of the probe station severely limits the maximum achievable link distance. In order to accommodate low-cost mass production of the envisioned system and provide maximum measurement flexibility, the high-frequency interface must be integrated on the PCB. For the wireless link, the interface takes the form of an antenna, while a DWG link needs a coupler to launch the signal in/from the waveguide. Optionally, these structures could be integrated on-chip. While this would avoid a lossy interface, the poor performance of on-chip antennas at sub-THz frequencies (low efficiency, low gain, limited bandwidth) due to the lossy silicon substrate and thin metals makes off-chip antennas the superior option.

Having to implement a 135 GHz antenna on PCB requires special attention when choosing the PCB technology. First of all, the relative permittivity, ϵ_r, is important as it determines the effective wavelength in the material and thus influences the size of high-frequency structures. Since the minimal physical pitch and spatial resolution of PCB technologies are rather limited, a low ϵ_r is ideal to avoid impractically small dimensions. Furthermore, a low-loss material (low $\tan(\delta)$) is desired in the substrate to minimize the insertion loss of the realized component. Lastly, the ϵ_r and $\tan(\delta)$ must remain as stable as possible over frequency and temperature to ensure robust performance. From the numerous available high-frequency substrate materials, the Tachyon 100G substrate from Isola was eventually chosen for this work, with an ϵ_r of 3.02 and a $\tan(\delta)$ of only 0.0021. It must be noted that the permittivity and loss tangent are measured at 10 GHz, which is not necessarily valid at 135 GHz. Nevertheless, measuring dielectric properties above 100 GHz is quite challenging so these values are adopted during the design.

An overview of the different layers and types of vias is depicted in Fig. 7.55. A four-layer stack with a minimum trace pitch of 160 μm is chosen to provide adequate high-frequency performance and routing flexibility without making the

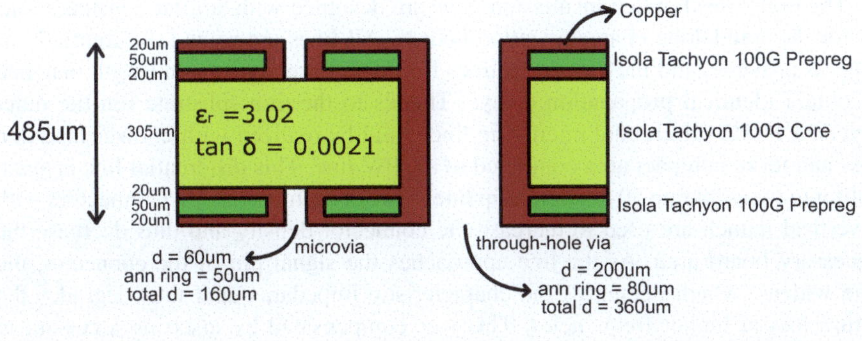

Fig. 7.55 Layer build-up of the PCB technology employed for the high-frequency board

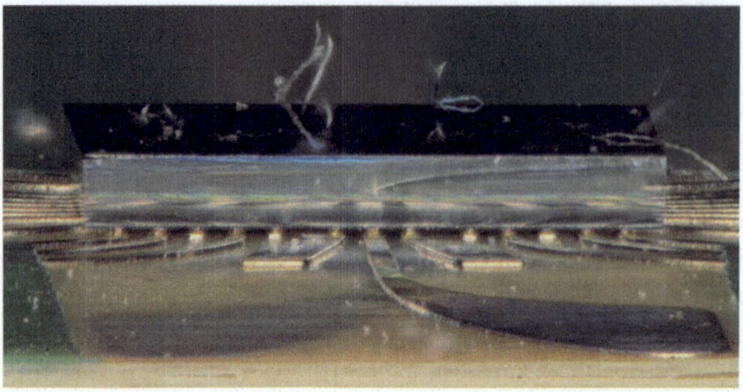

Fig. 7.56 Sideways microscope picture of a silicon die connected to the board via flip-chip bumps. For scale, the silicon die has a thickness of about 300 μm. The Vivaldi structure is also clearly visible

board overly expensive. Four layers also enable the PCB to have thin outer dielectric layers, necessary for >100 GHz operation, while guaranteeing a minimum board thickness with a thick inner layer for mechanical robustness.

To enable a low-loss electrical interface at 135 GHz the chips are connected to the PCB via flip-chip bumps, drastically reducing the large series inductance of the bondwire (>500 pH) to about 50 pH. This rather small inductance can easily be absorbed in the matching network connecting the GSG input/output of the chips to the EM structure on PCB. Additionally, the lower flip-chip inductance enhances the bandwidth at the data interface considerably, as this is mostly determined by the ESD capacitance combined with this inductance. However, compared to a rather simple bondwire interface, the flip-chip process is quite delicate and requires additional material considerations. For example, to avoid the chip from detaching from the board while being heated and pressed on the board, the PCB substrate needs sufficient mechanical stiffness and minimal thermal expansion relative to the silicon substrate. Both of these specifications are met with the chosen material. As an illustration, Fig. 7.56 shows a sideways depiction through a microscope of a silicon die connected to the board via flip-chip bumps.

The high-speed data transmission lines are designed with similar considerations as for the standalone characterization boards. All lines are routed differentially as long as possible, and the data input lines for the TX are once again length matched to obtain identical propagation delays. Thanks to the thin substrate (on the outer layer), a 100 Ω differential microstrip line could be realized with a lower insertion loss and more compact area compared to a CPW line. This differential line is again split into two separate 50 Ω microstrip lines toward each connector. Connectors with a vertical launch are used to increase the connector density and thus decrease the necessary board area. As the line approaches the signal pin of the connector, the line widens, which decreases the characteristic impedance and thus degrades the return loss at higher frequencies. This was compensated by inserting a cut-out in the ground plane near the signal pin, which leads to a slight impedance increase and improvement of the return loss. The lines are simulated in HFSS and demonstrate sufficient bandwidth (the exact value depends on the length of the line) and return loss performance (< −20 dB over 30 GHz).

7.5.2.1 Vivaldi Antenna Design

The most challenging part of this board design is the 135 GHz antenna structure. Antipodal Vivaldi antennas are employed to function as wireless antennas [164] and DWG couplers [130]. The Vivaldi antenna is a nonresonant tapered slot antenna exhibiting high bandwidth with end-fire radiation [203]. The nonresonant operation makes this antenna particularly robust to manufacturing variations, but with a rather large footprint (significantly larger than the wavelength). The antipodal Vivaldi was implemented in the top two metal layers on the board, and microvias were carefully placed to perform single-ended to differential signal conversion. The entire transition from on-chip GSG pads to the off-chip antenna structure was thoroughly simulated to ensure a broadband, low-loss interconnect from the chip to the antenna. Two different Vivaldi designs were developed to support wireless transmission as well as guided transmission over a DWG, as is depicted in Fig. 7.57.

To function as an antenna, the Vivaldi was optimized for far-field radiation performance. An elliptic tapering was employed, and the corners were rounded to avoid unwanted radiation from sharp edges. At the beginning of the antenna, the electric field mainly has a vertical component between the two conductors. This field rotates along the antenna into the horizontal plane and detaches at the end of the antenna radiating outward. The antenna achieves a simulated realized gain around 8 dBi with less than 2 dB variation across the entire D-band, as shown in Fig. 7.58a.

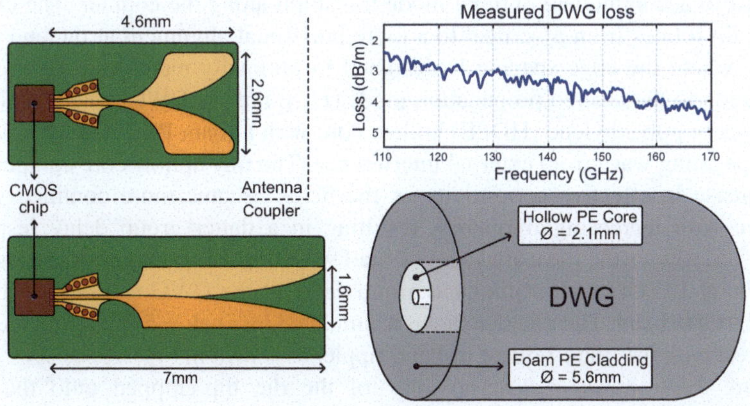

Fig. 7.57 Overview of the antipodal Vivaldi structure as a free-space antenna and DWG coupler, along with the cross-section and measured loss of the DWG

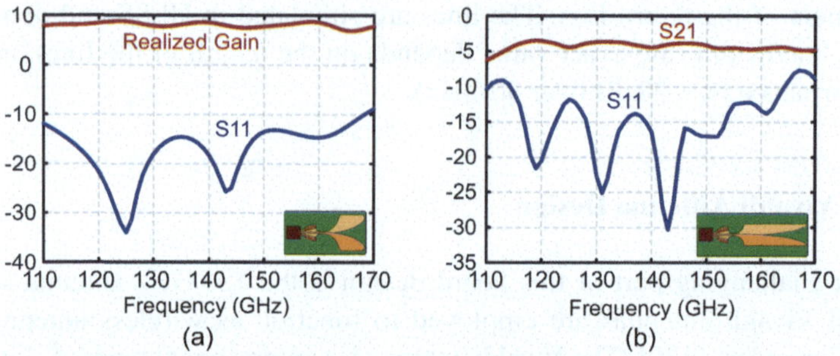

Fig. 7.58 Simulated performance of the Vivaldi antenna (**a**) and Vivaldi DWG coupler (**b**)

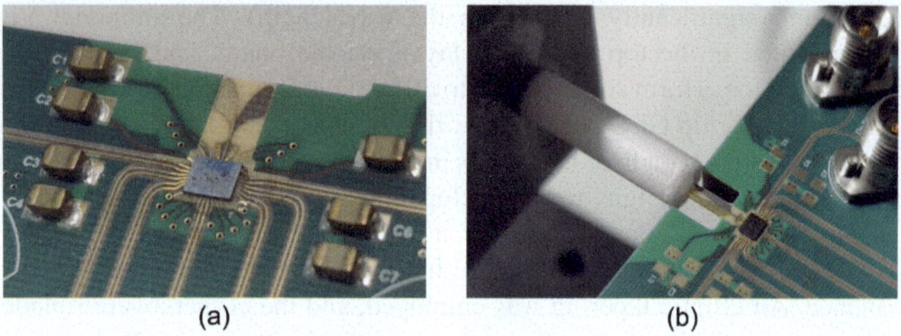

Fig. 7.59 Close-up view of the flip-chipped die connected to the Vivaldi antenna (**a**) and Vivaldi coupler launching into the DWG (**b**)

While implementing the DWG coupler, the Vivaldi was designed to efficiently couple the RF signal into the DWG with minimal insertion loss and a broadband characteristic. The slot is tapered elliptically again, and the length of the coupler is designed to accommodate optimal mode transition along the coupler. The electric field again rotates from a vertical to a more horizontal alignment at the end of the coupler where the edge spacing is designed to optimally match the electric field distribution to the desired propagation mode (HE_{11}) of the DWG. The DWG has a high-density polyethylene (HDPE) hollow core with a foam PE cladding to isolate the propagating wave from external interference. The tiny hollow core compensates the increase in effective permittivity as the fields become more confined within the core with increasing frequency, resulting in a flatter group delay response. Figure 7.57 shows the measured loss of the DWG over frequency, which is around 3.5 dB/m at 135 GHz. The simulated group delay of the DWG is around 100 ps/m over a 20 GHz band. The coupler shows a simulated loss below 5 dB across the band of interest with less than 1 dB of in-band ripple, as shown in Fig. 7.58b.

Figure 7.59 shows a close-up view of the die flip-chipped onto the PCB and connected to the Vivaldi antenna or DWG coupler. To allow for precise alignment between the coupler and DWG, a 3D-printed holder was designed and

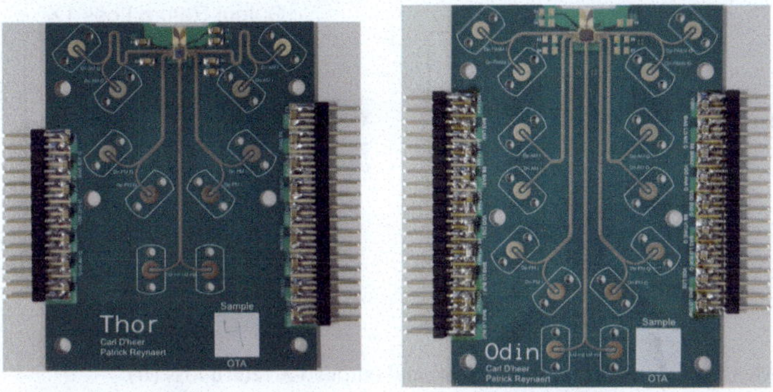

(a) (b)

Fig. 7.60 Pictures of the PCBs designed for link measurements: TX PCB (**a**) and RX PCB (**b**)

manufactured, as shown in the picture. The two designed PCBs (TX and RX for wireless measurements) are shown in Fig. 7.60.

7.6 Measurement Results

The transmitter (Fig. 7.30) and receiver (Fig. 7.50) chips, fabricated in a standard 28 nm bulk CMOS technology, occupy an area of 2.29 mm^2 and 3.62 mm^2, respectively. Thanks to the custom designed on-chip GSG pads, both chips can be interfaced to off-chip antennas via flip-chip bumps for full link measurements while also being compatible with GSG probes for standalone characterizations without having to de-embed off-chip structures.

7.6.1 Standalone Probed Measurements

Initially, standalone probe-based measurements were performed to verify and characterize the performance of the TX and RX chips. The chips were mounted on a two-layer Rogers 4350B PCB as discussed in the previous section. The RF input/output signals were probed using 50 μm GSG probes, while all other signals, including the low-frequency LO input, were wire-bonded to the PCB.

To validate the design of the GSG pads, 1-port VNA measurements were performed on the RX input pads and TX output pads. Figure 7.61 shows the measured return loss at the input of the RX and output of the TX, demonstrating a well-matched 50 Ω interface with the custom GSG pads on both chips.

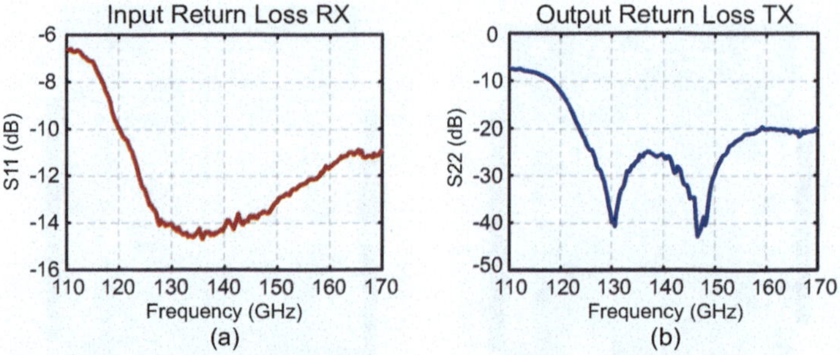

Fig. 7.61 Measured input RX return loss (**a**) and output TX return loss (**b**)

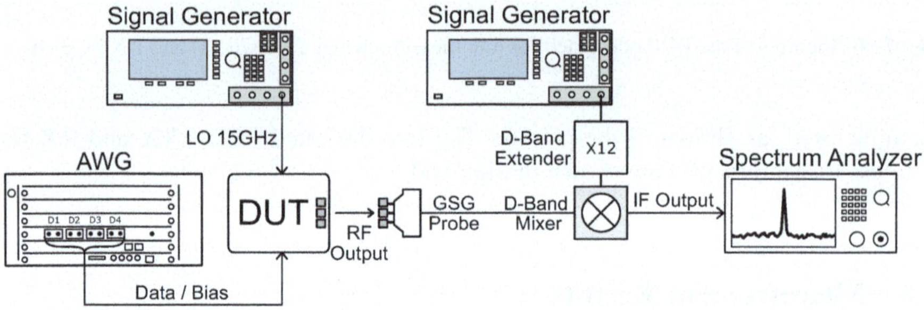

Fig. 7.62 Measurement setup for standalone characterization of the transmitter

7.6.1.1 TX Characterization

To characterize the TX, the TX output was downconverted to an intermediate frequency (IF) by an external D-band mixer and analyzed with a spectrum analyzer. The conversion loss of the mixer is calibrated using a PM5 power meter. The low-frequency LO references for the TX and D-band extender are generated using external signal generators. The measurement setup to perform the standalone characterization of the TX is shown in Fig. 7.62 (Fig. 7.63).

The measured CW TX output power at 135 GHz versus LO input power is shown in Fig. 7.64a, demonstrating a saturated output power of 0 dBm at 135 GHz with less than 3 dB variation over a 10 GHz band. The significantly lower saturated output power compared to simulations (Fig. 7.16) can be explained by larger than expected spurious tones generated by the LO chain. Indeed, Fig. 7.64a also shows the output power at 120 GHz, which reaches a value 2 dB larger than the desired tone at 135 GHz. The consequences of this large spur will be discussed more in detail at the end of this section. Figure 7.64b shows the measured output power of the TX at $9f_0$ versus frequency for various input powers at f_0. While this plot demonstrates the wideband performance of the LO and amplifier chain (P_{sat} 3-dB bandwidth around 12 GHz), it does not reflect the modulation bandwidth as this response includes

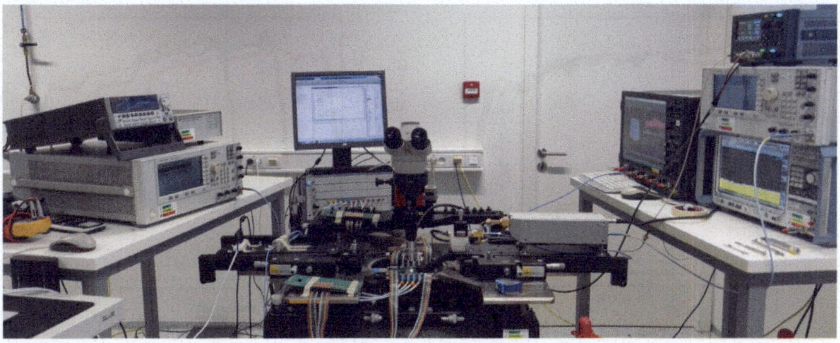

Fig. 7.63 Picture of the real-life measurement setup for TX measurements

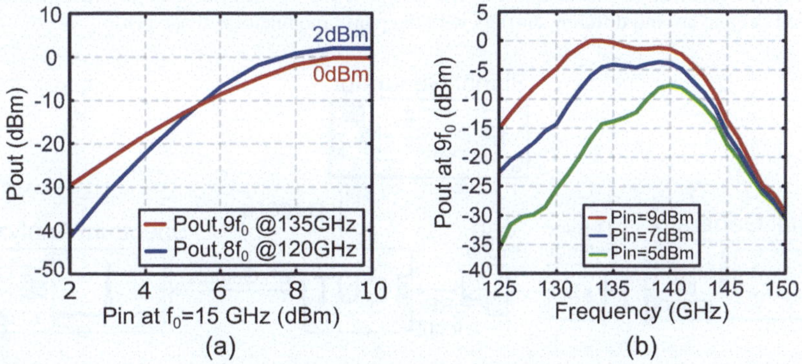

Fig. 7.64 Measured TX output power at 135 GHz and 120 GHz vs. LO input power at 15 GHz (**a**) and TX output power at $9f_0$ versus frequency for various input powers at f_0 (**b**)

bandwidth-limiting stages that do not confine the maximum modulation speed (e.g., the triplers and buffers in the LO chain).

To characterize the upconversion response of the TX, a 50% duty cycle square wave is used to drive one or more modulators, while the other modulators are kept constant. The power of the fundamental sidebands around the 135 GHz carrier was then observed while sweeping the frequency of the square wave input. Figure 7.65a depicts the upconversion response normalized to the value at the center frequency. Driving one of the phase modulators, amplitude modulators, or both modulators simultaneously yields a similar response. Furthermore, employing different carrier frequencies, as shown in Fig. 7.65b, reveals information about the frequency response of the TX as experienced by the modulated signal. The sharp drop-off in the response is most likely caused by the undesired spur, as discussed later in this section.

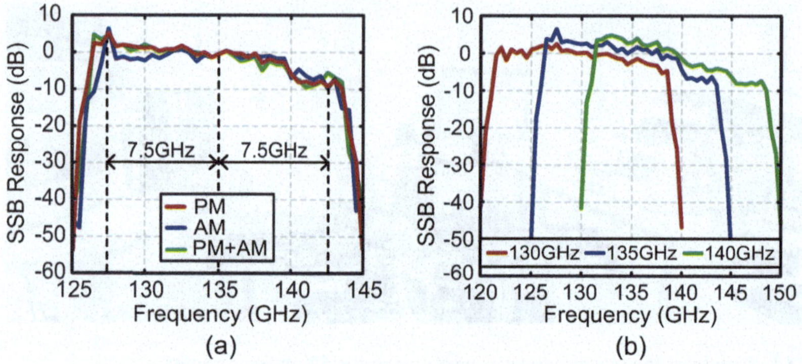

Fig. 7.65 Normalized upconversion response around a 135 GHz carrier with different modulators activated (**a**) and around different carriers with the phase modulator activated (**b**)

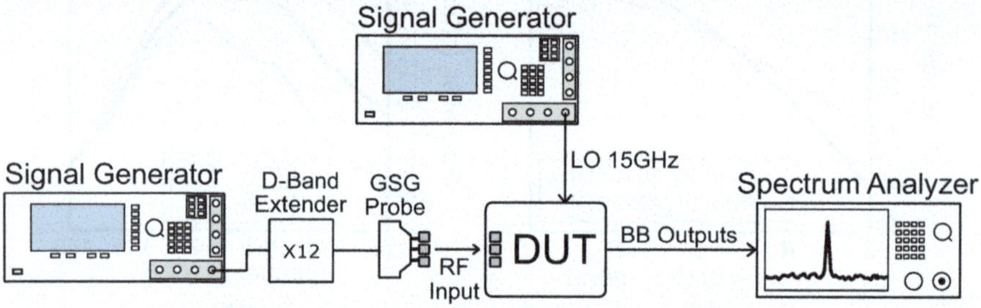

Fig. 7.66 Measurement setup for standalone characterization of the receiver

7.6.1.2 RX Characterization

To perform standalone RX characterization, an RF input signal is generated by an external D-band extender and calibrated using a PM5 power meter. The RX chip downconverts the D-band input into an IF which is further analyzed with a spectrum analyzer. The measurement setup for standalone RX characterization is shown in Fig. 7.66 (Fig. 7.67).

Figure 7.68a shows the measured RX conversion gain (to the PAM-4 test outputs) versus input frequency for a fixed LO frequency of 135 GHz and various CTLE tuning voltages. The RX demonstrates a maximum conversion gain of 25 dB with a 3-dB bandwidth of 16 GHz. By tuning the CTLE, the 3-dB bandwidth can be extended to 20 GHz at a cost of lower conversion gain. Figure 7.68b shows the measured conversion gain using slightly different LO frequencies, showing a clear asymmetry in the response to either side when offsetting the LO from its optimal value at 135 GHz. The input-referred 1-dB compression point (P1dB) is measured around −29 dBm, as illustrated in Fig. 7.69a. This compression point was measured with the CTLE tuned at its highest gain state. As most of the compression is caused by the final baseband stages, tuning down the CTLE to decrease the gain

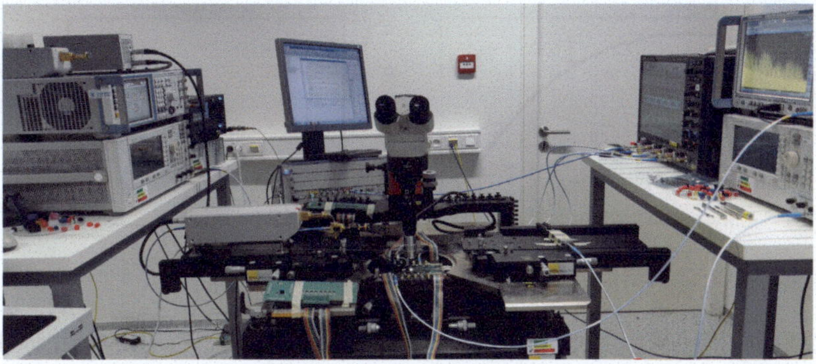

Fig. 7.67 Picture of the real-life measurement setup for RX measurements

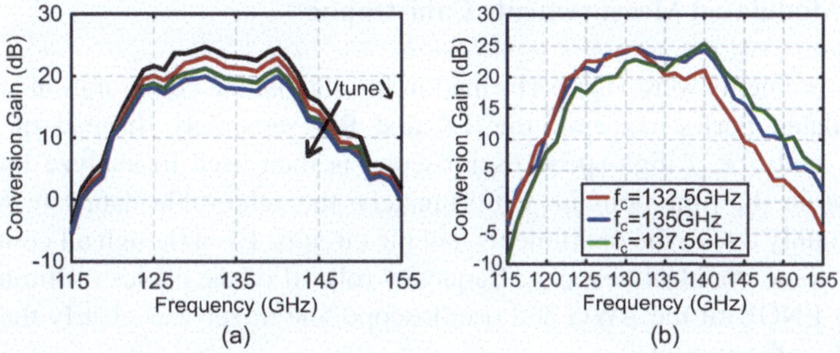

Fig. 7.68 Measured RX conversion gain for various CTLE tuning voltages with a 135 GHz LO (**a**) and conversion gain for different LO frequencies (**b**)

in the baseband chain would further improve the 1-dB compression point. This was unfortunately not measured.

The receiver noise figure (NF) is characterized using the cold-source method, as illustrated using the definition of the noise factor:

$$F = \frac{\frac{S_{in}}{N_{in}}}{\frac{S_{out}}{N_{out}}} = \frac{N_{out}}{G \cdot N_{in}} = \frac{N_{out}}{G \cdot kT}, \tag{7.13}$$

where $S_{out} = G \cdot S_{in}$ and the noise floor at the input $N_{in} = kt = -174$ dBm/Hz. By measuring the output noise floor of the receiver (with the receiver turned on) and referring it to the input (by dividing by the measured RX conversion gain G), the noise factor can be calculated by comparing that to an ideal 50 Ω noise source. The double-sideband (DSB) NF is depicted in Fig. 7.69b and shows a minimum value of 6 dB. The average NF integrated over a 15 GHz band is 9.8 dB.

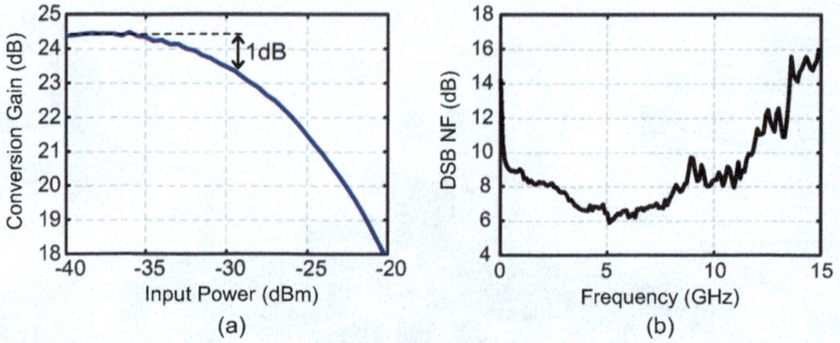

Fig. 7.69 Measured RX conversion gain versus input power with a 135 GHz LO (**a**) and DSB noise figure (**b**)

7.6.1.3 Modulated Measurements Catastrophe

Data measurements were also performed in the standalone configurations to assess the modulated performance of the TX and RX separately. Instead of using a spectrum analyzer, a high-speed oscilloscope is then used to analyze the output waveforms in the time domain. Unfortunately, the achievable data rate was very limited, mainly due to the instruments, not the circuits. Even though all components introduce some nonidealities, e.g., frequency roll-off of the cables or limited high-frequency ENOB of the AWG and oscilloscope, the mixer was clearly the biggest bottleneck in the system.

A D-band mixer was employed to convert the RF signal into a baseband (BB) or intermediate frequency (IF) signal, or the other way around. However, this mixer turned out not to work properly (possibly because of an ESD event), indicated by an excessive power consumption, which made measuring high data rates considerably more challenging. Although no good results were obtained because of this, interesting effects were observed, which could provide some insight and will be discussed for this reason. Also otherwise months of valuable time would have been wasted for no reason.

A “broken” mixer results in the degradation of all relevant electrical specifications of the component that influence the system in different ways. Most obvious is the deterioration in conversion gain, output power, and linearity, which results in a much lower linear output power or OP1dB of the mixer. For TX measurements, the mixer is able to handle much lower RF input powers and can thus output much smaller modulated output signals for the oscilloscope to demodulate. For RX measurements, the mixer can barely generate a modulated signal with sufficient output power to enable demodulation in the receiver. Next, the frequency response of the mixer is seriously degraded, leading to significant in-band ripple, lower bandwidth, and more asymmetry in the response, all impairing the modulated signal and introducing considerable ISI.

Another problem while using the malfunctioning mixer was the limited isolation between the RF and LO ports, leading to signal leakage between those ports. During

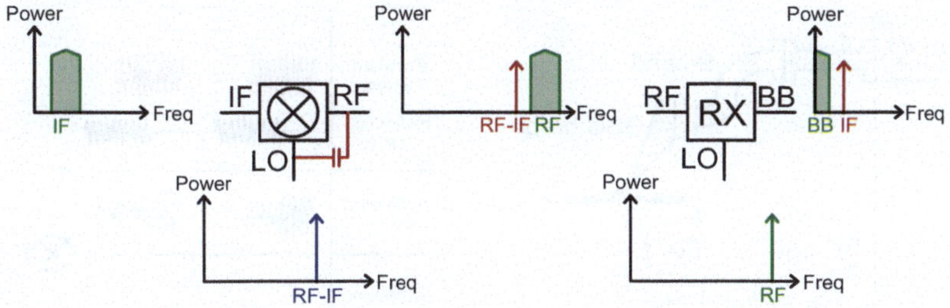

Fig. 7.70 Conceptual demonstration of how limited LO-RF isolation in the upconversion mixer with modulated IF input can lead to a problematic IF tone in the downconverted RX output

TX measurements, the modulated RF output of the TX leaks to the LO port of the mixer and mixes with itself, leading to an additional self-mixed signal on top of the desired baseband signal at the mixer output. If this leaked RF signal has a similar magnitude to the LO signal, this self-mixed component causes severe signal distortion. When measuring the RX, the leaked LO signal appears dangerously close to the upconverted IF signal at the output of the mixer and reveals itself as an unwanted IF tone very close to the modulated baseband after downconversion by the RX circuits, as illustrated in Fig. 7.70. This uncovers a trade-off when choosing the IF input of the mixer. While a large IF is desired to more easily filter out the IF tone at the baseband output, a larger IF decreases the modulated output power of the mixer and increases the loss/distortion induced by the cables and mixer frequency response. This trade-off fundamentally limits the achievable data rate during RX-only complex modulated measurements. By choosing the IF equal to zero, the above issue disappears, but only single-channel modulations (ASK, BPSK, APSK) can then be generated by the mixer, as complex modulations (M-QAM) require either I/Q baseband inputs or a nonzero complex modulated IF input.

Finally, the impaired characteristics of the mixer can lead to problematic reflections. Typically, when the return loss of all ports is sufficiently low, these reflections are of little concern. When the return loss is impaired, however, reflections can become a problem, depending on the channel. Figure 7.71a depicts the situation occurring in the RX-only measurement setup. The malfunctioning mixer provides a rather large reflection coefficient, Γ_{mixer}, while the receiver chip is fairly well matched, $\Gamma_{RX} \approx -14$ dB (Fig. 7.61a). Since the waveguide channel has relatively low loss ($\approx$2 dB), the reflections can propagate almost undisturbed and cause problems depending on the delay through the channel. Figure 7.71b illustrates the potential problem when the total delay of the reflected pulse (traveling twice the length of the channel) is similar to the symbol period, causing transitions in the signal level whenever one of the reflections arrives back at the output.

If only two reflections are considered (the magnitude of the reflections quickly declines every additional round trip through the channel), the in-phase, $I_{\mathrm{out}}(t)$, and quadrature, $Q_{\mathrm{out}}(t)$, baseband outputs resulting from an in-phase modulated input signal, $V_{\mathrm{in}}(t)$, can be expressed as follows:

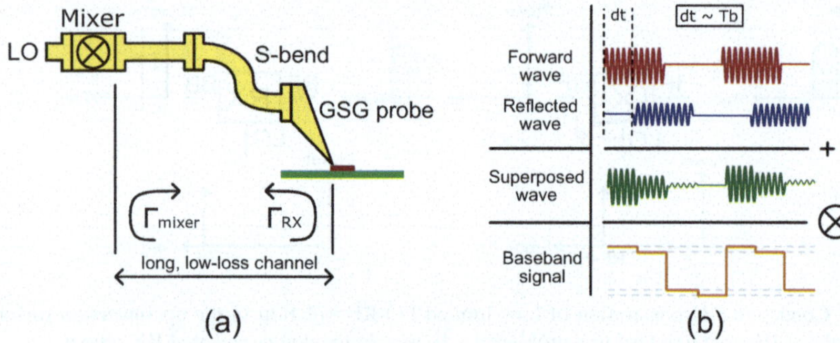

Fig. 7.71 Diagram visualization of the occurrence of reflections (**a**) and graphical representation of the influence of reflections on the output waveform (**b**)

$$V_{\text{in}}(t) = d(t)\cos(\omega_c t) \tag{7.14}$$

$$I_{\text{out}}(t) = [1 + \Gamma_L]\, d(t - l/v_p) + \Gamma_L \Gamma_S\, d(t - 3lv_p)\cos(\omega_c 2l/v_p) \tag{7.15}$$

$$Q_{\text{out}}(t) = -\Gamma_L \Gamma_S\, d(t - 3l/v_p)\sin(\omega_c 2lv_p), \tag{7.16}$$

where l is the length of the channel, Γ_S/Γ_L are the reflection coefficients of the source/load, and v_p is the propagation velocity of the wave through the channel. A full derivation of this expression, as well as a more detailed investigation and description of this phenomenon, can be found in Appendix A.4.

From this expression it can be seen that the original signal gets attenuated/amplified by Γ_L, distorted by a delayed version of the data, and even couples to the orthogonal channel distorting the signal present on that channel. All of these effects have indeed been observed during measurements. Depending on the exact values of Γ_S/Γ_L and the propagation delay relative to the symbol period, the resulting signal will be distorted in a certain way.[4] Also note that if the propagation delay would be larger than the symbol period, the reflections would distort subsequent symbols, effectively introducing ISI. Furthermore, in reality, multiple reflections occur resulting in specific distorted waveforms depending on the exact data sequence. Figure 7.72 shows examples of distorted constellations from TX-only data measurements and distorted baseband output signals from RX-only data measurements. A custom data sequence was used in Fig. 7.72b to identify the appearance of the reflection, which leads to additional signal levels in the eye diagram and additional constellation points in the constellation plot (Fig. 7.72a).

The most straightforward method to solve these issues, without repairing the mixer, is to suppress the reflections, for example, by inserting an attenuator between

[4] Note that, assuming $\Gamma_{mixer} = 0$ dB, $\Gamma_{RX} = -14$ dB, and 2 dB of channel loss, the reflected wave at the output would have an amplitude approximately equal to 16% of the desired signal, which is very significant, despite an excellent return loss at the RX input.

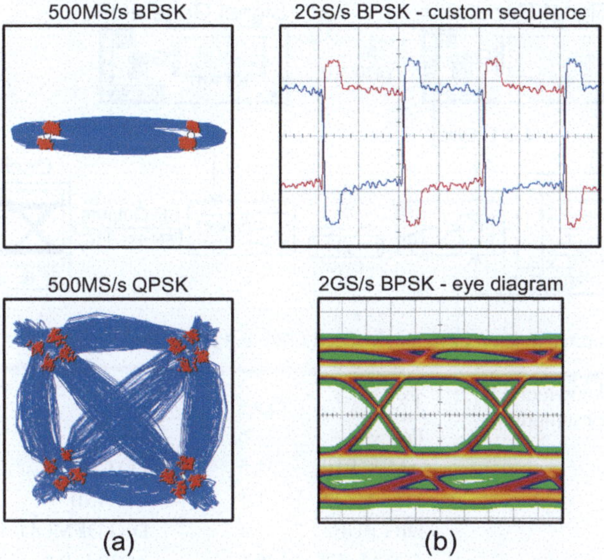

Fig. 7.72 Distorted waveforms caused by reflections: 500 MS/s BPSK and QPSK constellations (**a**) and 2 GS/s BPSK waveform and eye diagram (**b**)

the mixer and the probes. This adds additional loss to the channel and thus dampens the reflections. However, the attenuator also attenuates the desired signal, which is not ideal. A better option is to use an isolator component. An isolator is a nonreciprocal device which is designed to transmit power in only one direction. As a result, the forward wave only experiences little insertion loss ($\approx$3 dB), while the reflected wave is dampened significantly ($\approx$30 dB). The isolator is thus able to suppress the reflections without severely degrading the SNR of the measurement. Due to the unavailability of the component, an attenuator had to be used during the TX measurements, while an isolator could be used for the RX measurements. However, even without the reflections, RX standalone complex modulated measurements were not possible due to the LO-RF leakage in the mixer, as described before.

To conclude, due to a malfunctioning mixer component, high-speed modulated measurements were hampered in the standalone setups, mainly due to limited LO-RF isolation and reflections. Particularly the reflections lead to interesting observations which are more extensively investigated in Appendix A.4. To perform modulated measurements without having to rely on external components for important blocks, a full link should be realized using the implemented TX and RX chips instead.

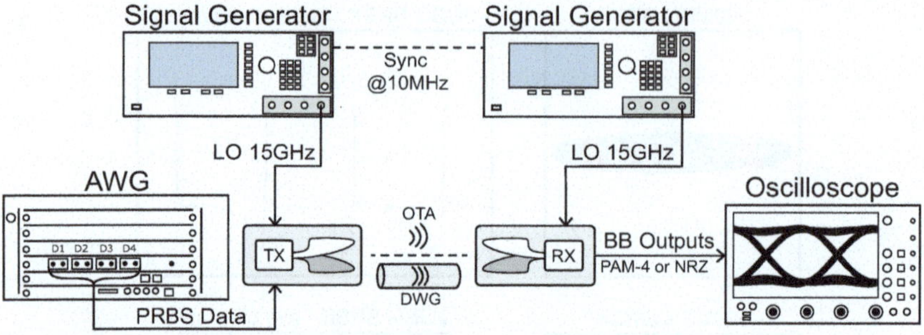

Fig. 7.73 Measurement setup of the high-speed OTA and DWG link

Table 7.1 Breakdown of TX and RX power consumption

TX	P_{DC}[a] (mW)	RX	P_{DC}[a] (mW)
PA	21.6	LNA	27.9
AM & BUF	31.5	LO CHAIN	43.2
LO CHAIN	71.1	BB AMP	63.0
DIG BUF	5.4	DIG DEMOD	23.4
TOTAL TX	129.6	TOTAL RX	157.5

[a] From a 0.9 V supply

7.6.2 *Modulated Link Measurements*

After TX/RX characterization, high-speed modulated link measurements were performed by linking the TX and RX chips over-the-air (OTA) and over a dielectric waveguide (DWG) channel using off-chip antenna structures. As discussed in the previous section, the dies are flip-chip packaged onto a four-layer PCB with high-frequency Isola Tachyon 100G substrate. Two distinct antipodal Vivaldi antennas are designed to function as wireless antennas and as DWG couplers.

Figure 7.73 shows the setup for the high-speed link measurements. The data inputs for the transmitter are four rectangular, rail-to-rail binary PRBS7 sequences generated with an arbitrary waveform generator and supplied to the chip with a set of phase-matched cables. The low-frequency LO references for both chips are generated using external signal generators which are locked through a 10 MHz locking signal. The LO phases are manually tuned to obtain optimal coherent detection in the receiver. The downconverted baseband outputs of the receiver, either PAM-4 or NRZ signals, are sampled and analyzed by a high-speed oscilloscope.

The TX and RX chips consume a combined total of 287 mW from a 0.9 V supply while transmitting and demodulating high-speed data. Table 7.1 summarizes the measured power consumption of the different TX and RX blocks.

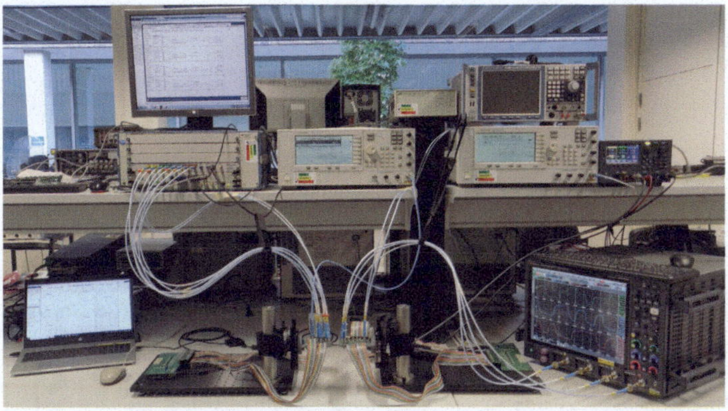

Fig. 7.74 Picture of the real-life measurement setup for wireless link measurements

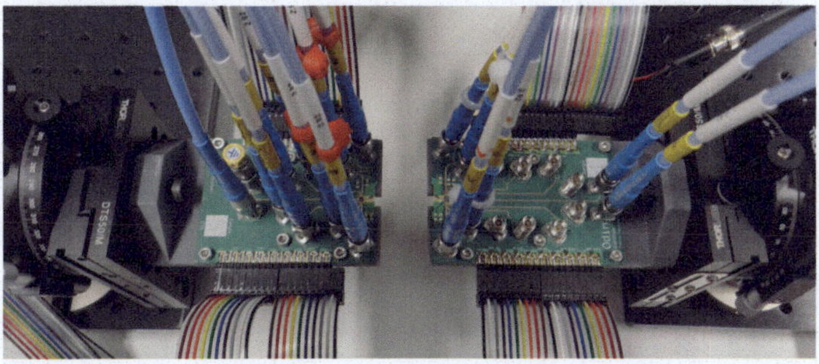

Fig. 7.75 Close-up picture of the boards in the wireless link setup

7.6.2.1 Wireless Link

A picture of the real-life measurement setup of the wireless link is shown in Fig. 7.74 with a close-up view of the two boards depicted in Fig. 7.75. First, the PAM-4 test outputs are observed to assess the performance of the transmitter and receiver RF front-end with baseband chain, without the demodulation circuitry. The PAM-4 signals in I and Q channels are captured and used to reconstruct the constellation and calculate the EVM. To maximize the SNR of the measurement, the PAM-4 signals are filtered in the oscilloscope to reduce the integrated noise power affecting the signal. A Bessel filter response is preferred due to its superior phase linearity and flat magnitude response. Furthermore, using the model presented in [54], the minimum required EVM value can be calculated to achieve a certain BER given a specific modulation scheme. An overview of the required EVM values as a function of BER and modulation scheme is summarized in Table 7.2. This equation assumes that the SNR is mainly limited by white Gaussian noise, which is not entirely accurate, but serves as a decent estimation to evaluate performance.

Table 7.2 Table summarizing the minimal required EVM vs. desired BER

BER	1e−3	1e−6	1e−12
BPSK	−6.8 dB	−10.5 dB	−13.9 dB
QPSK	−9.8 dB	−13.6 dB	−17.0 dB
4-APSK	−10.5 dB	−14.5 dB	−17.9 dB
Star-QAM	−15.3 dB	−19.2 dB	−22.7 dB
16-QAM	−16.6 dB	−20.4 dB	−23.9 dB

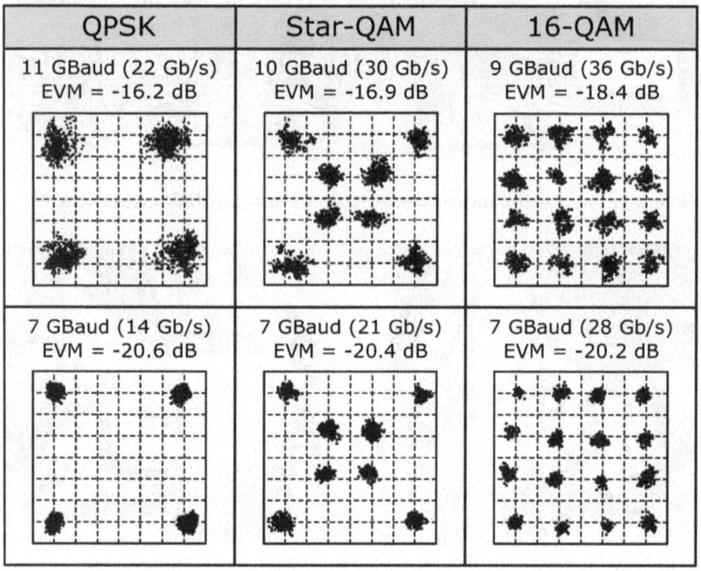

Fig. 7.76 Measured constellations for various modulation schemes over a 2 cm OTA link

These link measurements have been performed using various modulation schemes at multiple distances. Figure 7.76 shows the measured constellation diagrams of the three supported IQ modulations over a 2 cm OTA link. The highest speed constellations are shown in the top row, while the bottom row shows the signals at 7 GBaud to demonstrate the linearity of the TX and RX. A maximum data rate of 36 Gb/s is achieved using 16-QAM with an EVM of −18.4 dB, equivalent to a BER below 10^{-4}. Using star-QAM and QPSK leads to maximum data rates of 30 Gb/s and 22 Gb/s, respectively.

To get a better understanding of how the performance of the link changes with frequency, data rate, distance, and modulation scheme, the EVM of the output constellation is measured and plotted versus various parameters. Figure 7.77a gives an overview of the measured EVM versus data rate for all measured modulations in a 2 cm OTA link. To verify the robustness of the link over frequency, the EVM was measured for 20 Gb/s 16-QAM transmission at a distance of 2 cm while sweeping the carrier frequency, as shown in Fig. 7.77b. The EVM stays below −16.6 dB (BER $< 10^{-3}$) across an 8 GHz LO range. Figure 7.78a shows the evaluated EVM

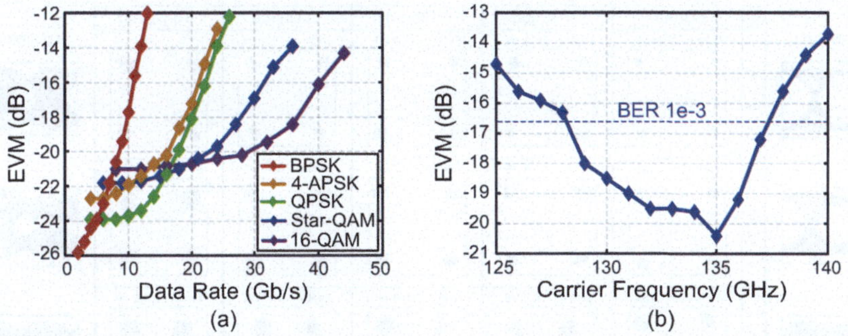

Fig. 7.77 Measured EVM versus data rate of a 2 cm OTA link for all modulation schemes (**a**) and measured 16-QAM EVM versus carrier frequency over a 20 Gb/s 2 cm OTA link (**b**)

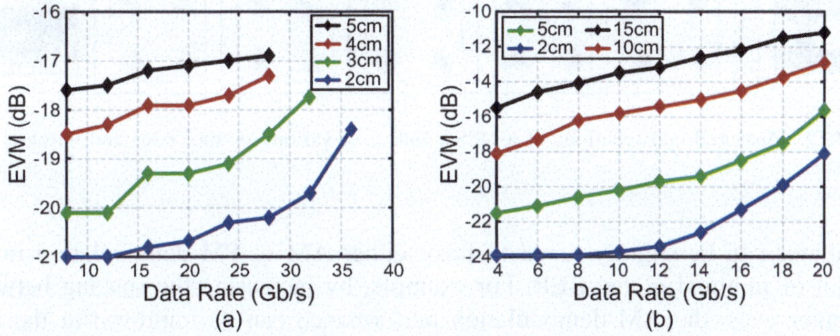

Fig. 7.78 Measured EVM versus data rate and link distance for 16-QAM (**a**) and QPSK (**b**) transmission over an OTA link

for 16-QAM transmission versus data rate for various wireless link distances. At 5 cm distance a 16-QAM data rate of 28 Gb/s is supported with −16.9 dB EVM. Similarly, the measured EVM versus data rate and distance for QPSK transmission is shown in Fig. 7.78b, demonstrating 20 Gb/s over up to 15 cm distance. Figure 7.79 gives an overview of the measured QPSK constellations over various symbol rates and link distances. From left to right, the symbol rate increases, and as a result, the constellation and eye diagram degrade accordingly due to ISI. From top to bottom, the wireless distance increases and thus the SNR of the signal clearly deteriorates. It is important to remember that these constellations and EVM values are all obtained without using any digital predistortion, pulse shaping, or equalization, thus demonstrating the linearity and robustness of the designed system.

Secondly, to verify the functionality of the full link including the PAM-4 decoders, the demodulated AM and PM outputs bits were observed and compared to the input data supplied to the transmitter to perform BER measurements, at a link distance of 2 cm. The operation settings for the TX and RX are pretty much identical for optimal demodulation performance compared to minimal 16-QAM EVM performance. As an additional tool, the bias voltage of the amplitude

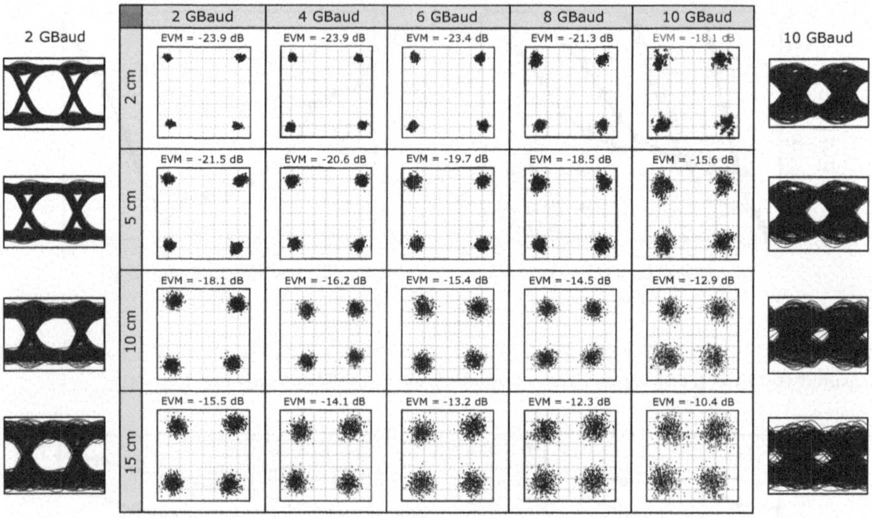

Fig. 7.79 Measured constellations of a QPSK link over various symbol rates and wireless link distances

modulator can be slightly tuned to favor either AM or PM demodulation in the pursuit of minimal global BER. For example, by increasing the spacing between the outer eyes, the AM demodulation performance can be improved at the cost of a smaller inner eye height and thus worse PM demodulation performance. The measured BER versus symbol rate is shown in Fig. 7.80 for a 16-QAM input as well as a 4-APSK input (meaning only the I or Q channel contains modulated data). A 28 Gb/s (7 GBaud) 16-QAM signal was successfully demodulated with a BER below 10^{-6} (no errors were observed in practice, but the minimum measurable BER was limited by the memory of the oscilloscope). At 32 Gb/s (8 GBaud) a BER of $1.5 \cdot 10^{-3}$ was achieved, which is lower than the $4.4 \cdot 10^{-3}$ BER threshold needed for a CI-BCH forward error correcting (FEC) code to achieve a post-FEC BER $<10^{-15}$ with 7% overhead [59]. It is clear that the demodulator can process higher symbol rates in case only a single channel is modulated (4-APSK) instead of both channels (16-QAM), indicating that coupling between the I and Q channels leads to non-negligible degradation of the PAM-4 signals which affects the demodulation performance.

Figure 7.81 shows the demodulated AM and PM demodulated waveforms at 5 GBd, 7 GBd, and 8 GBd. It can be observed that the demodulated AM output eye contains slightly more jitter than the demodulated PM output. This can be mainly attributed to the performance of the rectifier which is the bottleneck of the demodulation performance. Furthermore, a significant degradation of the horizontal eye width is seen when increasing the symbol rate from 5 to 8 GBd. This degradation is caused by the more serious ISI which is converted into jitter by the comparators and drivers, as discussed earlier in this chapter.

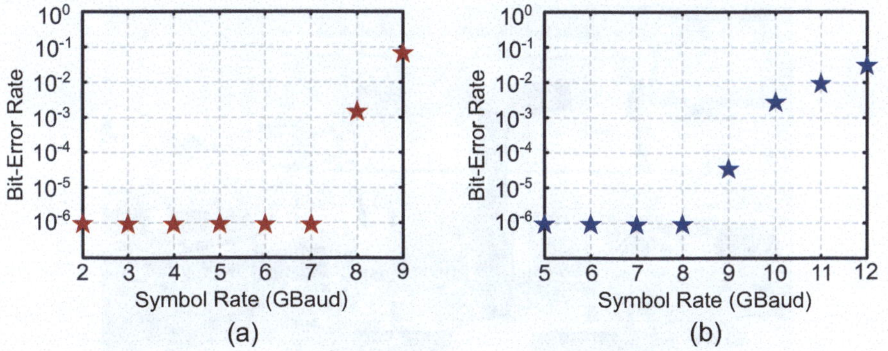

Fig. 7.80 Measured BER versus symbol rate of the 2 cm OTA link for 16-QAM (**a**) and 4-APSK (**b**) modulation

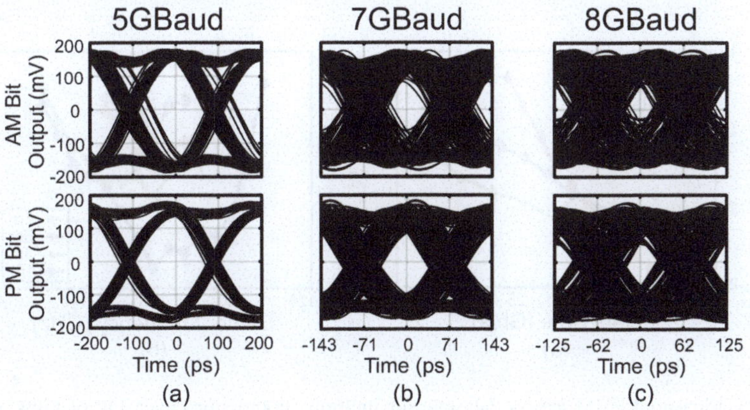

Fig. 7.81 Measured eye diagrams of the AM and PM demodulated NRZ outputs at 5 GBaud (**a**), 7 GBaud (**b**), and 8 GBaud (**c**) 16-QAM input

7.6.2.2 Dielectric Waveguide Link

Modulated link measurements were also performed using a DWG channel with various lengths. A picture of the real-life measurements setup for a 1 m DWG link is shown in Fig. 7.82. Again, measurements were first performed without the demodulation circuitry by reconstructing the constellations from the PAM-4 test outputs of the RX. Figure 7.83a shows the measured EVM versus data rate for various modulations at 1 m distance, achieving a maximum 16-QAM data rate of 32 Gb/s at 1 m distance with an equivalent BER below 10^{-3}. The 20 Gb/s 16-QAM EVM versus carrier frequency is depicted in Fig. 7.83b for multiple DWG lengths, showing robust performance over frequency. The measured EVM versus data rate and distance is summarized in Fig. 7.84a for 16-QAM and in Fig. 7.84b for QPSK transmission. Over 3 m and 5 m of DWG a maximum data rate of 32 Gb/s and

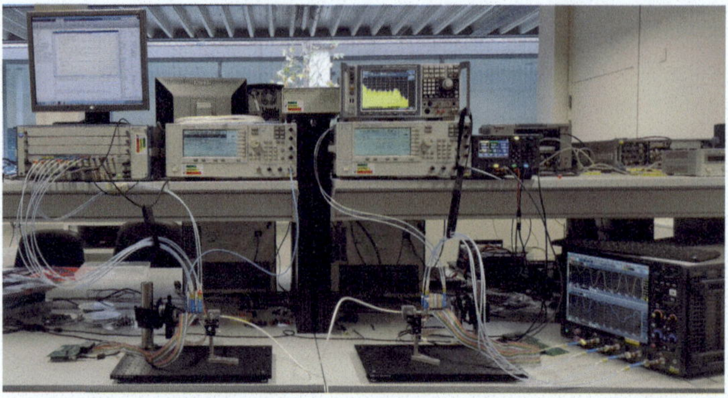

Fig. 7.82 Picture of the real-life measurement setup for DWG link measurements

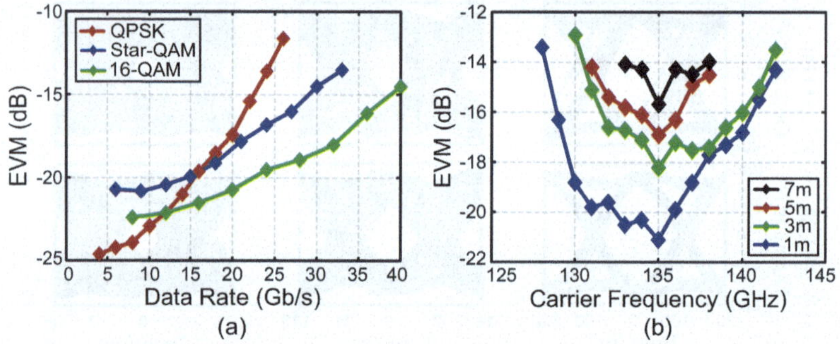

Fig. 7.83 Measured EVM versus data rate for multiple modulations over 1 m of DWG (**a**) and measured EVM over carrier frequency for 16-QAM transmission over various DWG lengths (**b**)

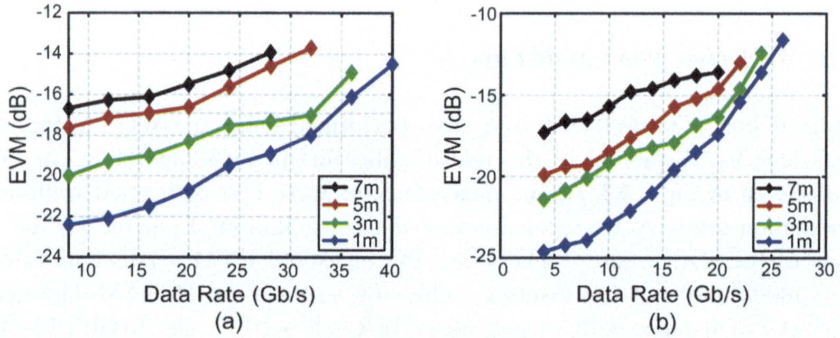

Fig. 7.84 Measured EVM versus data rate over various DWG lengths for 16-QAM (**a**) and QPSK (**b**) transmission

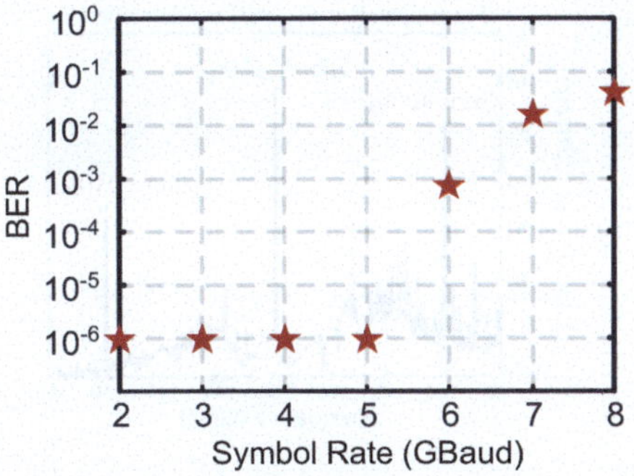

Fig. 7.85 Measured BER versus symbol rate of the 1 m DWG link for 16-QAM modulation

20 Gb/s is achieved with 16-QAM, respectively. The link distance can be increased further by switching to QPSK modulation, still achieving a data rate of 20 Gb/s at 7 m.

Furthermore, demodulation was also verified over a 1 m DWG link, as shown in Fig. 7.85. At 24 Gb/s 16-QAM, a BER below 10^{-3} was achieved, while a BER below 10^{-6} (zero errors observed) was demonstrated at 20 Gb/s.

7.6.3 Performance Limitations

Since the ambitious goal of 100 Gb/s has not quite been met, it is insightful to investigate the main performance limiting factors. Even though the data rate can be bound by a number of unpredictable factors, the reported link performance in this chapter is largely limited by the spurs in the LO chain of the TX. This subsection is thus split into one part analyzing the spur problems and one part discussing other nonspur related factors.

7.6.3.1 Spur Related Limitations

From standalone TX measurements, it was observed that the TX output spectrum contained a significantly larger spur at $8\,f_0$ than predicted from rigorous simulations (Fig. 7.27). Figure 7.86a shows the CW output spectrum of the TX, without de-embedding the downconversion gain of the employed mixer. Calibrated power measurements, plotted in Fig. 7.64a, showed that the output power of the $8\,f_0$ tone is even 2 dB larger than the desired tone at the output of the TX. This increased spur level is likely due to insufficient isolation of the fundamental tone (f_0) to the input of the second tripler leading to a larger intermodulation product at $8\,f_0$ relative to the desired tone at $9\,f_0$. Thanks to an additional LO buffer added between both triplers, this problem did not occur in the RX. Furthermore, this $8\,f_0$ LO spur is

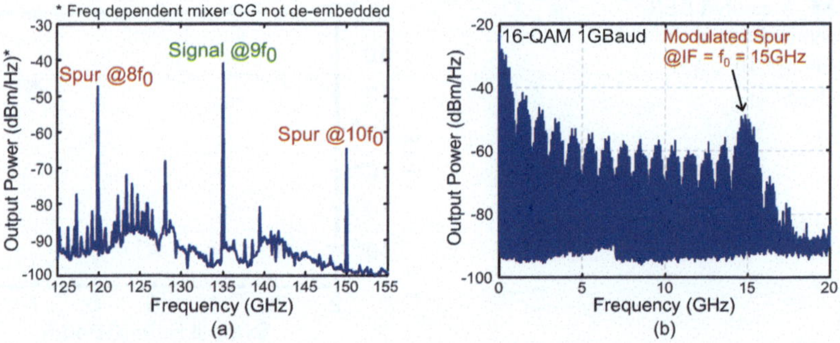

Fig. 7.86 Measured CW TX output spectrum (**a**) and measured downconverted 1 GBaud 16-QAM spectrum at a PAM-4 test output (**b**), both uncalibrated

poorly rejected by the filtering stages in the TX and RX due to their broadband frequency responses.

By adapting some of the TX/RX bias voltages, the spur levels could be reduced slightly. The bias of the frequency triplers in the TX was re-optimized to maximize third harmonic power generation relative to fundamental power generation, at the cost of a lower overall LO output power. Next, the bias of the buffers in the LO chain was lowered to shift the frequency response to higher frequencies, thus attenuating the $8f_0$ spur more relative to the desired $9f_0$. However, this again reduces the power gain and output power of the LO chain at 135 GHz. The same bias lowering technique was applied to the PA in the TX and LNA in the RX. Unfortunately, this not only decreases the power gain of these blocks, but also degrades the frequency response and so aggravates the ISI when amplifying a large bandwidth modulated signal. Decreasing the level of the spurs thus comes with an inevitable sacrifice of link SNR and ISI. Hence, unfortunately, the level of the spur could not be suppressed sufficiently without seriously degrading the link performance in several ways.

Firstly, this large spur causes the LO buffers to compress early, therefore limiting the maximum output power of the LO chain and consequently that of the transmitter. As the PA was not able to reach compression, the TX operated at maximum output power for the 16-QAM link measurements instead of at a power back-off. The decrease in TX output power (0 dBm compared to a simulated Psat around 10 dBm) results in significantly lower link SNR than predicted from simulations.

Second, the link performance is seriously impacted by the modulated spur in the TX output as this leads to a modulated IF signal at $9f_0 - 8f_0 = f_0 =$ 15 GHz after downconversion in the RX, as shown in Fig. 7.86b. This large IF tone severely degrades the sensitivity and demodulation speed of the PAM-4 decoders and cannot be filtered out on-chip. Furthermore, the signal integrity becomes corrupted as the spur inevitably overlaps with the desired baseband spectrum. As the symbol rate increases, this overlap becomes increasingly more problematic, even when measuring the PAM-4 test outputs as the spur becomes progressively more challenging to filter out. By plotting the EVM values from Fig. 7.77a versus symbol

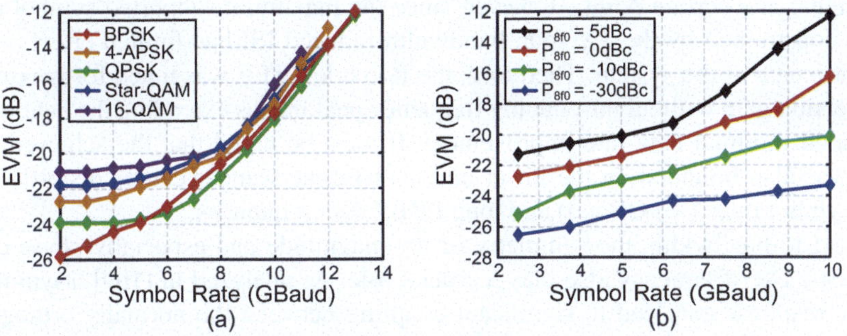

Fig. 7.87 Measured EVM values versus symbol rate of the 2 cm wireless link (**a**) and simulated TX 16-QAM EVM values versus symbol rate for various $8f_0$ spur levels (**b**)

rate, attaining Fig. 7.87a, it can be observed that the EVM for all modulations starts degrading rapidly from 7-8 GBaud onward, which is when the null-to-null (desired and spur) signal bands start to overlap. This overlap is also observed in the TX upconversion responses from Fig. 7.65, where the response steeply drops around 7.5 GHz ($f_0/2$) from the carrier, independent of the center frequency.

The effects of the $8f_0$ spur were also verified in full TX simulations. First, the measured output powers at 135 GHz and 120 GHz from Fig. 7.64a were reproduced by varying the amount of f_0 coupling to the second tripling stage in the LO chain. It was found that an $8f_0$ spur of 5 dBc relative to the desired tone at $9f_0$ at the output of the LO chain corresponded to the measured spur level at the output of the TX, while also predicting the observed 10 dB decrease in $9f_0$ TX output power. Second, modulated 16-QAM TX simulations were performed over symbol rate with various levels of the $8f_0$ spur at the LO chain output, including the estimated spur level in measurements (i.e., 5 dBc), as shown in Fig. 7.87b. While low spur levels show a small expected increase in EVM with symbol rate, high spur levels lead to a serious EVM degradation from 7-8 GBaud, similar to what was observed in modulated measurements.

This spur problem thus explains the large discrepancy in maximum data rate compared to simulations and can readily be solved in future iterations by adding an LO buffer between both triplers as in the RX LO chain where no problematic spurious tones were observed.

7.6.3.2 Non-spur Related Limitations

Even if, in an ideal world, the spurs in the TX were not an issue, various other factors would limit the obtainable data rate in the link, for example, caused by the designed circuits and prototype boards. The most obvious limit would be the bandwidth of the circuits, PCB antenna/coupler, flip-chip interface, and PCB transmission lines. As the symbol rate is increased to very high values, ISI would completely distort the signal at a certain point. However, since the maximum supported symbol rates were constrained by the TX spur, bandwidth-induced ISI had limited effect on the performance of the system. Therefore, the RX-side CTLE was tuned for maximum conversion gain without introducing frequency peaking, as extending the bandwidth provided minimal data rate improvement. It must be noted that the achieved data rate was also bounded on the lower end, as baseline wander causes distortion for low symbol rates (<1 GBaud) and long PRBS data sequences.

Next to bandwidth, the symmetry of the magnitude and especially phase characteristic over frequency also play a crucial role. As analyzed in [109], asymmetry in the response can lead to significant coupling between the normally orthogonal I and Q channels of the 16-QAM signal. In the designed system, all RF blocks have an inevitable asymmetry due to the significant gain roll-off of the transistors in the sub-THz range. Without explicitly compensating this roll-off, which is quite challenging and comes with other trade-offs, IQ coupling will cause serious signal distortion, especially at higher symbol rates. The degrading effect of IQ coupling is illustrated in the measurement results by the higher EVM values for 16-QAM compared to 4-APSK in Fig. 7.77a and the larger BER for 16-QAM compared to 4-APSK demodulation in Fig. 7.80.

Next to the circuit-level limitations, the wireless setup was limited by the specific antenna choice. Due to the large antenna size, the far-field distance of the antenna is quite large (about 5 cm). At 5 cm distance, the free-space path loss is already 49 dB, significantly limiting the link margin. Reducing the distance in order to increase the link SNR places both antennas in each other's near-field, where the antenna characteristics (radiation pattern, directivity, etc.) deviate from the designed far-field values. As a result, no close-proximity wireless measurements could be performed to assess the performance without SNR limitations. The best performance was achieved at 2 cm distance, but it is unclear how the antennas behave in this setup. In a real-life application, multiple antenna cells would be needed to obtain a link over a practical distance (i.e., more than a few centimeters) at sub-THz frequencies.

Lastly, specific elements in the DWG setup also limited the maximum achievable data rate over a guided channel. Over short DWG distances, the available link budget was significantly larger than in the wireless setup. However, even with a short DWG length, the coupler–DWG interface leads to small reflections along the coupler caused by the small, but nonzero return loss of the coupler. These reflections intensify when the coupler is misaligned with the waveguide and always lead to slight pulse distortions limiting the minimal achievable EVM compared to the wireless link despite the higher SNR. As the DWG length increases, waveguide dispersion becomes more prominent and problematic for higher symbol rates. Furthermore, due to practical considerations, longer waveguides have to be bent. Depending on the curvature, these bends can lead to additional dispersion and power loss. So although larger link distances can be achieved over a DWG compared to wireless transmission, the maximum achievable data rate at small distances is lower mainly due to the reflections of the coupler–DWG interface and dispersion of the DWG channel.

7.6.4 Performance Survey

When evaluating the reported results of this chapter, it is important to note that all reported data rates have been achieved without using any offline digital signal processing, such as predistortion, pulse shaping, or equalization, as would be the case in the ideal direct-digital modulation system as discussed in Sect. 7.1. The results thus include all impairments (e.g., limited bandwidth and dispersion) caused by the circuits, Vivaldi antennas, PCB traces, data cables, and the channel (especially the DWG).

Table 7.3 compares the presented wireless link to other state-of-the-art high-speed wireless sub-THz links in silicon-based technologies: TMTT 2022 [109], TMTT 2020 [156], IMS 2020 [160], CICC 2020 [162], MWCL 2019 [163], and ISSCC 2017 [47]. Table 7.4 compares the demonstrated DWG link to other state-of-the-art high-speed sub-THz silicon-based links over cladded dielectric waveguides reaching 1 m and above: SSCL 2022 [104], ESSCIRC 2021 [130], and JSSC 2019 [52]. The leftmost part of both tables shows the conventional architectures which (partly) operate with a multi-bit data interface and thus necessitate high-speed multi-bit DACs and/or ADCs to perform the (de)modulation. Furthermore, most works employ some form of DSP which further exacerbates the stringent specifications of the required data converters. This part of the table also includes the performance at the PAM-4 test outputs, which was demonstrated in this chapter. In this case, the consumption of the PAM-4 decoders, which are unused in this case, is excluded. In the absence of BER measurements, EVM values are shown approximately equivalent to a BER below 1e−3 [54] to keep the comparison reliable.

Compared to the conventional architectures, this chapter achieves significantly lower data rates. However, due to the challenges associated with the design of the data converters and the digital processing units, these works employ bulky and power-hungry arbitrary waveform generators and oscilloscopes to perform the data conversion and execute all of the signal processing offline. Moreover, equalization applied to the modulated results not only de-embeds channel and board impairments, but also inevitably compensates circuit nonidealities making it difficult to directly compare circuit performance independent of DSP efficacy. Finally, comparing the energy efficiency among these works is not straightforward as the power consumption of the external data converters and DSP is not included in the power budget of the system.

In order to more realistically compare the feasibility of the reported system and raw chip performance, the rightmost part of both tables only includes direct-digital (de)modulation architectures which employ a truly binary data interface without any offline DSP. Compared to these works, the presented system achieves better or similar data rates while supporting more spectrally efficient modulation schemes.

This chapter presents the first integrated direct-digital 16-QAM link achieving high-speed high-order modulation and demodulation with binary data interfaces. The link has demonstrated high-speed performance even without offline signal processing with a great energy efficiency thanks to the presented direct-digital

Table 7.3 Performance comparison with state-of-the-art high-speed sub-THz wireless links in silicon-based technologies

Architecture	Conventional architectures							Direct-digital architectures		
Reference	TMTT 2022 [13]	TMTT 2020 [35]	IMS 2020 [16]	CICC 2020 [5]	MWCL 2019 [20]	This work		ISSCC 2017 [24]	This work	
Technology	130 nm SiGe	130 nm SiGe	130 nm SiGe	28 nm CMOS	130 nm SiGe	28 nm CMOS		55 nm SiGe	28 nm CMOS	
Carrier frequency (GHz)	225	230	240	113	230	135		130	135	
Modulation scheme	16-QAM	DP-QPSK	16-QAM	16-QAM	16-QAM	16-QAM		OOK	16-QAM	
TX architecture	Direct-conversion	Direct-conversion	Direct-conversion	RF-DAC	Direct-Conversion	Direct-digital cartesian		Swtched PA	Direct-digital cartesian	
RX architecture	Direct-conversion	Direct-conversion	Direct-conversion	Direct-conversion	Direct-conversion	Direct-conversion		Envelope detector	Direct-conversion + PAM-4 decoders	
TX pout (dBm)	7	7.5	12^{c}	-2^{d}	5	0		9.3	0	
RX NF (dB)	17.5 (SSB)	18 (SSB)	15.5 $(DSB)^{c}$	–	14 (SSB)	6 (DSB)		–	6 (DSB)	
Distance (cm)	60	100	80	10	100	2	5	500	2	
Link interface	On-chip antenna + silicon lens	On-chip antenna + Silicon lens	On-chip antenna + HDPE lens	Off-chip antenna	On-chip antenna + silicon lens	Off-chip antenna		Off-chip antenna + 3D-printed lens	Off-chip antenna	
Data rate (Gb/s)	80	110	60	72	95	36	28	12.5	32	28

Bit-error rate/EVM	EVM = −17.7 dB	EVM = −9.9 dB	EVM = −16 dB	EVM = −16.1 dB	EVM = −16.4 dB	EVM = −18.4 dB	EVM = −16.9 dB	BER = 1e−6	BER = 1.5e−3	BER < 1e−6
Offline processing	TX: DPD + RRC PS	TX: DPD + RRC PS	TX: RRC PS	RX: Blind EQ	TX: RRC PS	None		None	None	
	RX: 51-tap FFE	RX: 61-tap FFE	RX: RRC PS + ch. EQ		RX: 17-tap FFE					
Data interface >2b @ fs ≥baudrate needed?	Yes	Yes	Yes	No (TX)	Yes	No (TX)		No	No	
				Yes (RX)		Yes (RX)				
Power consumption (mW)	2600[a,b]	2859[a,b]	2087[a,b]	470[a,b]	1410[a,b]	264[b]		38	287	
Link efficiency (pJ/bit)	32.5[a,b]	26[a,b]	34.7[a,b]	6.5[a,b]	14.8[a,b]	7.3[b]	9.4[b]	3.04	9	10.3

[a] Not including Pdc of offline DSP
[b] Not including Pdc of external multi-bit DACs and/or ADCs necessary for (de)modulation
[c] Simulated value
[d] Deduced from reported EIRP and antenna gain

Table 7.4 Performance comparison with state-of-the-art high-speed sub-THz DWG links in silicon-based technologies

Architecture	Conventional architectures							Direct-digital architectures							
Reference	SSCL 2022 [11]		ESSCIRC 2021 [12]		This work			ESSCIRC 2021 [12]			JSSC 2019 [30]			This work	
Technology	16nm FinFET		28 nm CMOS		28 nm CMOS			28 nm CMOS			28 nm CMOS			28 nm CMOS	
Carrier frequency (GHz)	109/135		135		135			135			140			135	
Modulation scheme	Dual-band 16-QAM		4-ASK		16-QAM			2-ASK			CP-FSK			16-QAM	
TX architecture	DB superheterodyne		Direct-conversion		Direct-digital cartesian			Direct-conversion			DCO			Direct-digital cartesian	
RX architecture	DB superheterodyne		Envelope detector		Direct-conversion			Envelope detector			Quadricorrelator			Direct-conversion + PAM-4 Decoders	
TX pout (dBm)	−5		−3		0			−3			6			0	
RX NF (dB)	–		–		6 (DSB)			–			–			6 (DSB)	
Distance (m)	3	4	3	5	1	3	5	1	4	6	1	2	4	1	
Link interface	Off-chip coupler		Off-chip coupler		Off-chip coupler			Off-chip coupler			Off-chip coupler			Off-chip coupler	
Data rate (Gb/s)	120	80	50	24	32	32	20	27	18	10	12	10	7	24	20
Bit-error rate/EVM	EVM = −16.9 dB	EVM = −16.4 dB	BER = 1e−6	BER = 1e−6	EVM = −18 dB	EVM = −17 dB	EVM = −16.6dB	BER = 1e −12	BER = 1e−12	BER = 1e−12	BER = 1e−12	BER = 1e−12	BER = 1e−12	BER = 1e−3	BER < 1e−6

Offline processing	RX: 2x2 oversampled 20-tap FFE		TX: DPD + RRC PS RX: nonlinear EQ		None			None			None			None	
Data interface > 2b @ fs ≥ baudrate needed?	Yes		Yes		No (TX) Yes (RX)			No			No			No	
Power consumption (mW)	1116[a,b,c]		130[a,b]		264[b]			130			230			287	
Link efficiency (pJ/bit)	9.3[a,b,b]	14[a,b,b]	2.6[a,b]	5.4[a,b]	8.25[b]	8.25[b]	13.2[b]	4.8	7.2	13	19.2	23	32.9	12	14.4

[a] Not including Pdc of offline DSP
[b] Not including Pdc of external multi-bit DACs and/or ADCs necessary for (de)modulation
[c] Not including 50-Ohm output buffers

architecture and circuit techniques. By addressing the primary limitations imposed by TX spurs, improving the spur rejection in the TX LO chain could facilitate data rates exceeding 50 Gb/s, as indicated by simulation results.

7.7 Conclusion

This chapter described the design of a 135 GHz direct-digital 16-QAM link, suitable for the next generation of wireless and wireline connectivity. First, the challenges of conventional architectures were discussed, and the direct-digital architecture was introduced as a possible solution. Next, the proposed 16-QAM direct-digital system was described with the most important system-level considerations. The subsequent sections provided a detailed explanation of the design of the transmitter circuits, receiver circuits, and prototype boards. Finally, standalone as well as modulated link measurement results were discussed and compared to the state-of-the-art.

The proposed Cartesian direct-digital modulation transmitter is capable of generating a low-EVM 16-QAM constellation without needing high-resolution DACs or offline predistortion. The presented direct-digital demodulation receiver performs 16-QAM demodulation in the analog domain after downconversion by means of unclocked PAM-4 decoders, obviating the need for high-resolution ADCs or any extensive DSP. The transmitter and receiver prototypes were implemented in 28 nm CMOS and linked over a wireless and dielectric waveguide channel via off-chip Vivaldi antennas. Data rates up to 32 Gb/s were demonstrated while performing high-order modulation and demodulation on-chip with a link efficiency of 9 pJ/b without any offline signal processing. The high-speed performance combined with the absence of any extensive DSP or external high-resolution data converters shows that the proposed system is a promising solution for future high-speed energy-efficient communication systems.

Chapter 8
Conclusion

This chapter aims to give an overview of the discussed topics in this book with the most important conclusions.

8.1 Overview and General Conclusions

The ever-expanding array of data-intensive applications in our interconnected world has led to an exponential surge in global data consumption. To meet the demands of this growing trend, the development of higher-capacity telecommunication systems has become essential. This thirst for higher data rates has led research into the realm of THz and sub-THz frequencies where bandwidth is abundantly available. Thanks to the recent advancements in CMOS technology, high-speed circuits at such high frequencies can be realized at large volumes. This thesis investigated new architectures and circuits at THz and sub-THz frequencies in CMOS for high-speed energy-efficient telecommunication.

Chapter 1 outlined the history and recent trends in telecommunication and introduced the THz and sub-THz part of the spectrum. While this frequency range provides a large bandwidth to support high-speed telecommunication, the high propagation loss and poor transceiver performance limit the effectiveness and range of high-frequency communication links. CMOS was chosen as the preferred IC technology for its integration potential and low-cost high-volume production capability, despite the sub-optimal performance at higher frequencies above 100 GHz. Based on the discussed benefits and challenges, the primary research goals of this chapter were outlined.

Chapter 2 covered the most important concepts necessary to analyze telecommunication systems. First, a concise mathematical groundwork was described, covering the basics of information theory, frequency-domain analysis, and digitization. Digital modulation was then introduced with a discussion of the most prevalent

C. D'heer, P. Reynaert, *THz and Sub-THz CMOS Electronics for High-Speed Telecommunication*, Analog Circuits and Signal Processing,
https://doi.org/10.1007/978-3-031-64439-9_8

modulation techniques and implementations. Finally, an extensive overview was given of the general framework of telecommunication systems along with the main performance affecting phenomena, including noise, nonlinearity, and bandlimiting filtering. The chapter concludes with a performance comparison of the different modulation techniques, revealing the trade-off between spectral efficiency, SNR requirements, and implementation complexity.

Chapter 3 detailed the fundamental physics and circuit theory that form the basis of high-frequency circuit design. Starting from Maxwell's equations and electromagnetism, the basics of circuit theory and finally transmission line theory were outlined. The most important active and passive electronic components were then introduced, describing their behavior at low and higher frequencies, as well as various implementation aspects. It was concluded that the performance of both the active and passive components generally degrades with increasing frequency. Furthermore, the scaling in advanced CMOS nodes does not seem to benefit the performance at frequencies above 100 GHz.

Chapter 4 combines the described components from the previous chapter into electronic circuits at high frequencies. The fundamentals of impedance matching were discussed along with stability verification methods within complex networks. The two primary building blocks for high-frequency amplifiers above 100 GHz in this chapter were then detailed: the neutralized pseudo-differential pair and transformer-based matching network. The pseudo-differential pair with capacitive neutralization achieves superior gain and noise performance above 100 GHz while ensuring unconditional stability. Transformer-based matching networks enable broadband impedance matching with a compact footprint and convenient circuit biasing. The behavior of both blocks was described along with a design methodology to optimize the performance above 100 GHz. Finally, using these building blocks, the design of several amplifier-based circuits was discussed.

Chapter 5 focused on the high-level design choices and performance trade-offs in mm-Wave communication systems. First, the primary degrees of freedom were discussed that determine the obtainable capacity: the number of channels, bandwidth, and available SNR. After a qualitative discussion, the link capacity was quantitatively modeled over frequency based on Shannon's capacity equation and published state-of-the-art transceiver performance. A single-cell wireless link was first described, including the most important system impairments, such as nonlinearity, phase noise, and inter-symbol interference. This model was then extended toward phased-array wireless links and DWG links. Finally, in addition to capacity, the energy efficiency of single-cell wireless and DWG links was also modeled and used to analyze various case studies. Based on these models, 16-QAM emerged as the optimal modulation scheme for speed and efficiency in most cases, providing a good trade-off between spectral efficiency and SNR requirements. Furthermore, while wireless links are constrained by noise, DWG links are mostly limited by dispersion. The system-level conclusions from this chapter can help make better architectural and circuit-level design decisions and point research efforts toward the topics that are most worthwhile.

Chapter 6 detailed the design of a tripler-last 390 GHz BPOOK transmitter in 28 nm CMOS. First, the most effective THz transmitter architectures above f_{max} for high-speed communication were discussed, concluding that a tripler-last transmitter can provide a large output power while supporting various modulation schemes. Then, after studying the rather unconventional BPOOK modulation scheme, the design of the individual circuit blocks was described. Finally, the continuous wave and modulated measurement results of the 390 GHz transmitter were discussed, and the important performance limitations were identified. A maximum data rate of 28 Gb/s was achieved using BPOOK modulation, while 18 Gb/s was obtained with OOK and BPSK modulation. Despite the high data rate, it was concluded that a lower carrier frequency may improve the link performance, as the large available bandwidth at the THz carrier cannot be adequately exploited due to the need for a nonlinear output stage to generate an above-f_{max} signal.

Chapter 7 presented a complete description of a 135 GHz direct-digital 16-QAM wireless and DWG link in 28 nm CMOS. First, the challenges of conventional transceiver architectures were discussed, concluding that the data converter and DSP blocks largely limit the efficiency and performance of the system. To circumvent these bottlenecks, the direct-digital system architecture was introduced, which obviates the need for high-speed multi-bit data converters and complex DSP by performing high-order modulation and demodulation directly in the RF/analog domain. The proposed direct-digital 16-QAM system was then described along with the system-level specifications. The design of each circuit block in the transmitter and receiver, as well as the prototype boards with antenna structures, was extensively detailed. Finally, standalone as well as modulated link measurement results were discussed and compared to the state-of-the-art. Data rates up to 32 Gb/s were demonstrated while performing high-order modulation and demodulation on-chip with a link efficiency of 9 pJ/b without any offline signal processing. The high achieved data rate combined with the absence of any extensive DSP or external high-resolution data converters shows that the proposed direct-digital system is a promising solution for high-speed energy-efficient communication at sub-THz frequencies.

8.2 Afterword

As we conclude our in-depth exploration of THz and sub-THz CMOS electronics, it is evident that we are on the cusp of a transformative era in high-speed telecommunication. The road ahead is fraught with challenges, yet these are dwarfed by the immense potential they hold. This book aimed not only to contribute to the advancement of integrated telecommunication but primarily to lay the groundwork for the revolutionary high-speed transceivers of tomorrow. It is my sincere hope that this work has ignited your passion to innovate. Good luck and may you accelerate global connectivity like never before!

Appendix A
Mathematical Derivations

A.1 Derivation of Bit-Error Rate Expressions

This section gives more reasoning and explanation for the bit-error rate expressions given in Sect. 2.3.2. The derivation and reasoning are mostly based on the work in [53].

First of all, as it is more convenient to reason in terms of symbol/bit energy E_s, E_b and noise density N_0 instead of received signal power P_r, the SNR can be rewritten as follows:

$$\mathrm{SNR} = \frac{P_r}{N_0 B} = \frac{E_s}{N_0 B T_s} = \frac{E_b}{N_0 B T_b} . \tag{A.1}$$

E_s/N_0 is thus equivalent to SNR if the signal bandwidth is equal to $B = 1/T_s$. In some texts, B is set to $2/T_s$ as would be the case for passband signaling with a rectangular pulse shape, and then the noise PSD is often defined as $N_0/2$, yielding the same relation. Also, in case of binary signaling, E_b/N_0 is also equivalent to the SNR.

Furthermore, assuming the symbol energy is equally divided among all bits, we can write:

$$E_b = \frac{E_s}{\log_2(M)} , \tag{A.2}$$

where the number of points in the constellation is equal to M. Furthermore, assuming Gray coding is used, one symbol error leads to exactly one bit error (at reasonable SNRs), leading to the following relation between bit error P_b and symbol error P_s:

C. D'heer, P. Reynaert, *THz and Sub-THz CMOS Electronics for High-Speed Telecommunication*, Analog Circuits and Signal Processing,
https://doi.org/10.1007/978-3-031-64439-9

$$P_b = \frac{P_s}{\log_2(M)} . \tag{A.3}$$

Knowing this, the expression for the BER of a BPSK signal can be expressed as [53]

$$P_{b,\text{BPSK}} = Q\left(\sqrt{\frac{2E_b}{N_0}}\right) , \tag{A.4}$$

which is identical to the previously derived equation 2.69 for binary signaling, as expected.

QPSK modulation consists of BPSK modulation on both orthogonal components of the signal and ideally does not influence each other. Hence, the bit-error rate on both branches is the same as for BPSK. However, compared to BPSK, the total symbol energy is divided among the symbols: $E_s = 2E_b$. Furthermore, compared to BPSK, each symbol now has two neighboring symbols at the same minimum distance[1] and hence has twice the probability of an error. With these observations, the symbol-error rate (SER) can be expressed as

$$P_{s,\text{QPSK}} = 2\,Q\left(\sqrt{\frac{2E_b}{N_0}}\right) = 2\,Q\left(\sqrt{\frac{E_s}{N_0}}\right) , \tag{A.5}$$

which, combined with the assumption that $P_b = P_s/2$, leads to the expression of the BER for a QPSK signal as a function of the energy per symbol (which is more convenient than the energy per bit for multi-bit modulations):

$$P_{b,\text{QPSK}} = Q\left(\sqrt{\frac{E_s}{N_0}}\right) . \tag{A.6}$$

Compared to BPSK, QPSK suffers from a reduction in SNR by a factor of $\sqrt{2}$ due to the spread in signal power over the two carriers.

This reasoning can be extended to general M-PAM and M-QAM modulations [53, 54]. First, an M-PAM signal, consisting of M different signal levels over a single dimension, has a BER that can be expressed in function of the SNR per symbol E_s/N_0:

$$P_{b,M\text{-PAM}} = \frac{2(1-1/M)}{\log_2(M)}\,Q\left(\sqrt{\frac{6}{M^2-1}\frac{E_s}{N_0}}\right) , \tag{A.7}$$

[1] Other neighbors at a larger distance contribute very marginally to the error rate and thus be neglected.

which is a generalization of the BER expression for BPSK. In front of the Q-function, the numerator equals the average number of neighbors at minimum distance, and the denominator takes care of the conversion from symbol-error rate to bit-error rate. Within the Q-function, the factor in front of E_s/N_0 denotes the division of total symbol energy among the different symbols.

Finally, let us consider M-QAM modulations with a square constellation of size $M = L^2$. Such a signal can be viewed as two M-PAM signals with constellations of size L transmitted over the two quadrature signal components, with each half the energy of the original M-QAM signal. To find the BER expression for M-QAM signals starting from Eq. A.7, replace the number of levels per dimension M by $L = \sqrt{M}$ and reduce the SNR by a factor of two (since the total energy is now spread over two carriers), arriving at the following expression:

$$P_{b,M\text{-QAM}} = \frac{4(1-1/\sqrt{M})}{\log_2(M)} Q\left(\sqrt{\frac{3}{M-1}\frac{E_s}{N_0}}\right). \tag{A.8}$$

The different factors in the expression have the same meaning as for the M-PAM expression. For instance, for 16-QAM, the factor $4(1-1/\sqrt{M})$ equals three, which is the average number of neighbors of the symbols in the constellation at minimum distance.[2] The factor $\log_2(M) = 4$ bit illustrates that each symbol error leads, in certain situations, to only one bit error thus making the BER a factor of 4 smaller than the SER. Lastly, the factor $3/(M-1) = 1/5$ describes how the symbol energy is spread over the 16 symbols modulated onto two carriers.

Finally, to make calculations more convenient, a general expression for the bit error can be found, where L is the number of levels per dimension and M the total number of constellation points:

$$P_b = \frac{2(1-1/L)}{\log_2(L)} Q\left(\sqrt{\frac{\log_2(L)}{\log_2(M)}\frac{6}{L^2-1}\frac{E_s}{N_0}}\right), \tag{A.9}$$

which is equivalent to Eq. 2.72 as an explicit function of SNR. For M-PAM, $L = M$, while for rectangular M-QAM, $L = \sqrt{M}$. In the general expression, the factor two in front of E_s/N_0 is taken care of by the factor $\log_2(L)/\log_2(M)$, which is equal to 1 or 1/2 in case of M-PAM or M-QAM, respectively.

[2] This can be quickly verified by noticing that the corner points have two neighbors, the other edge points three, and all inner points four. Hence: $(4\cdot 2 + 8\cdot 3 + 4\cdot 4)/16 = 3$.

A.2 Intermodulation Distortion

Assuming an input $x(t) = A_1 \cos \omega_1 t + A_2 \cos \omega_2 t$, intermodulation can be described by analyzing the response of a nonlinear system $(a_1 x + a_2 x^2 + a_3 x^3)$ to this input:

$$y(t) = a_1 (A_1 \cos \omega_1 t + A_2 \cos \omega_2 t) + a_2 (A_1 \cos \omega_1 t + A_2 \cos \omega_2 t)^2 \tag{A.10}$$

$$+ a_3 (A_1 \cos \omega_1 t + A_2 \cos \omega_2 t)^3 . \tag{A.11}$$

First, by expanding this equation the fundamental components can be found [60]:

$$y(t)|_{\omega=\omega_1,\omega_2} = \left(a_1 A_1 + \frac{3}{4} a_3 A_1^3 + \frac{3}{2} a_3 A_1 A_2^2 \right) \cos \omega_1 t \tag{A.12}$$

$$+ \left(a_2 A_2 + \frac{3}{4} a_3 A_2^3 + \frac{3}{2} a_3 A_2 A_1^2 \right) \cos \omega_2 t , \tag{A.13}$$

where gain compression can occur in both components caused by either components, if $a_1 \cdot a_3 < 0$. Next, by discarding the DC terms, harmonics, and components at $\omega_1 \pm \omega_2$ (as they both fall outside of the band of interest), the intermodulation products can be determined [60]:

$$y(t)|_{\omega=2\omega_1 \pm \omega_2} = \frac{3 a_3 A_1^2 A_2}{4} \cos(2\omega_1 + \omega_2)t + \frac{3 a_3 A_1^2 A_2}{4} \cos(2\omega_1 - \omega_2)t \tag{A.14}$$

$$y(t)|_{\omega=2\omega_2 \pm \omega_1} = \frac{3 a_3 A_1 A_2^2}{4} \cos(2\omega_2 + \omega_1)t + \frac{3 a_3 A_1 A_2^2}{4} \cos(2\omega_2 - \omega_1)t . \tag{A.15}$$

The amplitudes of the different components can then be used to calculate the relative intermodulation strength and third intercept point in Sect. 2.3.3.3.

A.3 Analysis of Duobinary Encoding

This section provides a derivation of the duobinary encoding scheme and its characteristics. Duobinary encoding is employed to generate a BPOOK signal which has half the bandwidth of a OOK or BPSK signal modulated at the same speed, as described in Chap. 6. Since BPOOK is the passband equivalent of duobinary signaling, the analysis will be performed on the baseband version with all advantages (particularly the bandwidth compression) extending to BPOOK signals as well.

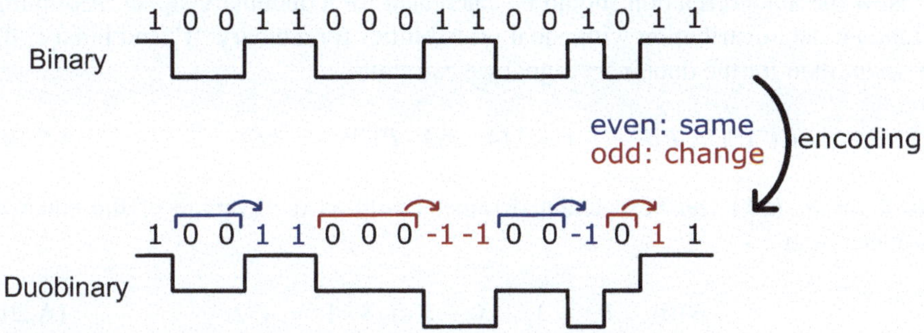

Fig. A.1 Visual representation of the duobinary encoding scheme

First it should be understood how binary data can be encoded into duobinary data, as originally described by Lender in 1963 [144]. Let a_n be the nth symbol of the binary data and b_n the nth symbol of the corresponding duobinary data. A binary "0" will always lead to a duobinary "0," so $b_n = 0$ if $a_n = 0$. Then suppose k zeros are present between a_n and a_{n+k}. If k is even, $b_{n+k} = b_n$. If k is odd, $b_{n+k} = -b_n$. So the sign in the duobinary data changes when an odd number of zeros precede a nonzero value. This encoding schema is visualized in Fig. A.1.

Now the question is how this encoding results in bandwidth reduction. For this the notion of power spectral density (PSD) should be introduced. The PSD describes the normalized power of a waveform versus frequency and can be expressed as follows [41]:

$$\mathrm{PSD}(f) = \frac{|F(f)|^2}{T} \sum_{-\infty}^{\infty} R(k) e^{j2\pi k f T}, \tag{A.16}$$

where $F(f)$ is the spectrum of the pulse shape, $R(k)$ is the autocorrelation of the data, and $T = 1/R$ is the bit period. If a rectangular pulse shape is used, $|F(f)|$ becomes

$$|F(f)| = \left(\frac{\sin(\pi f T)}{\pi f T}\right)^2. \tag{A.17}$$

The autocorrelation can be expressed as [41]

$$R(k) = \sum_{i=1}^{I} (a_n \cdot a_{n+k})_i \cdot P_i, \tag{A.18}$$

where a_n is the amplitude of the nth symbol and P_i the probability that combination i occurs.

Now the autocorrelation should be calculated for a duobinary signal. Assuming a random data distribution with equal probabilities for a binary “1” and binary “0” to occur, then for the duobinary signal we can state:

$$P("1") = P("-1") = 1/4 \quad \text{and} \quad P("0") = 1/2. \tag{A.19}$$

For $k = 0$, $R(k)$ can be calculated rather easily as the average of the squared duobinary data:

$$R(0) = \overline{b_n^2} = 1/2 \cdot 0 + 2 \cdot 1/4 \cdot 1^2 = 1/2. \tag{A.20}$$

For $k > 0$, it becomes more complicated. It can already be noticed that only the terms without a 0 in the product will contribute to $R(k)$. As a result, only patterns of $k+1$ symbols that start and end with “1” or “−1” must be looked at. Or, looking at the binary stream, only sequences starting and ending with “1” are relevant. For each sequence of $k+1$ symbols there are 2^{k-1} possibilities[3] each with a probability of $(1/2)^{k+1}$ since a “1” and “0” have equal probabilities in the binary stream. Furthermore, the value of $(b_n \cdot b_{n+k})$ is either 1 or −1 each with an equal probability of 1/4. So $R(k)$ can be written as

$$R(k) = \Delta \cdot \frac{1}{2^{k+1}} \cdot 1, \tag{A.21}$$

where Δ is the number of patterns (a_n, a_{n+k}) with $(b_n \cdot b_{n+k}) > 1$ minus the number of patterns with $(b_n \cdot b_{n+k}) < 1$. But how big is Δ?

If $k = 1$, $\Delta = 1$ because there is only one possible pattern (1,1) which always results in $(b_n \cdot b_{n+k}) > 1$.

For $k > 1$, Δ will be calculated by induction. For $k + 1$, the sequences of $k + 2$ symbols are created by adding a “0” or a “1” after the first “1.” If a “0” is added, the sign of $(b_n \cdot b_{n+k})$ is changed, and in case a “1” is added, the sign remains unchanged. Thus, $\Delta = 0$ for every $k > 1$.

In conclusion, $R(0) = 1/2$, $R(1) = 1/4 = R(-1)$, and $R(k > 1) = 0$. The power spectral density of a duobinary signal can thus be written as

$$\text{PSD}_{\text{duobin}}(f) = \frac{1}{T} \, |F(f)|^2 \, [1/2 + 2 \cdot 1/4 \cdot e^{\pm j2\pi fT}] \tag{A.22}$$

$$= \frac{1}{2T} \, |F(f)|^2 \, [1 + 1/2 \cdot e^{j2\pi fT} + 1/2 \cdot e^{-j2\pi fT}] \tag{A.23}$$

$$= \frac{1}{2T} \, |F(f)|^2 \, [1 + \cos(2\pi fT)]. \tag{A.24}$$

[3] Looking at the binary stream, this can be understood as a permutation of $k + 1$ binary numbers with 2 numbers fixed, so actually a permutation of $k - 1$ binary symbols.

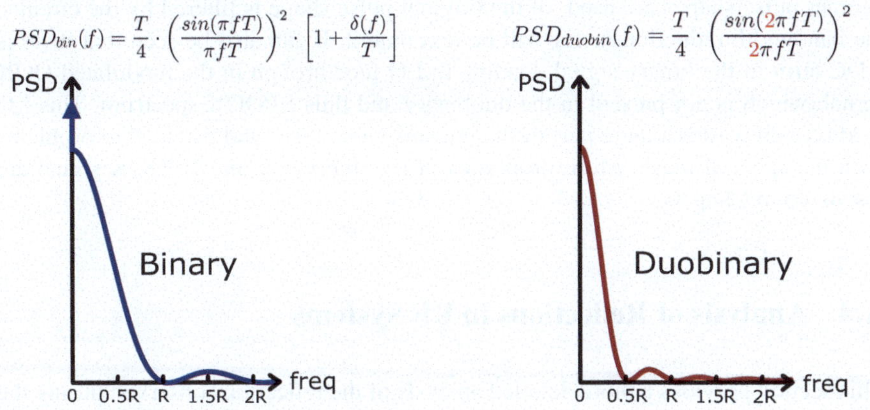

Fig. A.2 Comparison of the spectra of binary and duobinary data

Using trigonometry,[4] the sinc function of the pulse shape and cosine term can be combined into

$$\mathrm{PSD}_{\mathrm{duobin}}(f) = T \cdot \left(\frac{\sin(2\pi f T)}{2\pi f T}\right)^2. \tag{A.25}$$

Note that the amplitude of the signal is chosen to be ± 1, so NOT normalized for average power of unity. To get the PSD equal to that of a binary sequence at DC, the amplitude must be scaled by a factor of 1/2, yielding the following PSD:

$$\mathrm{PSD}_{\mathrm{duobin}}(f) = \frac{T}{4} \cdot \left(\frac{\sin(2\pi f T)}{2\pi f T}\right)^2. \tag{A.26}$$

For NRZ (non-return to zero) binary signaling, the PSD can be derived to be [41]

$$\mathrm{PSD}_{\mathrm{bin}}(f) = \frac{T}{4} \cdot \left(\frac{\sin(\pi f T)}{\pi f T}\right)^2 \cdot \left[1 + \frac{\delta(f)}{T}\right]. \tag{A.27}$$

Figure A.2 shows the PSD of the calculated binary and duobinary baseband signals. The PSD of the OOK and BPOOK signals can be derived by translating the PSD of the binary and duobinary baseband signal to the carrier frequency, leading to similar conclusions.

The first clear observation is the null-to-null bandwidth reduction by a factor of 2 for duobinary signaling compared to binary signaling, for the same bit rate R. It is important to note that this result is only valid in case of a rectangular pulse shape.

[4] Using the following property: $\frac{\sin(\pi f T)^2}{(\pi f T)^2} \cdot (1 + \cos(2 \cdot \pi f T)) = \frac{\sin(2 \cdot \pi f T)^2}{2 \cdot (\pi f T)^2}$.

If other pulse shapes are used, of the original pulse shape is filtered by the circuits, the bandwidth reduction factor will be less than 2. It can also be seen that there is a DC term in the binary signal, leading to LO feedthrough in the modulated OOK signal, which is not present in the duobinary and thus BPOOK spectrum. This LO feedthrough wastes a large part of the transmission power and can lead to problems with the spectral mask. More properties of the duobinary and BPOOK signal are discussed in Chap. 6.

A.4 Analysis of Reflections in RF Systems

This section provides a more detailed analysis of the reflections observed during the standalone measurements of the 135 GHz TX and RX discussed in Chap. 7.

Figure A.3a depicts the situation in which the reflections were first observed. A mixer with a large return loss is connected to a well-matched GSG probe over a long, low-loss waveguide channel. A simplified schematic, shown in Fig. A.3b, is constructed to model the effect of these reflections on a modulated waveform. The mixer is modeled as a voltage source transmitting a modulated signal over the channel with reflection coefficient Γ_S toward the GSG probe, modeled as a load with reflection coefficient Γ_L. The channel can be modeled as a (lossless) transmission line with characteristic impedance Z_0 and propagation constant β.

When a wave, V_{in}, is transmitted by the source, it propagates through the channel until it arrives at the load. At that point, part of the wave is reflected, $V_{\text{in}}\Gamma_L$, and the remaining part is absorbed in the load. This reflected wave propagates back and is again partly reflected at the source, $V_{\text{in}}\Gamma_L\Gamma_S$, and partly absorbed in the source impedance. This process is visualized in Fig. A.4a. An alternate way of viewing the progress of a wave propagating in time and position along a channel is with a bounce diagram [65], as shown in Fig. A.4b. It can also show the contribution of all components to the signal at a certain position and time. This bounce diagram will

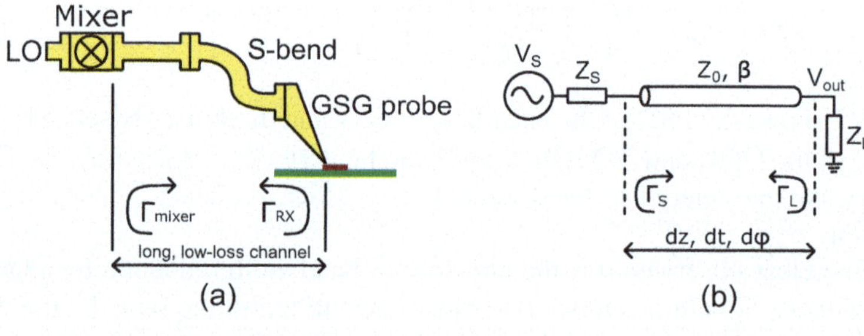

Fig. A.3 Configuration where reflections occurred during measurements (**a**) and simplified diagram to model the reflections (**b**)

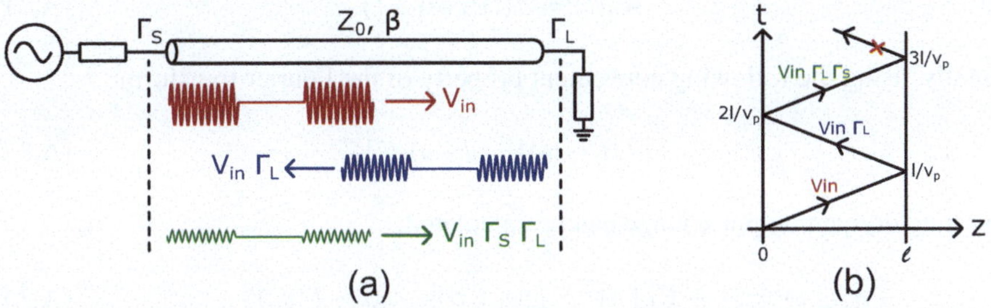

Fig. A.4 Depiction of the forward and reflected waves propagating along the channel (**a**) and bounce diagram to visualize the wave propagation in time and position (**b**)

now be used to express the evolution of the output signal $V(t, z = l)$ through time in the presence of reflections.

Initially, at $t = 0$, no wave has arrived at the output so the output equals zero. At $t = l/v_p$, the forward traveling wave arrives at the load and gets reflected back, both leading to a certain output voltage. At $t = 3l/v_p$, the initially reflected wave has traveled all the way to the source, is reflected back to the load, and thus contributes to the output signal. In this analysis only two reflections are considered to avoid overly complicated expressions, but all insights can easily be extended to situations with more reflections. Let us now try to express this more mathematically.

Assuming the forward traveling wave coming from the source is equal to V_{in} at the input of the transmission line, the output signal is composed of three components:

$$\begin{aligned} V_{\text{out}}(\omega) &= V_{\text{in}}(\omega)\, e^{-j\beta l} \\ &+ V_{\text{in}}(\omega)\, e^{-j\beta l}\, \Gamma_L \\ &+ V_{\text{in}}(\omega)\, e^{-3j\beta l}\, \Gamma_L\, \Gamma_S, \end{aligned} \tag{A.28}$$

which can be simplified to

$$V_{\text{out}}(\omega) = V_{\text{in}}(\omega) e^{-j\beta l}\, [1 + \Gamma_L + e^{-2j\beta l}\Gamma_L\Gamma_S]. \tag{A.29}$$

To convert this frequency-domain expression to the time domain, an inverse Fourier transform must be applied. For this, we first observe that the propagation delay over the transmission line can be expressed as

$$dt = \frac{d\phi}{\omega} = \frac{\beta l}{\omega} = \frac{l}{v_p}. \tag{A.30}$$

Furthermore, we consider an input signal which only has an in-phase component and can thus be written as

$$V_{\text{in}}(t) = d(t)\ \cos(\omega_c t). \tag{A.31}$$

Finally, using the following convenient property of the Fourier transform:

$$\mathcal{F}\{f(t - t_0)\} = F(\omega)e^{-j\omega t_0}, \tag{A.32}$$

the time-domain output voltage can be calculated:

$$\begin{aligned} V_{\text{out}}(t) = [1 + \Gamma_L]\ d(t - l/v_p)\ \cos(\omega_c(t - l/v_p)) \\ + \Gamma_L\Gamma_S\ d(t - 3l/v_p)\ \cos(\omega_c(t - 3l/v_p)). \end{aligned} \tag{A.33}$$

This passband output signal can then be downconverted coherently with the forward traveling wave (by multiplying with 2 $\cos[\omega_c(t - l/v_p)]$) and low-pass filtered to extract the baseband outputs:

$$I_{\text{out}}(t) = [1 + \Gamma_L]\ d(t - l/v_p) + \Gamma_L\Gamma_S\ d(t - 3lv_p)\ \cos(\omega_c 2l/v_p) \tag{A.34}$$

$$Q_{\text{out}}(t) = -\Gamma_L\Gamma_S\ d(t - 3l/v_p)\ \sin(\omega_c 2lv_p). \tag{A.35}$$

The first term of Eq. A.34 is the desired output, delayed by the expected propagation delay and attenuated (or amplified) by Γ_L. The second term of Eq. A.34 is the reflected signal which arrives two propagation delays later and distorts the original signal. As shown in Eq. A.35, the reflected wave can couple to the other channel, possibly distorting the data modulated on the quadrature component of the transmitted wave.

The nature of the distortion depends on the relation of the propagation delay, dt, compared to the carrier period, T_C, and symbol period, T_S. Three different situations are shown in Fig. A.5 for generic OOK modulated signal with $T_C \ll T_S$ and various propagation delays.

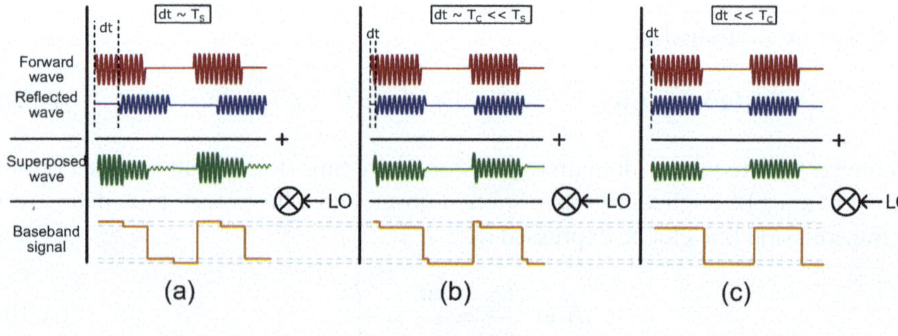

Fig. A.5 Three different situations where a reflected wave can cause signal distortions: a large propagation delay similar to the symbol period (**a**), a propagation delay much smaller than the symbol period but similar to the carrier period (**b**), and a propagation delay much smaller than the carrier period (**c**)

Figure A.5a shows the situation where the propagation delay is a significant part of the symbol period. In this case, the transition in the output signal is clearly visible and thus distorts the signal pulse. Note that if the propagation delay is larger than the symbol period, the reflected wave will distort the subsequent symbol and thus introduce significant inter-symbol interference.

If the propagation delay is much smaller than the symbol period, but larger than the carrier period (Fig. A.5b), the transition caused by the reflected wave happens very quickly. This might cause less visual distortion of the pulse but can still seriously degrade the resulting signal amplitude depending on the exact value of the delay.

Finally, if the propagation delay is extremely small, distinct transitions during the symbol period are not visible. All reflections happen instantaneously and accumulate to determine the resulting voltage. This situation represents, for example, on-chip matching networks where the physical distance is negligible compared to the wavelength of the signal. As reflections will not arise, the reflection coefficient can be optimized for other metrics, such as maximal power gain. Indeed, conjugate matching often leads to nonzero reflection coefficients, which is not problematic in terms of reflections, as proven by the executed analysis.

References

1. Etymology of Telecommunication. https://www.etymonline.com/word/telecommunication. Accessed: April 2023
2. Wikipedia: Telecommunications. https://en.wikipedia.org/wiki/Telecommunications. Accessed: April 2023
3. 1956 Nobel Prize in Physics. https://www.bell-labs.com/about/awards/1956-nobel-prize-physics/. Accessed: April 2023
4. G. Moore, Cramming more components onto integrated circuits. Electron. Mag. **38**, 114–117 (1965)
5. Moore's law: The number of transistors per microprocessor. https://ourworldindata.org/technological-change. Accessed: April 2023
6. S. Cherry, Edholm's Law of bandwidth. IEEE Spect. **41**(7), 58–60 (2004)
7. T.H. Lee, *The Design of CMOS Radio-Frequency Integrated Circuits*, 2nd edn. (Cambridge University Press, Cambridge, 2003)
8. S. Mirabbasi, L.C. Fujino, K.C. Smith, Through the looking glass-the 2022 edition: trends in solid-state circuits from ISSCC. IEEE Solid-State Circ. Mag. **14**(1), 54–72 (2022)
9. E. Khorov, I. Levitsky, I.F. Akyildiz, Current status and directions of IEEE 802.11be, the future Wi-Fi 7. IEEE Access **8**, 88664–88688 (2020)
10. Wikipedia: Comparison of mobile phone standards
11. IEA: Data Centres and Data Transmission Networks. https://www.iea.org/reports/data-centres-and-data-transmission-networks. Accessed: April 2023
12. Worldwide IDC Global DataSphere Forecast. https://www.idc.com/getdoc.jsp?containerId=US49018922. Accessed: April 2023
13. C.E. Shannon, A mathematical theory of communication. Bell Syst. Tech. J. **27**(3), 379–423 (1948)
14. The ITU radio regulations in 2016. https://www.itu.int/pub/R-REG-RR-2016. Accessed: August 2023
15. ETSI guidelines on millimetre Wave Transmission in the W-band. https://www.etsi.org/deliver/etsi_gr/mWT/001_099/018/01.01.01_60/gr_mWT018v010101p.pdf. Accessed: August 2023
16. ETSI: millimetre Wave Transmission (mWT); Analysis of Spectrum, License Schemes and Network Scenarios in the D-band. https://www.etsi.org/deliver/etsi_gr/mWT/001_099/008/01.01.01_60/gr_mWT008v010101p.pdf. Accessed: August 2023

C. D'heer, P. Reynaert, *THz and Sub-THz CMOS Electronics for High-Speed Telecommunication*, Analog Circuits and Signal Processing,
https://doi.org/10.1007/978-3-031-64439-9

17. ECC Recommendation on radio frequency channel/block arrangements for Fixed Service systems operating in the the W-band. https://docdb.cept.org/document/6037. Accessed: August 2023
18. ECC Recommendation on radio frequency channel/block arrangements for Fixed Service systems operating in the the D-band. https://docdb.cept.org/document/2012. Accessed: August 2023
19. IEEE Standard for High Data Rate Wireless Multi-Media Networks–Amendment 2: 100 Gb/s Wireless Switched Point-to-Point Physical Layer. *IEEE Std 802.15.3d-2017 (Amendment to IEEE Std 802.15.3-2016 as amended by IEEE Std 802.15.3e-2017)* (2017), pp. 1–55
20. International Telecommunication Union (ITU) Recommendation: Attenuation by atmospheric gases. https://www.itu.int/dms_pubrec/itu-r/rec/p/R-REC-P.676-9-201202-S!!PDF-E.pdf. Accessed: April 2023
21. Power Amplifiers Performance Survey 2000-Present. https://ideas.ethz.ch/Surveys/pa-survey.html. Accessed: April 2023
22. Low-Noise Amplifiers Performance Survey 2018-Present. https://ideas.ethz.ch/Surveys/lna-survey.html. Accessed: April 2023
23. V. Petrov, A. Pyattaev, D. Moltchanov, Y. Koucheryavy, Terahertz band communications: Applications, research challenges, and standardization activities, in *2016 8th International Congress on Ultra Modern Telecommunications and Control Systems and Workshops (ICUMT)* (2016), pp. 183–190
24. Apple Watch Series 7 Supports 60.5GHz Wireless Data Transfer. https://www.macrumors.com/2021/09/23/apple-watch-60-5-ghz-wireless-data-transfer/. Accessed: April 2023
25. H. Elayan, O. Amin, R.M. Shubair, M.-S. Alouini, Terahertz communication: The opportunities of wireless technology beyond 5G, in *2018 International Conference on Advanced Communication Technologies and Networking (CommNet)* (2018), pp. 1–5
26. V. Petrov, D. Moltchanov, Y. Koucheryavy, Applicability assessment of terahertz information showers for next-generation wireless networks, in *2016 IEEE International Conference on Communications (ICC)* (2016), pp. 1–7
27. W. Tong, P. Zhu, *6G: The Next Horizon: From Connected People and Things to Connected Intelligence* (Cambridge University Press, Cambridge, 2021)
28. Microwave Backhaul Evolution - Reaching Beyond 100GHz. https://www.ericsson.com/en/reports-and-papers/ericsson-technology-review/articles/microwave-backhaul-evolution-reaching-beyond-100ghz. Accessed: April 2023
29. M. Civas, O.B. Akan, Terahertz wireless communications in space. ITU J. Future Evol. Technol. **2**(7), 1–8 (2021)
30. J.W. Holloway, G.C. Dogiamis, R. Han, 11.9 A 105Gb/s dielectric-waveguide link in 130nm BiCMOS using channelized 220-to-335GHz signal and integrated waveguide coupler, in *2021 IEEE International Solid- State Circuits Conference (ISSCC)*, vol. 64 (2021), pp. 196–198
31. The Data Deluge: What do we do with the Data Generated by AVs?. https://blogs.sw.siemens.com/polarion/the-data-deluge-what-do-we-do-with-the-data-generated-by-avs/. Accessed: April 2023
32. Wiring harness development in today's automotive world. https://blogs.sw.siemens.com/ee-systems/2020/07/28/wiring-harness-development-in-todays-automotive-world/. Accessed: April 2023
33. Wikipedia: Information Theory. https://en.wikipedia.org/wiki/Information_theory. Accessed: April 2023
34. T.M. Cover, J.A. Thomas, *Elements of Information Theory*, 2nd edn. (Wiley, Hoboken, 2005)
35. R.V.L. Hartley, Transmission of information. Bell Syst. Tech. J. **7**, 535–563 (1928)
36. S. Höst, *Information and Communication Theory*, 1st edn. (Wiley, Hoboken, 2019)
37. H. Nyquist, Certain factors affecting telegraph speed. Bell Syst. Tech. J. **3**(2), 324–346 (1924)
38. H. Nyquist, Certain topics in telegraph transmission theory. Trans. Amer. Inst. Electr. Eng. **47**(2), 617–644 (1928)
39. Wikipedia: Shannon–Hartley theorem. https://en.wikipedia.org/wiki/Shannon-Hartley_theorem. Accessed: April 2023

40. Wikipedia: Noisy-channel coding theorem. https://en.wikipedia.org/wiki/Noisy-channel_coding_theorem. Accessed: April 2023
41. W.L. Couch, *Digital & Analog Communication Systems*, 8th edn. (Pearson, London, 2012)
42. Wikipedia: Fourier transform. https://en.wikipedia.org/wiki/Fourier_transform. Accessed: April 2023
43. M.J.M. Pelgrom, *Analog-to-Digital Conversion*, 4th edn. (Springer, Berlin, 2022)
44. Wikipedia: Nyquist-Shannon sampling theorem. https://en.wikipedia.org/wiki/Nyquist%E2%80%93Shannon_sampling_theorem. Accessed: April 2023
45. A.T. Ramkaj, M.J.M. Pelgrom, M.S.J. Steyaert, F. Tavernier, *Multi-Gigahertz Nyquist Analog-to-Digital Converters* (Springer, Berlin, 2023)
46. E. McCune, *Practical Digital Wireless Signals* (Cambridge University Press, Cambridge, 2010)
47. N. Dolatsha, B. Grave, M. Sawaby, C. Chen, A. Babveyh, S. Kananian, A. Bisognin, C. Luxey, F. Gianesello, J. Costa, C. Fernandes, A. Arbabian, 17.8 a compact 130GHz fully packaged point-to-point wireless system with 3D-printed 26dBi lens antenna achieving 12.5Gb/s at 1.55pJ/b/m, in *2017 IEEE International Solid-State Circuits Conference (ISSCC)* (2017), pp. 306–307
48. M.H. Eissa, A. Malignaggi, R. Wang, M. Elkhouly, K. Schmalz, A.C. Ulusoy, D. Kissinger, Wideband 240-GHz transmitter and receiver in BiCMOS technology with 25-Gbit/s data rate. IEEE J. Solid-State Circ. **53**(9), 2532–2542 (2018)
49. C. D'heer, P. Reynaert, A high-speed 390GHz BPOOK transmitter in 28nm CMOS, in *2020 IEEE Radio Frequency Integrated Circuits Symposium (RFIC)* (2020), pp. 223–226
50. S. Kang, S.V. Thyagarajan, A.M. Niknejad, A 240 GHz fully integrated wideband QPSK transmitter in 65 nm CMOS. IEEE J. Solid-State Circ. **50**(10), 2256–2267 (2015)
51. S. Lee, S. Hara, R. Dong, K. Takano, S. Amakawa, T. Yoshida, M. Fujishima, A 272-GHz CMOS analog BPSK/QPSK demodulator for IEEE 802.15.3d, in *ESSCIRC 2021 - IEEE 47th European Solid State Circuits Conference (ESSCIRC)* (2021), pp. 415–418
52. M. De Wit, Y. Zhang, P. Reynaert, Analysis and design of a foam-cladded PMF link with phase tuning in 28-nm CMOS. IEEE J. Solid-State Circ. **54**(7), 1960–1969 (2019)
53. A. Goldsmith, *Wireless Communications* (Cambridge University Press, Cambridge, 2005)
54. R.A. Shafik, M.S. Rahman, A.R. Islam, On the extended relationships among EVM, BER and SNR as performance metrics, in *2006 International Conference on Electrical and Computer Engineering* (2006), pp. 408–411
55. A.K. Gupta, J.F. Buckwalter, Linearity considerations for low-EVM, millimeter-wave direct-conversion modulators. IEEE Trans. Microw. Theory Techn. **60**(10), 3272–3285 (2012)
56. H. Friis, Noise figures of radio receivers. Proc. IRE **32**(7), 419–422 (1944)
57. Wikipedia: Error correcting code. https://en.wikipedia.org/wiki/Error_correction_code. Accessed: May 2023
58. R.W. Hamming, Error detecting and error correcting codes. Bell Syst. Tech. J. **29**(2), 147–160 (1950)
59. F. Chang, K. Onohara, T. Mizuochi, Forward error correction for 100 G transport networks. IEEE Commun. Mag. **48**(3), S48–S55 (2010)
60. B. Razavi, *RF Microelectronics* (Prentice Hall, Hoboken, 2012)
61. N. Dolatsha, C. Chen, A. Arbabian, Loss and dispersion limitations in mm-Wave dielectric waveguides for high-speed links. IEEE Trans. Terahertz Sci. Technol. **6**(4), 637–640 (2016)
62. H. Friis, A note on a simple transmission formula. Proc. IRE **34**(5), 254–256 (1946)
63. J.C. Maxwell, A treatise on electricity and magnetism. Nature **7**, 478–480 (1873)
64. C. Yeh, F.I. Shimabukuro, *The Essence of Dielectric Waveguides*, 1st edn. (Springer, Berlin, 2010)
65. D.M. Pozar, *Microwave Engineering*, 4th edn. (Wiley, Hoboken, 2011)
66. A. Standaert, P. Reynaert, Analysis of hollow circular polymer waveguides at millimeter wavelengths. IEEE Trans. Microw. Theory Techn. **64**(10), 3068–3077 (2016)
67. A.M. Niknejad, *Electromagnetics for High-Speed Analog and Digital Communication Circuits* (Cambridge University Press, Cambridge, 2007)

68. The Feynman Lectures on Physics. https://www.feynmanlectures.caltech.edu/. Accessed: May 2023
69. W.M.C. Sansen, *Analog Design Essentials* (Springer, Berlin, 2006)
70. A.M. Niknejad, H. Hashemi, *mm-Wave Silicon Technology* (Springer, Berlin, 2008)
71. B. Razavi, R.-H. Yan, K. Lee, Impact of distributed gate resistance on the performance of MOS devices. IEEE Trans. Circ. Syst. I Fundam. Theory Appl. **41**(11), 750–754 (1994)
72. D. Zhao, P. Reynaert, A 60-GHz dual-mode class AB power amplifier in 40-nm CMOS. IEEE J. Solid-State Circ. **48**(10), 2323–2337 (2013)
73. Intel's Revolutionary 22nm transistor technology. https://download.intel.com/newsroom/kits/22nm/pdfs/22nm-Details_Presentation.pdf. Accessed: June 2023
74. B. Philippe, P. Reynaert, 24.7 A 15dBm 12.8%-PAE compact D-band power amplifier with two-way power combining in 16nm FinFET CMOS, in *2020 IEEE International Solid- State Circuits Conference - (ISSCC)* (2020), pp. 374–376
75. I. Aoki, S. Kee, D. Rutledge, A. Hajimiri, Distributed active transformer-a new power-combining and impedance-transformation technique. IEEE Trans. Microw. Theory Techn. **50**(1), 316–331 (2002)
76. N. Deferm, P. Reynaert, *CMOS Front Ends for Millimeter Wave Wireless Communication Systems* (Springer, Berlin, 2015)
77. K. Kurokawa, Power waves and the scattering matrix. IEEE Trans. Microw. Theory Techn. **13**(2), 194–202 (1965)
78. D. Zhao, P. Reynaert, *CMOS 60-GHz and E-band Power Amplifiers and Transmitters* (Springer, Berlin, 2015)
79. H.W. Bode, *Network Analysis and Feedback Amplifier Design*. Bell Telephone Laboratories Series (D. Van Nostrand Company, New Jersy, 1945)
80. R.M. Fano, Theoretical limitations on the broadband matching of arbitrary impedances. J. Franklin Inst. **249**(2), 139–154 (1955)
81. M. Vigilante, P. Reynaert, On the design of wideband transformer-based fourth order matching networks for E-band receivers in 28-nm CMOS. IEEE J. Solid-State Circ. **52**(8), 2071–2082 (2017)
82. C.R. Chappidi, K. Sengupta, Globally optimal matching networks with lossy passives and efficiency bounds. IEEE Trans. Circ. Syst. I Regular Papers **65**(1), 257–269 (2018)
83. S. Ooms, *Millimeter Wave Circuits for Directive Communication Links with Dielectric Fibers*. PhD Dissertation, KU Leuven, 2021
84. J. Rollett, Stability and Power-gain invariants of linear twoports. IRE Trans. Circ. Theory **9**(1), 29–32 (1962)
85. D. Woods, Reappraisal of the unconditional stability criteria for active 2-port networks in terms of S parameters. IEEE Trans. Circ. Syst. **23**(2), 73–81 (1976)
86. M. Edwards, J. Sinsky, A new criterion for linear 2-port stability using a single geometrically derived parameter. IEEE Trans. Microw. Theory Techn. **40**(12), 2303–2311 (1992)
87. H. Barkhausen, *Lehrbuch der Elektronenrohr* (1921)
88. H. Nyquist, Regeneration theory. Bell Syst. Techn. J. **11**(1), 126–147 (1932)
89. H.W. Bode, Relations between attenuation and phase in feedback amplifier design. Bell Syst. Techn. J. **19**(3), 421–454 (1940)
90. K. Kurokawa, Some basic characteristics of broadband negative resistance oscillator circuits. Bell Syst. Techn. J. **48**(6), 1937–1955 (1969)
91. A. Anakabe, J.-M. Collantes, J. Portilla, J. Jugo, A. Mallet, L. Lapierre, J.-P. Fraysse, 'Analysis and elimination of parametric oscillations in monolithic power amplifiers, in *2002 IEEE MTT-S International Microwave Symposium Digest (Cat. No.02CH37278)*, vol. 3 (2002), pp. 2181–2184
92. J. Jugo, J. Portilla, A. Anakabe, A. Suarez, J.M. Collantes, Closed-loop stability analysis of microwave amplifiers. Elec. Letters **37**(4), 226–228 (2001)
93. M. Vigilante, P. Reynaert, 20.10 A 68.1-to-96.4GHz variable-gain low-noise amplifier in 28nm CMOS, in *2016 IEEE International Solid-State Circuits Conference (ISSCC)* (2016), pp. 360–362

94. S.V. Thyagarajan, A.M. Niknejad, C.D. Hull, A 60 GHz linear wideband power amplifier using cascode neutralization in 28 nm CMOS, in *Proceedings of the IEEE 2013 Custom Integrated Circuits Conference* (2013), pp. 1–4
95. P. Reynaert, M. Steyaert, *RF Power Amplifiers for Mobile Communications* (Springer, Berlin, 2006)
96. M. Vigilante, P. Reynaert, A wideband class-AB power amplifier with 29–57-GHz AM-PM compensation in 0.9-V 28-nm bulk CMOS. IEEE J. Solid-State Circ. **53**(5), 1288–1301 (2018)
97. D. Frickey, Conversions between S, Z, Y, H, ABCD, and T parameters which are valid for complex source and load impedances. IEEE Trans. Microw. Theory Techn. **42**(2), 205–211 (1994)
98. B. Philippe, D. Zhao, P. Reynaert, *Mm-wave Circuit Design in 16nm FinFET for 6G Applications* (Springer, Berlin, 2023)
99. R. Kananizadeh, O. Momeni, High-power and high-efficiency millimeter-wave harmonic oscillator design, exploiting harmonic positive feedback in CMOS. IEEE Trans. Microw. Theory Techn. **65**, 3922–3936 (2017)
100. G. Guimarães, P. Reynaert, A 670-GHz 4 × 2 oscillator-radiator array achieving 7.4-dBm EIRP in 40-nm CMOS. IEEE J. Solid-State Circ. **56**(11), 3399–3411 (2021)
101. S. Lee, S. Hara, T. Yoshida, S. Amakawa, R. Dong, A. Kasamatsu, J. Sato, M. Fujishima, An 80-Gb/s 300-GHz-band single-chip CMOS transceiver. IEEE J. Solid-State Circ. **54**(12), 3577–3588 (2019)
102. E. Khorov, I. Levitsky, I.F. Akyildiz, Current status and directions of IEEE 802.11be, the future Wi-Fi 7. IEEE Access **8**, 88664–88688 (2020)
103. A. Karakuzulu, W.A. Ahmad, D. Kissinger, A. Malignaggi, A four-channel bidirectional D-band phased-array transceiver for 200 Gb/s 6G wireless communications in a 130-nm BiCMOS technology. IEEE J. Solid-State Circ. **58**(5), 1310–1322 (2023)
104. G.C. Dogiamis, T.W. Brown, N. Prabhu Gaunkar, Y.S. Nam, T.S. Rane, S. Ravikumar, V.B. Neeli, J.C. Chou, S. Rami, J. Swan, A 120-Gb/s 100–145-GHz 16-QAM dual-band dielectric waveguide interconnect with package integrated diplexers in Intel 16. IEEE Solid-State Circ. Lett. **5**, 178–181 (2022)
105. A. Singh, M. Sayginer, M.J. Holyoak, J. Weiner, J. Kimionis, M. Elkhouly, Y. Baeyens, S. Shahramian,A D-band radio-on-glass module for spectrally-efficient and low-cost wireless Backhaul, in *2020 IEEE Radio Frequency Integrated Circuits Symposium (RFIC)* (2020), pp. 99–102
106. C. D'heer, P. Reynaert, A 135 GHz 32 Gb/s direct-digital modulation 16-QAM transmitter in 28 nm CMOS, in *ESSCIRC 2022- IEEE 48th European Solid State Circuits Conference (ESSCIRC)* (2022), pp. 481–484
107. C. D'heer, P. Reynaert, A 135 GHz 24 Gb/s direct-digital demodulation 16-QAM receiver in 28 nm CMOS, in *ESSCIRC 2022- IEEE 48th European Solid State Circuits Conference (ESSCIRC)* (2022), pp. 485–488
108. S. Callender, A. Agrawal, A. Whitcombe, R. Bhat, M. Rahman, C.C. Lee, P. Sagazio, G.C. Dogiamis, B.R. Carlton, C. Hull, S. Pellerano, A fully integrated 160-Gb/s D-band transmitter achieving 1.1-pJ/b efficiency in 22-nm FinFET. IEEE J. Solid-State Circ. **57**(12), 3582–3598 (2022)
109. J. Grzyb, P. Rodríguez-Vázquez, S. Malz, M. Andree, U.R. Pfeiffer, A SiGe HBT 215–240 GHz DCA IQ TX/RX chipset with built-in test of USB/LSB RF asymmetry for 100+ Gb/s data rates. IEEE Trans. Microw. Theory Techn. **70**(3), 1696–1714 (2022)
110. K.K. Tokgoz, S. Maki, J. Pang, N. Nagashima, I. Abdo, S. Kawai, T. Fujimura, Y. Kawano, T. Suzuki, T. Iwai, K. Okada, A. Matsuzawa, A 120Gb/s 16QAM CMOS millimeter-wave wireless transceiver, in *2018 IEEE International Solid - State Circuits Conference - (ISSCC)* (2018), pp. 168–170
111. K. Chuang, H. Yektaii, D. McLaurin, C. Mayer, Radio challenges, architectures, and design considerations for wireless infrastructure: creating the core technologies that connect people around the world. IEEE Microw. Mag. **23**(12), 42–59 (2022)

112. J. Salinas, J. Cosson-Martin, M. Laghaei, H. Shakiba, A. Sheikholeslami, Performance comparison of baseband signaling and discrete multi-tone for wireline communication. IEEE Open J. Circ. Syst. **2**, 65–77 (2021)
113. C. Desset, P. Wambacq, Y. Zhang, M. Ingels, A. Bourdoux, A flexible power model for mm-wave and THz high-throughput communication systems, in *2020 IEEE 31st Annual International Symposium on Personal, Indoor and Mobile Radio Communications* (2020), pp. 1–6
114. D. Leeson, A simple model of feedback oscillator noise spectrum. Proc. IEEE **54**(2), 329–330 (1966)
115. H. Jia, R. Ma, W. Deng, Z. Wang, B. Chi, A 53.6-to-60.2GHz many-core fundamental oscillator with scalable mesh topology achieving -136.0dBc/Hz phase noise at 10MHz offset and 190.3dBc/Hz peak FoM in 65nm CMOS, in *2022 IEEE International Solid- State Circuits Conference (ISSCC)*, vol. 65 (2022), pp. 154–156
116. L. Iotti, A. Mazzanti, F. Svelto, Insights into phase-noise scaling in switch-coupled multi-core LC VCOs for E-band adaptive modulation links. IEEE J. Solid-State Circ. **52**(7), 1703–1718 (2017)
117. T. Siriburanon, S. Kondo, M. Katsuragi, H. Liu, K. Kimura, W. Deng, K. Okada, A. Matsuzawa, A low-power low-noise mm-wave subsampling PLL using dual-step-mixing ILFD and tail-coupling quadrature injection-locked oscillator for IEEE 802.11ad. IEEE J. Solid-State Circ. **51**(5), 1246–1260 (2016)
118. X. Yi, C.C. Boon, H. Liu, J.F. Lin, W.M. Lim, A 57.9-to-68.3 GHz 24.6 mW frequency synthesizer with in-phase injection-coupled QVCO in 65 nm CMOS technology. IEEE J. Solid-State Circ. **49**(2), 347–359 (2014)
119. S. Shahramian, A. Hart, A. Tomkins, A. Chan Carusone, P. Garcia, P. Chevalier, S.P. Voinigescu, Design of a dual W- and D-band PLL. IEEE J. Solid-State Circ. **46**(5), 1011–1022 (2011)
120. F. Säckinger, *Broadband Circuits for Optical Fiber Communication* (Wiley, Hoboken, 2005)
121. B. Sadhu, A. Paidimarri, D. Liu, M. Yeck, C. Ozdag, Y. Tojo, W. Lee, K.X. Gu, J.-O. Plouchart, C.W. Baks, Y. Uemichi, S. Chakraborty, Y. Yamaguchi, N. Guan, A. Valdes-Garcia, A 24–30-GHz 256-element dual-polarized 5G phased array using fast on-chip beam calculators and magnetoelectric dipole antennas. IEEE J. Solid-State Circ. **57**(12), 3599–3616 (2022)
122. S. Shahramian, M.J. Holyoak, A. Singh, Y. Baeyens, A fully integrated 384-element, 16-tile, W -band phased array with self-alignment and self-test. IEEE J. Solid-State Circ. **54**(9), 2419–2434 (2019)
123. B. Sadhu, X. Gu, A. Valdes-Garcia, The more (antennas), the merrier: a survey of silicon-based mm-Wave phased arrays using multi-IC scaling. IEEE Microw. Mag. **20**(12), 32–50 (2019)
124. M. Elkhouly, J. Ha, M. J. Holyoak, D. Hendry, M. Sayginer, R. Enright, I. Kimionis, Y. Baeyens, S. Shahramian, Fully integrated 2D scalable TX/RX chipset for D-band phased-array-on-glass modules, in *2022 IEEE International Solid- State Circuits Conference (ISSCC)*, vol. 65 (2022), pp. 76–78
125. P. Reynaert, K. Dens, C. D'heer, D. Simic, J. Vaes, B. Philippe, S. Ooms, Polymer microwave fiber: A new communication concept that blends wireless, wireline and optical communication, in *2019 26th IEEE International Conference on Electronics, Circuits and Systems (ICECS)* (2019), pp. 755–758
126. Q.J. Gu, Sub-THz/THz interconnect, complement to electrical and optical interconnects: addressing fundamental challenges related to communication distances. IEEE Solid-State Circ. Mag. **12**(4), 20–32 (2020)
127. A. Meyer, M. Schneider, Signal distortions in circular dielectric waveguides at mm-wave frequencies. Int. J. Microw. Wirel. Technol. **14**(1), 8–18 (2022)
128. S.J. Orfanidis, *Electromagnetic Waves and Antennas* (Sophocles, Piscataway, 2016)
129. G. Agrawal, P. Anthony, T.-M. Shen, Dispersion penalty for 1.3 mu m lightwave systems with multimode semiconductor lasers. J. Lightwave Technol. **6**(5), 620–625 (1988)

130. K. Dens, J. Vaes, S. Ooms, M. Wagner, P. Reynaert, A PAM4 dielectric waveguide link in 28 nm CMOS, in *ESSCIRC 2021 - IEEE 47th European Solid State Circuits Conference (ESSCIRC)* (2021), pp. 479–482
131. B. Murmann, Mixed-signal computing for deep neural network inference. IEEE Trans. Very Large Scale Integr. Syst. **29**(1), 3–13 (2021)
132. A.M.A. Ali, H. Dinc, P. Bhoraskar, S. Bardsley, C. Dillon, M. McShea, J.P. Periathambi, S. Puckett, A 12-b 18-GS/s RF sampling ADC with an integrated wideband track-and-hold amplifier and background calibration. IEEE J. Solid-State Circ. **55**(12), 3210–3224 (2020)
133. K. Sun, G. Wang, Q. Zhang, S. Elahmadi, P. Gui, A 56-GS/s 8-bit time-interleaved ADC with ENOB and BW enhancement techniques in 28-nm CMOS. IEEE J. Solid-State Circ. **54**(3), 821–833 (2019)
134. L. Kull, D. Luu, C. Menolfi, M. Braendli, P.A. Francese, T. Morf, M. Kossel, A. Cevrero, I. Ozkaya, T. Toifl, A 24-to-72GS/s 8b time-interleaved SAR ADC with 2.0-to-3.3pJ/conversion and >30dB SNDR at Nyquist in 14nm CMOS FinFET, in *2018 IEEE International Solid - State Circuits Conference - (ISSCC)* (2018), pp. 358–360
135. S. Su, M.S.-W. Chen, A 16-bit 12-GS/s single-/dual-rate DAC with a successive bandpass delta-sigma modulator achieving < - 67-dBc IM3 within DC to 6-GHz tunable passbands. IEEE J. Solid-State Circ. **53**(12), 3517–3527 (2018)
136. Y. Greshishchev, J. Aguirre, S. Aouini, M. Besson, R. Gibbins, C.Y. Gouk, J. Lam, D. McPherson, M. Parvizi, T. Wen, N. Ben-Hamida, A 60 GS/s 8-b DAC with >29.5dB SINAD up to Nyquist frequency in 7nm FinFET CMOS, in *2019 IEEE BiCMOS and Compound semiconductor Integrated Circuits and Technology Symposium (BCICTS)* (2019), pp. 1–4
137. Y.M. Greshishchev, D. Pollex, S.-C. Wang, M. Besson, P. Flemeke, S. Szilagyi, J. Aguirre, C. Falt, N. Ben-Hamida, R. Gibbins, P. Schvan, A 56GS/s 6b DAC in 65nm CMOS with 256 × 6b memory, in *2011 IEEE International Solid-State Circuits Conference* (2011), pp. 194–196
138. K. Katayama, K. Takano, S. Amakawa, S. Hara, A. Kasamatsu, K. Mizuno, K. Takahashi, T. Yoshida, M. Fujishima, A 300 GHz CMOS transmitter with 32-QAM 17.5 Gb/s/ch capability over six channels. IEEE J. Solid-State Circ. **51**(12), 3037–3048 (2016)
139. K. Takano, S. Hara, K. Katayama, S. Amakawa, T. Yoshida, M. Fujishima, Quintic mixer: A subharmonic up-conversion mixer for THz transmitter supporting complex digital modulation, in *2016 IEEE MTT-S International Microwave Symposium (IMS)* (2016), pp. 1–3
140. K. Takano, S. Amakawa, K. Katayama, S. Hara, R. Dong, A. Kasamatsu, I. Hosako, K. Mizuno, K. Takahashi, T. Yoshida, M. Fujishima, 17.9 A 105Gb/s 300GHz CMOS transmitter, in *2017 IEEE International Solid-State Circuits Conference (ISSCC)* (2017), pp. 308–309
141. G. Guimarães, P. Reynaert, A 670-GHz 4 × 2 oscillator-radiator array achieving 7.4-dBm EIRP in 40-nm CMOS. IEEE J. Solid-State Circ. **56**(11), 3399–3411 (2021)
142. Y. Wang, B. Liu, R. Wu, H. Liu, A.T. Narayanan, J. Pang, N. Li, T. Yoshioka, Y. Terashima, H. Zhang, D. Tang, M. Katsuragi, D. Lee, S. Choi, K. Okada, A. Matsuzawa, A 60-GHz 3.0-Gb/s spectrum efficient BPOOK transceiver for low-power short-range wireless in 65-nm CMOS. IEEE J. Solid-State Circ. **54**(5), 1363–1374 (2019)
143. Yu-Lun Chen, C. Kao, P.-J. Peng, J. Lee, A 94GHz duobinary keying wireless transceiver in 65nm CMOS, in *2014 Symposium on VLSI Circuits Digest of Technical Papers* (2014), pp. 1–2
144. A. Lender, The duobinary technique for high-speed data transmission. Trans. Amer. Inst. Electr. Eng. Part I Commun. Electr. **82**(2), 214–218 (1963)
145. J. Lee, M.-S. Chen, H.-D. Wang, Design and comparison of three 20-Gb/s backplane transceivers for duobinary, PAM4, and NRZ data. IEEE J. Solid-State Circ. **43**(9), 2120–2133 (2008)
146. W. Lee, S. Hong, Low-loss and small-size 28 GHz CMOS SPDT switches using switched inductor, in *2018 IEEE Radio Frequency Integrated Circuits Symposium (RFIC)* (2018), pp. 148–151

147. Q. Zhong, Z. Chen, N. Sharma, S. Kshattry, W. Choi, K.O. Kenneth, 300-GHz CMOS QPSK transmitter for 30-Gbps dielectric waveguide communication, in *2018 IEEE Custom Integrated Circuits Conference (CICC)* (2018), pp. 1–4
148. S. Hu, Y. Xiong, B. Zhang, L. Wang, T. Lim, M. Je, M. Madihian, A SiGe BiCMOS transmitter/receiver chipset with on-chip SIW antennas for terahertz applications. IEEE J. Solid-State Circ. **47**, 2654–2664 (2012)
149. Y. Liang, C.C. Boon, H.C. Zhang, X.-L. Tang, Q. Zhang, H. Yu, A 13.5-Gb/s 140-GHz silicon redriver exploiting metadevices for short-range OOK communications. IEEE Trans. Microw. Theory Techn. **70**(1), 239–253 (2022)
150. S. Ooms, P. Reynaert, A 120-GHz wireless link using 3D-printed lens with flexible dielectric fiber feed and 28-nm CMOS transceiver. IEEE Solid-State Circ. Lett. **3**, 142–145 (2020)
151. H.-I. Song, H. Choi, J.y. Yoo, H.-S. Won, C.M. Lee, H. Jin, T.y. Kim, W. Kwon, K. Lim, K. Kwon, C.-A. Kim, T. Kim, J.G. Jo, J. Eu, S. Park, H.-M. Bae, A 50Gb/s PAM-4 Bi-directional plastic waveguide link with carrier synchronization using PI-based costas loop, in *2022 IEEE International Solid- State Circuits Conference (ISSCC)*, vol. 65 (2022), pp. 1–3
152. Y. Kim, B. Hu, R. Huang, A. Tang, C. Joye, T. Itoh, M.-C.F. Chang, 150-GHz CMOS TX/RX with digitally predistorted PAM-4 modulation for terahertz contactless/plastic waveguide communications. IEEE Trans. Terahertz Sci. Technol. **10**(4), 370–382 (2020)
153. N. Van Thienen, Y. Zhang, P. Reynaert, Bidirectional communication circuits for a 120-GHz PMF data link in 40-nm CMOS. IEEE J. Solid-State Circ. **53**(7), 2023–2031 (2018)
154. N. Van Thienen, Y. Zhang, M. De Wit, P. Reynaert, An 18Gbps polymer microwave fiber (PMF) communication link in 40nm CMOS, in *ESSCIRC Conference 2016: 42nd European Solid-State Circuits Conference* (2016), pp. 483–486
155. Y. Yang, S. Zihir, H. Lin, O. Inac, W. Shin, G.M. Rebeiz, A 155 GHz 20 Gbit/s QPSK transceiver in 45nm CMOS, in *2014 IEEE Radio Frequency Integrated Circuits Symposium* (2014), pp. 365–368
156. P. Rodriguez-Vazquez, J. Grzyb, B. Heinemann, U.R. Pfeiffer, A QPSK 110-Gb/s Polarization-diversity MIMO wireless link with a 220–255 GHz tunable LO in a SiGe HBT technology. IEEE Trans. Microw. Theory Techn. **68**(9), 3834–3851 (2020)
157. S.V. Thyagarajan, S. Kang, A.M. Niknejad, A 240 GHz fully integrated wideband QPSK receiver in 65 nm CMOS. IEEE J. Solid-State Circ. **50**(10), 2268–2280 (2015)
158. G.C. Dogiamis, T.W. Brown, N.P. Gaunkar, Y.S. Nam, T.S. Rane, S. Ravikumar, V.B. Neeli, J.C. Chou, S. Rami, A 60-Gbps 108-GHz 16-QAM dielectric waveguide interconnect with package integrated filters, in *2022 IEEE/MTT-S International Microwave Symposium - IMS 2022* (2022), pp. 556–559
159. T.W. Brown, G.C. Dogiamis, Y.-S. Yeh, D. Correas-Serrano, T.S. Rane, S. Ravikumar, J.C. Chou, V.B. Neeli, J. Koo, M. Marulanda, N.P. Gaunkar, I. Huang, H. Chandrakumar, J.W. Bates, Z. Tuli, Q. Yu, M. Weiss, J. Rangaswamy, C.F. Nieva, D. Frolov, T. Kamgaing, Y.S. Nam, H. Braunisch, S. Rami, A 50-Gb/s 134-GHz 16-QAM 3-m dielectric waveguide transceiver system implemented in 22-nm FinFET CMOS. IEEE Solid-State Circ. Lett. **4**, 206–209 (2021)
160. M.H. Eissa, N. Maletic, E. Grass, R. Kraemer, D. Kissinger, A. Malignaggi, 100 Gbps 0.8-m wireless link based on fully integrated 240 GHz IQ transmitter and receiver, in *2020 IEEE/MTT-S International Microwave Symposium (IMS)* (2020), pp. 627–630
161. I. Abdo, T. Fujimura, T. Miura, K.K. Tokgoz, H. Hamada, H. Nosaka, A. Shirane, K. Okada, A 300GHz wireless transceiver in 65nm CMOS for IEEE802.15.3d using push-push sub-harmonic mixer, in *2020 IEEE/MTT-S International Microwave Symposium (IMS)* (2020), pp. 623–626
162. A. Townley, N. Baniasadi, S. Krishnamurthy, C. Sideris, A. Hajimiri, E. Alon, A. Niknejad, A fully integrated, dual channel, flip chip packaged 113 GHz transceiver in 28nm CMOS supporting an 80 Gb/s wireless link, in *2020 IEEE Custom Integrated Circuits Conference (CICC)* (2020), pp. 1–4
163. P. Rodríguez-Vázquez, J. Grzyb, B. Heinemann, U.R. Pfeiffer, A 16-QAM 100-Gb/s 1-M wireless link with an EVM of 17% at 230 GHz in an SiGe technology. IEEE Microw. Wirel. Comp. Lett. **29**(4), 297–299 (2019)

164. I. Abdo, C. da Gomez, C. Wang, K. Hatano, Q. Li, C. Liu, K. Yanagisawa, A.A. Fadila, T. Fujimura, T. Miura, K.K. Tokgoz, J. Pang, H. Hamada, H. Nosaka, A. Shirane, K. Okada, A Bi-directional 300-GHz-band phased-array transceiver in 65-nm CMOS with outphasing transmitting mode and LO emission cancellation. IEEE J. Solid-State Circ. **57**(8), 2292–2308 (2022)
165. V. Ariyarathna, A. Madanayake, J.M. Jornet, Toward real-time software-defined radios for ultrabroadband communication above 100 GHz [Application Notes]. IEEE Microw. Mag. **24**(8), 50–59 (2023)
166. H. Huang, J. Heilmeyer, M. Grözing, M. Berroth, J. Leibrich, W. Rosenkranz, An 8-bit 100-GS/s distributed DAC in 28-nm CMOS for optical communications. IEEE Trans. Microw. Theory Techn. **63**(4), 1211–1218 (2015)
167. Fujitsu Semiconductor, Factsheet LEIA 55 – 65 GSa/s 8-bit DAC (2012). Accessed: June 2022
168. A. Zandieh, P. Schvan, S.P. Voinigescu, Design of a 55-nm SiGe BiCMOS 5-bit time-interleaved flash ADC for 64-GBd 16-QAM fiberoptics applications. IEEE J. Solid-State Circ. **54**(9), 2375–2387 (2019)
169. L. Kull, J. Pliva, T. Toifl, M. Schmatz, P.A. Francese, C. Menolfi, M. Brändli, M. Kossel, T. Morf, T.M. Andersen, Y. Leblebici, Implementation of low-power 6–8 b 30–90 GS/s time-interleaved ADCs with optimized input bandwidth in 32 nm CMOS. IEEE J. Solid-State Circ. **51**(3), 636–648 (2016)
170. S. Palermo, S. Hoyos, S. Cai, S. Kiran, Y. Zhu, Analog-to-digital converter-based serial links: an overview. IEEE Solid-State Circ. Mag. **10**(3), 35–47 (2018)
171. H. Chandrakumar, T.W. Brown, D. Frolov, Z. Tuli, I. Huang, S. Rami, A 48dB-SFDR, 43dB-SNDR, 50GS/s 9-bit 2x-interleaved Nyquist DAC in Intel 16, in *2022 IEEE Custom Integrated Circuits Conference (CICC)* (2022), pp. 1–2
172. J. Cao, D. Cui, A. Nazemi, T. He, G. Li, B. Catli, M. Khanpour, K. Hu, T. Ali, H. Zhang, H. Yu, B. Rhew, S. Sheng, Y. Shim, B. Zhang, A. Momtaz, 29.2 a transmitter and receiver for 100Gb/s coherent networks with integrated 4 $\times$ 64GS/s 8b ADCs and DACs in 20nm CMOS, in *2017 IEEE International Solid-State Circuits Conference (ISSCC)* (2017), pp. 484–485
173. H. Wang, H. Mohammadnezhad, P. Heydari, Analysis and design of high-order QAM direct-modulation transmitter for high-speed point-to-point mm-Wave wireless links. IEEE J. Solid-State Circ. **54**(11), 3161–3179 (2019)
174. H. Mohammadnezhad, H. Wang, A. Cathelin, P. Heydari, A 115–135-GHz 8PSK receiver using multi-phase RF-correlation-based direct-demodulation method. IEEE J. Solid-State Circ. **54**(9), 2435–2448 (2019)
175. K.-L. Chiu, H.-W. Chan, W.-C. Lee, C.-T. Wu, H. Lopez, H.-C. Liu, M.-Y. Huang, C.-Y. Liu, T.-H. Lee, H.-T. Chang, C.-W. Jen, N.-H. Chang, P.-Y. Tsai, Y.-C. Kuan, S.-J. Jou, A millimeter wave digital CMOS baseband transceiver for wireless LAN applications, in *2019 IEEE Asian Solid-State Circuits Conference (A-SSCC)* (2019), pp. 275–278
176. N. Saito, T. Tsukizawa, N. Shirakata, T. Morita, K. Tanaka, J. Sato, Y. Morishita, M. Kanemaru, R. Kitamura, T. Shima, T. Nakatani, K. Miyanaga, T. Urushihara, H. Yoshikawa, T. Sakamoto, H. Motozuka, Y. Shirakawa, N. Yosoku, A. Yamamoto, R. Shiozaki, K. Takinami, A fully integrated 60-GHz CMOS transceiver chipset based on WiGig/IEEE 802.11ad with built-in self calibration for mobile usage. IEEE J. Solid-State Circ. **48**(12), 3146–3159 (2013)
177. Z. Toprak-Deniz, J.E. Proesel, J.F. Bulzacchelli, H.A. Ainspan, T.O. Dickson, M.P. Beakes, M. Meghelli, A 128-Gb/s 1.3-pJ/b PAM-4 transmitter with reconfigurable 3-Tap FFE in 14-nm CMOS. IEEE J. Solid-State Circ. **55**(1), 19–26 (2020)
178. B. Ye, K. Sheng, W. Gai, H. Niu, B. Zhang, Y. He, S. Jia, C. Chen, J. Yu, A 2.29pJ/b 112Gb/s wireline transceiver with RX 4-Tap FFE for medium-reach applications in 28nm CMOS, in *2022 IEEE International Solid- State Circuits Conference (ISSCC)*, vol. 65 (2022), pp. 118–120
179. N. Kocaman, U. Singh, B. Raghavan, A. Lyer, K. ThaSARi, S. Surana, J.W. Jung, J. Jeong, H. Zhang, A. Vasani, Y. Shim, Z. Huang, A. Garg, H.-b. Lee, B. Wu, F. Liu, R. Wang, M. Loh,

A. Wang, M. Caresosa, B. Zhang, A. Momtaz, An 182mW 1-60Gb/s configurable PAM-4/NRZ transceiver for large scale ASIC integration in 7nm FinFET technology, in *2022 IEEE International Solid- State Circuits Conference (ISSCC)*, vol. 65 (2022), pp. 120–122
180. Z. Guo, A. Mostafa, A. Elshazly, B. Chen, B. Wang, C. Han, C. Wang, D. Zhou, D. Visani, E. Hsiao, F. Chu, F. Lu, G. Cui, H. Zhang, H. Wang, H. Zhao, J. Lin, J. Gu, L. Luo, L. Jiang, M. Singh, M. Gambhir, M. Hasan, M. Wu, M.J. Yoo, P. Liu, S. Kollu, T. Ye, X. Zhao, X. Yang, X. Han, Y. Huang, Y. Sun, Z. Yu, Z.H. Jiang, Z. Adal, Z. Yan, A 112.5Gb/s ADC-DSP-based PAM-4 long-reach transceiver with >50dB channel loss in 5nm FinFET, in *2022 IEEE International Solid-State Circuits Conference (ISSCC)*, vol. 65 (2022), pp. 116–118
181. M.-A. LaCroix, E. Chong, W. Shen, E. Nir, F.A. Musa, H. Mei, M.-M. Mohsenpour, S. Lebedev, B. Zamanlooy, C. Carvalho, Q. Xin, D. Petrov, H. Wong, H. Ho, Y. Xu, S.N. Shahi, P. Krotnev, C. Feist, H. Huang, D. Tonietto, 8.4 A 116Gb/s DSP-based wireline transceiver in 7nm CMOS achieving 6pJ/b at 45dB loss in PAM-4/Duo-PAM-4 and 52dB in PAM-2, in *2021 IEEE International Solid-State Circuits Conference (ISSCC)*, vol. 64 (2021), pp. 132–134
182. L. Wang, Y. Fu, M. LaCroix, E. Chong, A.C. Carusone, A 64Gb/s PAM-4 transceiver utilizing an adaptive threshold ADC in 16nm FinFET, in *2018 IEEE International Solid-State Circuits Conference - (ISSCC)* (2018), pp. 110–112
183. M. Morsy-Osman, M. Sowailem, E. El-Fiky, T. Goodwill, T. Hoang, S. LesSARd, D.V. Plant, DSP-free 'coherent-lite' transceiver for next generation single wavelength optical intra-datacenter interconnects. Opt. Express **26**, 8890–8903 (2018)
184. N. Nambath, R.K. Raveendranath, D. Banerjee, A. Sharma, A. Sankar, S. Gupta, Analog domain signal processing-based low-power 100-Gb/s DP-QPSK receiver. J. Lightw. Technol. **33**(15), 3189–3197 (2015)
185. J.K. Perin, A. Shastri, J.M. Kahn, Design of low-power DSP-free coherent receivers for data center links. J. Lightw. Technol. **35**(21), 4650–4662 (2017)
186. K. Dasgupta, S. Daneshgar, C. Thakkar, S. Kang, A. Chakrabarti, S. Yamada, N. Narevsky, D. Choudhury, J.E. Jaussi, B. Casper, A 60-GHz transceiver and baseband with polarization MIMO in 28-nm CMOS. IEEE J. Solid-State Circ. **53**(12), 3613–3627 (2018)
187. C. Thakkar, L. Kong, K. Jung, A. Frappé, E. Alon, A 10Gb/s 45mW adaptive 60GHz baseband in 65nm CMOS, in *2011 Symposium on VLSI Circuits - Digest of Technical Papers* (2011), pp. 24–25
188. C. Thakkar, A. Chakrabarti, S. Yamada, D. Choudhury, J. Jaussi, B. Casper, A 42.2-Gb/s 4.3-pJ/b 60-GHz digital transmitter with 12-b/symbol polarization MIMO. IEEE J. Solid-State Circ. **54**(12), 3565–3576 (2019)
189. O.E. Mattia, M. Sawaby, K. Zheng, A. Arbabian, B. Murmann, A 10-Gbps continuous-time linear equalizer for mm-Wave dielectric waveguide communication. IEEE Solid-State Circ. Lett. **3**, 266–269 (2020)
190. M. Verplaetse, J. Lambrecht, M. Vanhoecke, L. Breyne, H. Ramon, P. Demeester, G. Torfs, Analog I/Q FIR filter in 55-nm SiGe BiCMOS for 16-QAM optical communications at 112 Gb/s. IEEE J. Solid-State Circ. **55**(7), 1935–1945 (2020)
191. O.E. Mattia, D. Guermandi, G. Torfs, P. Wambacq, An up to 36Gbps analog baseband equalizer and demodulator for mm-wave wireless communication in 28nm CMOS, in *2017 IEEE Custom Integrated Circuits Conference (CICC)* (2017), pp. 1–4
192. A. Atharav, B. Razavi, A 56-Gb/s 50-mW NRZ receiver in 28-nm CMOS. IEEE J. Solid-State Circ. **57**(1), 54–67 (2022)
193. P. Nazari, S. Jafarlou, P. Heydari, A CMOS two-element 170-GHz fundamental-frequency transmitter with direct RF-8PSK modulation. IEEE J. Solid-State Circ. **55**(2), 282–297 (2020)
194. S. Shopov, O.D. Gurbuz, G.M. Rebeiz, S.P. Voinigescu, A D-band digital transmitter with 64-QAM and OFDM free-space constellation formation. IEEE J. Solid-State Circ. **53**(7), 2012–2022 (2018)
195. Q. Pan, L. Wang, X. Luo, C.P. Yue, A low-power PAM4 receiver with an adaptive variable-gain rectifier-based decoder. IEEE Trans. Very Large Scale Integr. Syst. **28**(10), 2099–2108 (2020)

196. I. Petricli, H. Zhang, E. Monaco, G. Albasini, A. Mazzanti, A 112 Gb/s PAM-4 RX front-end with unclocked decision feedback equalizer. IEEE Trans. Circ. Syst. II: Express Briefs **68**(1), 256–260 (2021)
197. J.G. Proakis, M. Salehi, *Digital Communications*, 5th edn. (McGraw-Hill, New York, 2008)
198. M. Baert, W. Dehaene, A 5-GS/s 7.2-ENOB time-interleaved VCO-based ADC achieving 30.5 fJ/cs. IEEE J. Solid-State Circ. **55**(6), 1577–1587 (2020)
199. B. Philippe, P. Reynaert, A quadrature phase detector in 28nm CMOS for differential mm-Wave sensing applications using dielectric waveguides, in *ESSCIRC 2018—IEEE 44th European Solid State Circuits Conference (ESSCIRC)* (2018), pp. 114–117
200. S. Ooms, P. Reynaert, Stability analysis of parasitic coupling between on-chip antenna and mm-Wave front-end, in *2020 IEEE Radio and Wireless Symposium (RWS)* (2020), pp. 275–278
201. S. Shekhar, J. Walling, D. Allstot, Bandwidth extension techniques for CMOS amplifiers. IEEE J. Solid-State Circ. **41**(11), 2424–2439 (2006)
202. J.S. Walling, S. Shekhar, D.J. Allstot, Wideband CMOS Amplifier Design: time-domain considerations. IEEE Trans. Circ. Syst. I: Regular Papers **55**(7), 1781–1793 (2008)
203. C.A. Balanis, *Antenna Theory: Analysis and Design*, 4th edn. (Wiley, Hoboken, 2016)

Index

C. D'heer, P. Reynaert, *THz and Sub-THz CMOS Electronics for High-Speed Telecommunication*, Analog Circuits and Signal Processing,
https://doi.org/10.1007/978-3-031-64439-9

The manufacturer's authorised representative in the EU is Springer Nature Customer Service Centre GmbH, Europaplatz 3, 69115 Heidelberg, Germany. If you have any concerns regarding our products, please contact ProductSafety@springernature.com

Printed and bound by CPI Group (UK) Ltd, Croydon, CR0 4YY
01/10/2026
02225904-0016